Advances in the Monitoring, Diagnosis, and Optimisation of Water Systems

Advances in the Monitoring, Diagnosis, and Optimisation of Water Systems

Editors

Miquel À. Cugueró-Escofet
Vicenç Puig

MDPI

Basel • Beijing • Wuhan • Barcelona • Belgrade • Novi Sad • Cluj • Manchester

Editors
Miquel À. Cugueró-Escofet
Polytechnic University of Catalonia
(UPC-BarcelonaTech)
Terrassa
Spain

Vicenç Puig
Institut de Robòtica i Informàtica Industrial
CSIC-UPC
Barcelona
Spain

Editorial Office
MDPI
St. Alban-Anlage 66
4052 Basel, Switzerland

This is a reprint of articles from the Special Issue published online in the open access journal *Sensors* (ISSN 1424-8220) (available at: https://www.mdpi.com/journal/sensors/special_issues/Optimisation_Water_Systems).

For citation purposes, cite each article independently as indicated on the article page online and as indicated below:

Lastname, A.A.; Lastname, B.B. Article Title. *Journal Name* **Year**, *Volume Number*, Page Range.

ISBN 978-3-0365-9594-8 (Hbk)
ISBN 978-3-0365-9595-5 (PDF)
doi.org/10.3390/books978-3-0365-9595-5

Cover image courtesy of Miquel À. Cugueró-Escofet

Contents

About the Editors

Miquel À. Cugueró-Escofet

Miquel À. Cugueró-Escofet is Assistant Professor at the Automatic Control Department of the Universitat Politècnica de Catalunya (UPC-BarcelonaTech) and Researcher at the Advanced Control Systems Research Group (SAC-UPC). His research interests are on the field of systems fault diagnosis and control, with a strong focus on data analysis, including system modelling and supervision, optimisation, adaptive control, robust control, sensor placement, and sensor data validation, reconstruction, and forecasting in different applications such as water systems management or active noise control. He has co-authored over 45 publications including JCR-indexed journals, international conference papers, and book chapters, and participated in over 30 EU and national R&D projects in collaboration with multidisciplinary scientific teams in different countries. His current research area is related to supervision, fault diagnosis, and the optimisation of critical infrastructure systems such as water sanitation systems, with a special focus on trying to reduce the gap between theory and practice.

Vicenç Puig

Vicenç Puig is Full Professor at the Automatic Control Department of the Universitat Politècnica de Catalunya (UPC-BarcelonaTech) and a Researcher at the Institut de Robòtica i Informàtica Industrial (IRI), CSIC-UPC. He is the Director of the Automatic Control Department and Head of the research group on Advanced Control Systems (SAC) at UPC. He has made important scientific contributions in the areas of fault diagnosis and fault-tolerant control using interval and linear parameter-varying models adopting set-based approaches. He has also developed important scientific contributions in the areas of monitoring, control, and supervision of large-scale systems with particular applicability to systems in the water cycle (drinking water networks, sewage/drainage networks, irrigation networks, and waste treatment plants). He has participated in more than 20 European and national research projects in the last decade. He has also led many private contracts with several companies and has published more than 140 journal articles, as well as over 400 contributions in international conference/workshop proceedings. He has supervised over 25 PhD dissertations and over 50 master's theses/final projects. He is currently Chair of the IFAC Safeprocess TC Committee 6.4 and served as Vice Chair from 2016 to 2019. He was the general chair of the 3rd IEEE Conference on Control and Fault-Tolerant Systems (SysTol 2016) and IPC Chair of IFAC Safeprocess 2018.

sensors

MDPI

Editorial

Advances in the Monitoring, Diagnosis and Optimisation of Water Systems

Miquel Àngel Cugueró-Escofet [1,*] and Vicenç Puig [1,2]

1 Advanced Control Systems (SAC) Research Group, Polytechnic University of Catalonia (UPC-Barcelo-naTech), Terrassa Campus, Gaia Research Bldg, Rambla Sant Nebridi, 22, 08222 Terrassa, Spain
2 Institut de Robòtica i Informàtica Industrial (CSIC-UPC), 46 Llorens i Artigas Street, 08028 Barcelona, Spain
* Correspondence: miquel.angel.cuguero@upc.edu

In the context of global climate change, with the increasing frequency and severity of extreme events—such as draughts and floods—which will likely make water demand more uncertain and jeopardise its availability, those in charge of water system management face new operational challenges because of increasing resource scarcity, intensive energy requirements, growing populations (especially in urban areas), costly and ageing infrastructures, increasingly stringent regulations, and rising attention towards the environmental impact of water use. The shift from a linear to a circular economy and the need for a transition to a low-carbon production system represents an opportunity to address these emerging challenges related to water, energy, and the efficient use of resources. These challenges impel network managers to improve their methods and techniques for the monitoring, diagnosis, prognosis, supervision, and optimisation of the performance of water-related systems to adhere to the current sustainability agenda.

In this context, the increasing number of advanced installed sensors—and the corresponding increase in available data—allow for the implementation of Industry 4.0 (I4.0) techniques, which are strongly focused on interconnectivity, automation, artificial intelligence (AI), and real-time data acquisition, and will facilitate the development of intelligent tools to tackle such challenges. Within this framework, the successful implementation of I4.0 techniques in water-cycle-management facilities may prompt a breakthrough in improving the processes involved, drastically increasing their performance.

In this Special Issue, a selection of these techniques applied to the integral water cycle—i.e., water distribution and water sanitation—is introduced to address different current water-management challenges. These challenges may be classified as water-quantity challenges and water-quality challenges. On the water-distribution side, these challenges may include fault detection—namely, leak localisation—in water-distribution networks (WDNs), e.g., in [1], where a process prior to the actual leak localisation—i.e., sensor placement—is carried out using information-theory simulation-based methodology; or in [2], where a new data-driven method for leak location considering pressure measurements and network topological information is presented; or in [3], where simultaneous leak detection and isolation is applied to real data. All these methodologies contribute to reducing water loss due to leaks, which may account for up to 65% of the total water depending on the network [3] and, hence, impact water-quantity-management challenges. WDNs are also the focus in [4], where a challenge from the water-quality side—particularly, the water-disinfection process in water distribution—is addressed, providing a water-quality model by an online chlorine-decay-model calibration method, which has a strong impact on human health, since its correct concentration is paramount to ensure safe water disinfection.

Work presented in [5–8] discusses the water-sanitation side. In this field, there is a growing interest in the adaptation and use of technologies related to the circular economy which promote environmental sustainability, where resource recovery is a key issue for industrial and environmental processes and involves a wide spectrum of study possibilities.

Citation: Cugueró-Escofet, M.À.; Puig, V. Advances in the Monitoring, Diagnosis and Optimisation of Water Systems. *Sensors* **2023**, *23*, 3256. https://doi.org/10.3390/s23063256

Received: 1 March 2023
Revised: 8 March 2023
Accepted: 11 March 2023
Published: 20 March 2023

In water sanitation, wastewater treatment plants (WWTPs) offer a wide range of possibilities for resource recovery, mainly related to sludge-treatment processes such as biogas generation via the substrate codigestion process, which can be an alternative source for thermal and electrical energy production. This potential for biogas generation could become a source of renewable natural gas, which has specific composition requirements that demand high-tech sensors to assure its quality no matter its origin. Due to their potential for resource recovery and the further implications in the water–food–energy nexus, WWTPs have been a research focus in different areas of expertise: from modelling and engineering design to process dynamics, simulation, and integration. This line of work is introduced in [6], where resource recovery—namely biogas in the latter reference—is optimised by a centralised codigestion method considering real data from a WWTP network. Different nature-inspired optimisation algorithms are compared in the performance of this task, providing potential dramatic improvement when compared with actual nonoptimised operation. The improved operation of WWTP is also sought in [5,7] by means of improving the controllers involved in the operation of certain key processes of the WWTP, e.g., the aeration process of biological reactors. Classic proportional-integral (PI) controllers have been traditionally considered as the control strategy for such processes; however, improved performance may be achieved with more complex structures and techniques, e.g., model predictive-control (MPC) schemes or artificial neural network (ANN) approaches. In [5], an economic MPC (EMPC) considering a linear parameter-varying (LPV) model is proposed to control dissolved oxygen concentration in the WWTP biological reactors. Since the MPC technique requires a model of the process involved for its control, in the latter reference, a reduced model of the complex nonlinear plant is represented in a quasilinear parameter-varying (qLPV) form to reduce the computational burden—enabling the real-time operation—and applied in a real facility. This model, however, may be not available or may be difficult to obtain since the processes involved in the WWTP include nonlinear relations. ANN schemes may provide an alternative to this issue since they are well suited to deal with such processes. In this line of work, [7] considers transfer-learning (TL) methods to train ANN nets supporting control operations in WWTPs, and compares this approach with traditional control schemes, providing improved control performance while reducing control-design complexity and time invested in the ANN training process, which can be considerably time-demanding. Last but not least, in this Special Issue collection, a soft-sensing approach to predict key performance indicators (KPIs) in water-quality monitoring and control of WWTPs—such as effluent biochemical oxygen demand (BOD) or ammonia nitrogen (NH3-N)—is presented in [8]. Water-quality KPIs in WWTPs are traditionally subject to nonautomated lab-based offline monitoring approaches. Instead, in the latter reference, a method to perform accurate predictions of these KPIs, aiming for online operation, is introduced.

Further work in this area is included in the Special Issue, e.g., in [9], where remote sensing (RS) image-based time series are considered to obtain mass balances and estimate the unfiltered volumes in topographic depressions which are seasonally filled with water in a real area; or in [10], where a soil-moisture monitoring technique in precision agriculture—which is becoming key to providing food sustainably in the context of world's increasing population and natural resource scarcity—is provided using a low-cost wireless sensor network in order to help farmers optimise the irrigation process, and is tested in a real plot of land. Finally, a comprehensive review of AI and computer-vision methods for intelligent water monitoring—namely, water-body extraction and water-quality monitoring—using RS techniques is presented in [11], discussing the main challenges of using AI and RS for water-information extraction, as well as pointing out research priorities in this area. Hence, all the contributions in this Special Issue have an impact on the advances in the monitoring, diagnosis, and optimisation of water systems and, overall, cover a wide and complete sector of knowledge within this area.

Conflicts of Interest: The authors declare no conflict of interest.

References

1. Santos-Ruiz, I.; López-Estrada, F.R.; Puig, V.; Valencia-Palomo, G.; Hernández, H.R. Pressure Sensor Placement for Leak Localization in Water Distribution Networks Using Information Theory. *Sensors* **2022**, *22*, 443. [CrossRef] [PubMed]
2. Alves, D.; Blesa, J.; Duviella, E.; Rajaoarisoa, L. Robust data-driven leak localization in water distribution networks using pressure measurements and topological information. *Sensors* **2021**, *21*, 7551. [CrossRef] [PubMed]
3. Navarro-Díaz, A.; Delgado-Aguiñaga, J.A.; Begovich, O.; Besançon, G. Two simultaneous leak diagnosis in pipelines based on input–output numerical differentiation. *Sensors* **2021**, *21*, 8035. [CrossRef]
4. Pérez, R.; Martínez-Torrents, A.; Martínez, M.; Grau, S.; Vinardell, L.; Tomàs, R.; Martínez-Lladó, X.; Jubany, I. Chlorine Concentration Modelling and Supervision in Water Distribution Systems. *Sensors* **2022**, *22*, 5578. [CrossRef] [PubMed]
5. Nejjari, F.; Khoury, B.; Puig, V.; Quevedo, J.; Pascual, J.; de Campos, S. Economic Linear Parameter Varying Model Predictive Control of the Aeration System of a Wastewater Treatment Plant †. *Sensors* **2022**, *22*, 6008. [CrossRef] [PubMed]
6. Palma-Heredia, D.; Verdaguer, M.; Puig-Cayuela, V.; Poch, M.; Cugueró-Escofet, M.À. Comparison of optimisation algorithms for centralised anaerobic co-digestion in a real river basin case study in Catalonia. *Sensors* **2022**, *22*, 1857. [CrossRef] [PubMed]
7. Pisa, I.; Morell, A.; Vilanova, R.; Vicario, J.L. Transfer learning in wastewater treatment plant control design: From conventional to long short-term memory-based controllers. *Sensors* **2021**, *21*, 6315. [CrossRef] [PubMed]
8. Zhou, M.; Zhang, Y.; Wang, J.; Shi, Y.; Puig, V. Water Quality Indicator Interval Prediction in Wastewater Treatment Process Based on the Improved BES-LSSVM Algorithm. *Sensors* **2022**, *22*, 422. [CrossRef] [PubMed]
9. Fil, P.P.; Yurova, A.Y.; Dobrokhotov, A.; Kozlov, D. Estimation of infiltration volumes and rates in seasonally water-filled topographic depressions based on remote-sensing time series. *Sensors* **2021**, *21*, 7403. [CrossRef] [PubMed]
10. Lloret, J.; Sendra, S.; Garcia, L.; Jimenez, J.M. A wireless sensor network deployment for soil moisture monitoring in precision agriculture. *Sensors* **2021**, *21*, 7243. [CrossRef] [PubMed]
11. Yang, L.; Driscol, J.; Sarigai, S.; Wu, Q.; Lippitt, C.D.; Morgan, M. Towards Synoptic Water Monitoring Systems: A Review of AI Methods for Automating Water Body Detection and Water Quality Monitoring Using Remote Sensing. *Sensors* **2022**, *22*, 2416. [CrossRef] [PubMed]

sensors

MDPI

Article

Pressure Sensor Placement for Leak Localization in Water Distribution Networks Using Information Theory

Ildeberto Santos-Ruiz [1], Francisco-Ronay López-Estrada [1,*], Vicenç Puig [2], Guillermo Valencia-Palomo [3] and Héctor-Ricardo Hernández [1]

[1] TURIX-Dynamics Diagnosis and Control Group, Tecnológico Nacional de México, I.T. Tuxtla Gutiérrez, Carretera Panamericana km 1080 S/N, Tuxtla Gutierrez 29050, Mexico; ildeberto.dr@tuxtla.tecnm.mx (I.S.-R.); hector.hl@tuxtla.tecnm.mx (H.-R.H.)

[2] Institut de Robòtica i Informàtica Industrial, CSIC-UPC, Universitat Politècnica de Catalunya, C/. Llorens i Artigas 4-6, 08028 Barcelona, Spain; vicenc.puig@upc.edu

[3] Tecnológico Nacional de México, I.T. Hermosillo, Av. Tecnológico y Periférico Poniente S/N, Hermosillo 83170, Mexico; gvalencia@hermosillo.tecnm.mx

* Correspondence: frlopez@tuxtla.tecnm.mx; Tel.: +52-961-615-0461

Abstract: This paper presents a method for optimal pressure sensor placement in water distribution networks using information theory. The criterion for selecting the network nodes where to place the pressure sensors was that they provide the most useful information for locating leaks in the network. Considering that the node pressures measured by the sensors can be correlated (mutual information), a subset of sensor nodes in the network was chosen. The relevance of information was maximized, and information redundancy was minimized simultaneously. The selection of the nodes where to place the sensors was performed on datasets of pressure changes caused by multiple leak scenarios, which were synthetically generated by simulation using the EPANET software application. In order to select the optimal subset of nodes, the candidate nodes were ranked using a heuristic algorithm with quadratic computational cost, which made it time-efficient compared to other sensor placement algorithms. The sensor placement algorithm was implemented in MATLAB and tested on the Hanoi network. It was verified by exhaustive analysis that the selected nodes were the best combination to place the sensors and detect leaks.

Keywords: sensor placement; pressure monitoring; information theory; leak localization; water distribution network

Citation: Santos-Ruiz, I.; López-Estrada, F.-R.; Puig, V.; Valencia-Palomo, G.; Hernández, H.-R. Pressure Sensor Placement for Leak Localization in Water Distribution Networks Using Information Theory. *Sensors* **2022**, *22*, 443. https://doi.org/10.3390/s22020443

Academic Editor: Hossam A. Gabbar

Received: 12 December 2021
Accepted: 5 January 2022
Published: 7 January 2022

1. Introduction

Finding a suitable sensor placement is a fundamental problem for monitoring water distribution networks (WDNs) because it is impossible to install sensors at each point of the geographic area covered by the distribution system. A WDN comprises hundreds of nodes; however, only a few sensors can be installed in certain carefully selected nodes. Then, the main question is how to select the optimal sensor placement. Finding an answer to this problem is not trivial because the selected nodes must capture the most relevant information to estimate hydraulic variables at non-measured points and provide essential information for different supervision algorithms, e.g., for leak localization [1,2]. Often there are pressure and flow instruments at the supplying nodes of a WDN and in some cases at critical points (e.g., at the minimum pressure node). However, these measurements are not sufficient for an accurate leak localization, so additional sensors must be installed at other sites [3]. A practical solution is to install more pressure sensors, because they are cheaper and easier to install and maintain than flow sensors. In addition, node pressures are more sensitive to leaks than flow rates, which is why many localization algorithms are based primarily on pressure measurements. The problem of sensor placement is closely related to other WDN management problems, such as the state estimation of the network [4–6], model calibration [7,8], water quality monitoring such as detection of contaminants and cyberattacks [9–15], among others. Nevertheless, the present work focuses on the context of leak detection and localization as discussed in [16,17]. Regarding techniques for optimal sensor placement for leak/burst detection and localization in water distribution systems, a comprehensive review can be found at [18].

In a mathematical/computational context, the placement of pressure sensors is a mixed-integer programming problem. In this problem, for a network with N nodes, a sensor placement consists of a selection $[s_1, s_2, \ldots, s_N]$ where s_i are binary decision variables such that $s_i = 1$ indicates that a sensor will be placed on the i-th node, whereas $s_i = 0$ indicates that no sensor will be placed on that node.

Combinatorial analysis shows that there are $2^N - 1$ possible sensor placements when non-empty subsets with any number of sensors are considered. If the number of sensors is previously set to a fixed number S, then the number of possible sensor placements is reduced to $\binom{N}{S}$, which is still a very large number. Therefore, in medium-sized and large networks, it is not feasible to check all possible combinations. For example, in a network containing 500 nodes the number of different placements for 10 sensors is $\binom{500}{10} \approx 2.5 \times 10^{20}$. That is why it is important to find an optimal placement method without analyzing all the possible combinations.

Usually, sensor placement focused on leak localization is addressed with an optimization approach from synthetic pressure data obtained by simulation. Some authors have focused on minimizing the number of undetectable leaks [19,20], whereas others reduce the error in the leak location [16,21]. In [22], a min-max optimization algorithm that considers the isolation of the leaks from their signatures obtained through simulation is proposed. In [23], a multi-objective approach to mitigate errors both in the detection and localization of leaks, considering minimum night flow conditions, is presented. Regarding the optimization of the objective function, two approaches are usually used: deterministic methods (e.g., branch and bound [24]) and metaheuristic methods, (e.g., genetic algorithms [25–27] and particle swarm optimization [28]). Deterministic approaches guarantee an optimal solution, but the computation time increases exponentially with the number of nodes and possible leak scenarios. On the other hand, metaheuristic methods search for a near-optimal solution that only guarantees optimality when the number of candidate solutions evaluated (named "population size") tends to infinity. Furthermore, optimization-based sensor placement methods are linked to a specific leak localization method because the objective function is expressed in terms of a localization error or isolation index for that method [16,28,29]. Based on this, a sensor placement method may be optimal for one specific leak localization method but not as good for others. Furthermore, the method should be independent of the leak localization method since it is not feasible to change it for every method. Thus, an improved leak localization method could be proposed based on an ensemble of different machine learning algorithms using the information provided by the sensors.

The huge computing time in networks with hundreds and thousands of nodes using optimization-based methods and the high dependence on the selected leak localization method has motivated the present work. In this new proposal, it is not considered how specific leak localization methods will use the information provided by the sensors, but rather that the sensor placement method only focuses on the sensors capturing as much information related to the leaks as possible. The proposed method consists of a heuristic algorithm to select the subset of nodes where to place the sensors, seeking to maximize the relevance of the information captured by the sensors while minimizing the redundancy between the pressures in the selected nodes. Both metrics, relevance and redundancy, are defined in terms of information theory.

An important contribution of this work is the reduction in computing time for sensor placement, compared to methods based on metaheuristic optimization. Another relevant contribution is the nondependence of the sensor placement on the leak localization method used, which allows the use of the same sensor placement with different localization methods. Some aspects not yet covered in this work are the possible heterogeneity of the sensors (e.g., different errors and measurement ranges) and the influence of the measurement noise in the optimal placement, but they are considered as future work.

The rest of the document is organized as follows: in Section 2, the concepts of redundancy and relevance is presented in terms of mutual information, and the information quotient used as the basis of the method is also defined. In Section 3, the proposed method is formally described and some guidelines for its implementation are given. In Section 4, the results of the proposed method applied to a simplified version of the Hanoi network (case study) are presented. Finally, in Section 5, the conclusions are presented and future related works are proposed.

2. Information Theory Fundamentals

In Shannon's information theory (IT), the self-information of a random variable is defined according to the unexpectedness of its values [30]. Thus, the information contained in a constant random variable is zero. Mathematically, if an event E has probability P, its information content is defined by:

$$I(E) \stackrel{\text{def}}{=} -\log_b(P), \tag{1}$$

where the unit of measure of I is defined by the base of the logarithm, b, which is called "bit" if $b = 2$. In a discrete random variable X with probability function $p(x) = \Pr(X = x)$, the self-information for obtaining x as a result when measuring X is given by:

$$I(x) = -\log_b(p(x)) = \log_b(1/p(x)). \tag{2}$$

To quantify the average information that a random variable contains, considering all its possible values, the *entropy* is used:

$$H(X) \stackrel{\text{def}}{=} E(I(x)) = \sum_x -p(x)\log_b(p(x)), \tag{3}$$

which is the expected value of the information contained in the measurements of X, that is, the sum of the self-information of each of its possible values weighted by its probability of occurrence.

The mutual information of two random variables, sometimes called "information gain", measures the amount of information obtained from one of the random variables by observing the other one. For example, in a practical application of WDN monitoring, the mutual information between two node pressures would indicate how much information about the pressure at one node is gained by knowing the pressure at the other one. In probabilistic terms, the mutual information determines how different the joint distribution of (X, Y) is from the product of the marginal distributions of X and Y.

For two discrete variables X and Y, defined over the space $\mathcal{X} \times \mathcal{Y}$, the mutual information is computed as the double sum:

$$I(X,Y) = \sum_{y\in\mathcal{Y}}\sum_{x\in\mathcal{X}} p(x,y)\log\frac{p(x,y)}{p(x)p(y)}, \tag{4}$$

where $p(x,y) = \Pr(X = x, Y = y)$ is the joint probability function of X and Y, whereas $p(x)$ and $p(y)$ are the marginal probability functions of X and Y, respectively. The mutual information (4) is derived from the entropy and the conditional probability by the following equivalences:

$$I(X,Y) \equiv H(X) - H(X \mid Y) \equiv H(Y) - H(Y \mid X). \tag{5}$$

Furthermore, $I(X,X) = H(X)$, $I(X,Y) = I(Y,X)$ and $I(X,Y) \geq 0$, where $I(X,Y) = 0$ iff X and Y are independent.

For continuous random variables, the summations in (4) are replaced by integrals and the probability functions by probability densities:

$$I(X,Y) = \int_{\mathcal{Y}} \int_{\mathcal{X}} p(x,y) \log \frac{p(x,y)}{p(x)p(y)} \, dx \, dy. \tag{6}$$

Due to the difficulty in modeling the probability densities and subsequently evaluating the double integrals in (6), a simplification to calculate the mutual information in continuous variables is to discretize the variables with n bits, so that the domain of each variable is reduced to 2^n bins. For example, to compute the mutual information of two node pressures in a hydraulic network, the span of the pressure variables $[P_{\min}, P_{\max}]$ must be divided into a discrete 8-bit grid (256 different values) and then (4) is applied.

3. Sensor Placement Method

The proposed sensor placement method is based on a dataset of node pressures that collects typical variations due to leaks of different sizes in all network nodes. The pressure dataset is obtained from simulations with the hydraulic model of the network in [31]. Each pressure data point is labeled with a "leak class" (the node where the leak occurs) so that the proposed method can be classified as supervised.

In the context of machine learning, the placement of pressure sensors is a *feature selection* stage. To select the features (the subset of nodes where the sensors will be placed), an algorithm is proposed that seeks to maximize the relevance of the selected features (node pressures) for the response variable (leaky node), while each of them avoids capturing information already contributed by the others, that is, minimizing redundancy.

The following definitions of relevance and redundancy, proposed in [32], are used as a basis for defining the methodology:

Definition 1 (Relevance). *A metric of the relevance of the subset of node pressures $\mathcal{S}$ for the response variable y (leak node), is given by*

$$\mathrm{Rel}(\mathcal{S}) \stackrel{\text{def}}{=} \frac{1}{S} \sum_{x \in \mathcal{S}} I(x,y), \tag{7}$$

where x is any feature in $\mathcal{S}$, and $S = |\mathcal{S}|$ is the number of features in $\mathcal{S}$ (the cardinality).

Definition 2 (Redundancy). *A metric for information redundancy in a feature subset $\mathcal{S}$ is given by:*

$$\mathrm{Red}(\mathcal{S}) \stackrel{\text{def}}{=} \frac{1}{S^2} \sum_{x,x' \in \mathcal{S}} I(x,x'), \tag{8}$$

where x and x' are any features in $\mathcal{S}$.

To apply the above definitions to compute a pressure sensor placement, first, a dataset of node pressures is built covering different scenarios that consider leaks of different magnitude in all nodes of the network. Through simulation with the hydraulic model of the network, a series of samples of the node pressures is obtained, one sample for each different leakage scenarios. In this way, if M different leakage scenarios are simulated in a network containing N nodes, the result of the simulation is a collection of N M-dimensional vectors, x and x' in (7) and (8), corresponding to the N candidate nodes (initially, it is assumed that all nodes are potential sensing nodes). In addition, an output vector, y in (7), is generated containing integer labels to indicate the leaky node corresponding to each simulated scenario.

The exhaustive search for the optimal subset of sensors, $\mathcal{S}$, requires testing the $2^N - 1$ different combinations, which would require an impractical computation time in networks with many nodes. Therefore, the use of the method proposed in [32] was considered to rank the node pressures through an iterative forward scheme that only requires $O(NS)$

computations. In fact, with this proposal, it is possible to rank all the node pressures in order of importance with a computational cost of $O(N^2)$.

Next, a heuristic algorithm is proposed, which orders the node pressures according to their importance to explain the different leak classes (leaky nodes). The first node pressures in the output list correspond to the nodes with the highest importance for explaining the leak positions according to the information contained in the dataset. The sequential selection of nodes starts from an empty subset and, at each iteration, adds the best-ranked node among those that are still available to be selected. At each iteration, the relevance of each available feature (node pressure) with respect to the output (leaky node) and its redundancy with respect to the variables that have been previously selected is evaluated using the following equations, adapted from (7) and (8):

$$\mathrm{Rel}_y(x) = I(x,y), \tag{9}$$

$$\mathrm{Red}_{\mathcal{S}}(x) = \frac{1}{S}\sum_{x' \in \mathcal{S}} I(x,x'). \tag{10}$$

Since maximizing relevance and simultaneously minimizing redundancy represents a multiobjective problem, a combined relevance/redundancy index (RRI) is defined that increases with increasing relevance and also with decreasing redundancy, so the problem is expressed as a single objective to be maximized:

$$\mathrm{RRI} = \mathrm{Rel}_y(x) / \mathrm{Red}_{\mathcal{S}}(x). \tag{11}$$

The complete node ranking process is formally expressed in Algorithm 1. When the process finishes, the nodes where to place the sensors are taken from the first positions in the list $\mathcal{S}$. If it is not necessary to obtain the complete ranking of the nodes, but only to know the best-ranked positions, the process may stop prematurely when the subset $\mathcal{S}$ already contains the number of sensors to be placed.

Algorithm 1: Node ranking based on information theory.

Data: Set with all node pressures, $\mathcal{A}$. The nodes in $\mathcal{A}$ will be placed in the ordered list $\mathcal{S}$ according to their importance (relevance/redundancy). During the process, $\tilde{\mathcal{S}}$ denotes the elements of $\mathcal{A}$ not yet added in $\mathcal{S}$.
Result: Set with ordered node pressures, $\mathcal{S}$.
Initialization:
  $\mathcal{S} \leftarrow \left[\ \arg\max_{x \in \mathcal{A}} \mathrm{Rel}_y(x)\ \right]$
repeat
  if $\exists x \in \tilde{\mathcal{S}}, \mathrm{Rel}_y(x) \neq 0, \mathrm{Red}_{\mathcal{S}}(x) = 0$ **then**
    $\mathcal{S} \leftarrow \left[\mathcal{S},\ \arg\max_{x \in \tilde{\mathcal{S}}, \mathrm{Red}_{\mathcal{S}}(x)=0} \mathrm{Rel}_y(x)\right]$
  else
    break
  end
until $\forall x \in \tilde{\mathcal{S}}, \mathrm{Red}_{\mathcal{S}}(x) \neq 0$
repeat
  $\mathcal{S} \leftarrow \left[\mathcal{S},\ \arg\max_{x \in \tilde{\mathcal{S}}, \mathrm{Rel}_y(x) \neq 0} \mathrm{Rel}_y(x) / \mathrm{Red}_{\mathcal{S}}(x)\right]$
until $\forall x \in \tilde{\mathcal{S}}, \mathrm{Rel}_y(x) = 0$
$\mathcal{S} \leftarrow [\mathcal{S}, \tilde{\mathcal{S}}]$

The number of sensors to place for leak localization purposes is determined by the equipment available in most cases. The minimum number of sensors for a successful leak localization method will depend on how that method uses the available information, the measurement noise, as well as the quality, resolution and calibration of the sensors. If there are enough resources to intensively instrument the network, it must be taken into account that increasing the number of sensors does not always lead to better performance in locating leaks. To determine how many sensors should be placed, it is suggested to start from the ranking obtained by Algorithm 1, and run a marginal analysis with the leak localization method to be used. Starting from one sensor (the best ranked), the number of sensors is progressively increased and the leak localization performance is evaluated for each new set of sensors until adding a new sensor no longer represents a significant benefit for locating leaks.

It should be noted that Algorithm 1 does not take into account the geographical distribution of the nodes, since relevance and redundancy depend only on the mutual information between node pressures. This means that the network topology is what determines the amount of mutual information rather than the distance between sensors (i.e., two sensors can be geographically very close but have little mutual information).

4. Results and Discussion

Algorithm 1 was implemented in MATLAB and tested on the Hanoi network [33]. The model of the Hanoi network is composed of one reservoir, 31 consumer nodes, and 34 pipes, as shown in Figure 1. Due to its reduced topology, this network has been used as a standarized benchmark in different works [21,27,34].

In order to build the pressure dataset, leaks of different magnitude were simulated at each junction node using the EPANET 2 simulation program [35] through the EPANET/MATLAB Toolkit [36]. The procedure to generate the dataset using EPANET, the training and the predictive use of classifiers in locating leaks have been described in [37]. The dataset generated by simulation for this work considered leaks at all junction nodes with flow rates from 50 L/s. In order to simulate leaks at a node, the demand assigned to that node in the EPANET hydraulic model was modified by increasing this demand by an amount equal to the flow of the simulated leak. Because the Hanoi network contains few nodes, the optimality of the sensor placement calculated by Algorithm 1 was exhaustively verified.

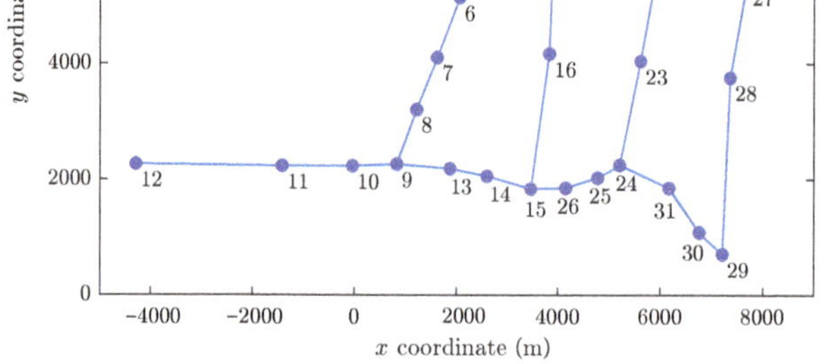

Figure 1. The Hanoi network.

To assess the optimality of the sensor placement obtained from Algorithm 1, leak localization tests were carried out using two machine learning methods that used the pressures in the selected nodes as features (input variables). The methods used were the *k*-nearest neighbors (*k*-NN) and quadratic discriminant analysis (QDA). These leak localization methods are based on classifiers that recognize directional patterns in pressure residuals using supervised learning techniques, as described in [38].

Through the marginal analysis, suggested at the end of Section 3, it was determined that $S = 3$ is an adequate number of sensors in the Hanoi network, because the addition of the fourth sensor does not produce a statistically significant improvement (with 0.95 confidence level) in leak location (considering that measurement noise may possibly increase the minimum number of sensors, but this discussion has been considered as future work). Because the Hanoi network contains few nodes, it was possible to comprehensively analyze all 4495 possible combinations of three sensor nodes. For each triplet of nodes (three-sensor placement), 50 leak localization tests were carried out with flow rates $q_{\text{leak}} = 1, 2, \ldots, 50$ L/s at each node of the network. Finally, the overall performance of both methods was evaluated for each candidate triplet using the classification accuracy (Acc) and the average topological distance (ATD) as performance metrics, as defined in [39]. The Acc is the fraction of exactly located leaks considering all leak scenarios in the test dataset, where Acc $= 1$ means that all leaks were correctly located, whereas Acc $= 0$ means that no leaks were correctly located. The ATD is a measure of how far from the true leaky node the classifier locates the leak, counting the number of separation links between the true leaky node and the estimated leaky node, averaged across all scenarios in the test dataset. Therefore, the best sensor placements are the ones that lead to the highest Acc values and the lowest ATD values.

The results in Table 1 show that the node triplet $\{12, 21, 28\}$ computed by Algorithm 1 is among the best ranked, since it presents the highest accuracy and the lowest average topological distance.

Table 1. Better positions to place three sensors in the Hanoi network, obtained by exhaustive analysis. The shaded selection is the one obtained by Algorithm 1.

Rank	Nodes	Location Method	
		k-NN	QDA
1	{12, 21, 28}	0.9974	0.9948
1	{12, 21, 27}	0.9974	0.9948
1	{12, 21, 31}	0.9974	0.9948
2	{7, 12, 21}	0.9961	0.9936
2	{12, 17, 21}	0.9961	0.9936
3	{3, 12, 21}	0.9961	0.9923
3	{4, 12, 21}	0.9961	0.9923
3	{6, 12, 21}	0.9961	0.9923
3	{5, 12, 21}	0.9961	0.9923

(a) Metric: classification accuracy

Rank	Nodes	Location Method	
		k-NN	QDA
1	{12,21,28}	0.0026	0.0052
1	{12, 21, 27}	0.0026	0.0052
1	{12, 21, 31}	0.0026	0.0052
2	{12, 13, 21}	0.0065	0.0065
3	{7, 12, 21}	0.0065	0.0090
3	{12, 17, 21}	0.0065	0.0090
4	{3, 12, 21}	0.0039	0.0129
4	{4, 12, 21}	0.0039	0.0129
5	{6, 12, 21}	0.0065	0.0129

(b) Metric: average topological distance

Figure 2 shows the geographic location of the three-sensor placement obtained considering the three nodes best ranked by Algorithm 1. Figure 3 shows the complete ranking considering the 31 nodes of the network.

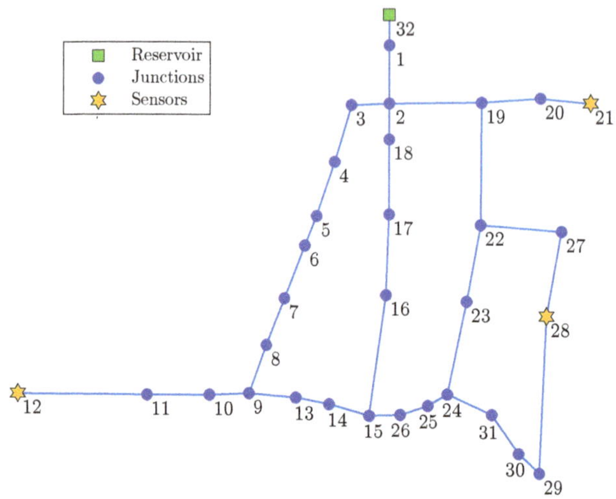

Figure 2. Computed three-sensor placement in the Hanoi network.

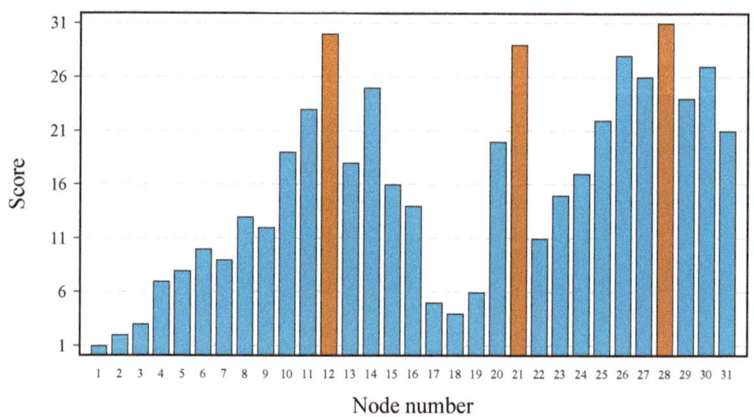

Figure 3. Node ranking in the Hanoi network.

Table 2 shows the sensor placements obtained for two, three and four sensors in the Hanoi network, and they are compared with the results obtained by metaheuristic methods reported in the literature [28]. The nodes selected by these methods are quite similar and produce very close results in terms of accuracy in locating leaks based on the pressures of the selected nodes. However, there is an important difference in the computation time of the IT-based method (Algorithm 1) compared with the metaheuristic methods. On a personal computer with an Intel 64-bit processor and 8 GB of RAM, the computation time for the IT-based method was around one second with the synthetic data from the Hanoi network, whereas it was 24 min for the genetic algorithm (it may be larger, depending on the initial population size) and about one hour for the exhaustive analysis.

Table 2. Optimal three-sensor placement in the Hanoi network using different methods.

S	IT [a]	GA [b]	PSO [c]	SE [d]
2	$\{12, 28\}$	$\{12, 21\}$	$\{12, 21\}$	$\{12, 21\}$
3	$\{12, 21, 28\}$	$\{12, 21, 27\}$	$\{12, 14, 21\}$	$\{12, 21, 29\}$
4	$\{12, 21, 26, 28\}$	$\{1, 12, 21, 29\}$	$\{1, 12, 21, 24\}$	$\{1, 12, 21, 29\}$

[a] Algorithm 1. [b] Genetic algorithm, reported in [28]. [c] Particle swarm optimization, reported in [28]. [d] Semi-exhaustive search, reported in [28].

Further tests were made on larger networks, e.g., in some midsize sectors of the Madrid network. Figure 4 shows a 10-sensor placement obtained using Algorithm 1 in a sector of the Madrid network containing one reservoir, 312 junction nodes and around 14 km of pipes. In this case, optimality was not exhaustively tested due to the vast number of possible placements to compare. However, it was found that the average accuracy in leak localization with sensor placements obtained by Algorithm 1 was at least better than that obtained with an existing placement (previously obtained by a genetic algorithm) for different leak scenarios.

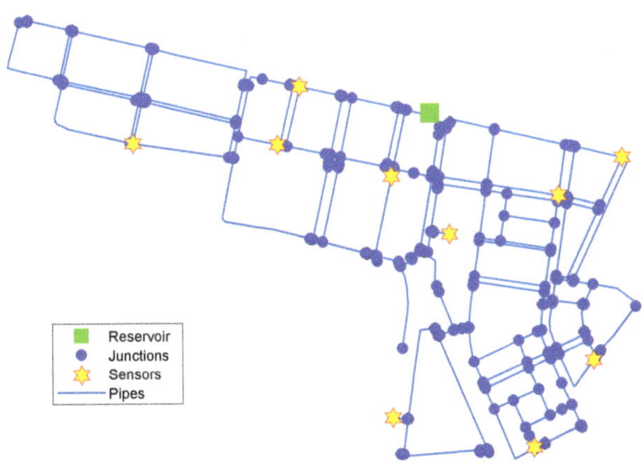

Figure 4. The optimal 10-sensor placement in a sector of the Madrid network.

Figures 2 and 4 show that the computed sensor placements do not show geometric regularity (i.e., the sensors do not appear equally spaced), since geometric or spatial criteria are not used to distribute the sensors in the network. However, regardless of geometric irregularity, leak location tests with these placements demonstrated that pressure measurements at these nodes provided the most useful information for discerning between different leak scenarios. In fact, when the placement of sensors obtained by Algorithm 1 is compared with the results reported by other authors using metaheuristics, sometimes very close performances can be found even though the sensors are distributed in different nodes, because the proposed algorithm does not optimize the position of each sensor individually but the entire set of sensors. This can be explained with an informal analogy: two soccer teams can achieve similar performances using different players.

Although, as noted above, there may be different sensor placements that lead to a good performance in locating leaks, the one obtained by Algorithm 1 has the advantage of being calculated in less time than the methods based on metaheuristics and that it is not linked to a specific leak location method, so changing the leak location method does not imply changing the location of the sensors, which would be impractical.

5. Conclusions

This paper has presented a technique for finding optimal sensor placements from information theory using a sequential forward selection, maximizing the relevance and minimizing the redundancy of the selected node subset. The proposed technique is computationally less expensive than other methods reported in the literature because the proposed technique operates directly on the values of node pressures without performing calculations for leak localization in the implementation of the algorithm. The optimality of the sensor placement obtained with the proposed method was extensively tested by simulation with the Hanoi network. It was found that the selection of nodes where to place sensors using information theory produced the best combination of pressure variables to locate leaks using different machine learning methods.

An implicit assumption in the proposed algorithms is that all network nodes have the same availability to place the sensors. However, in practice, some specific nodes may have placement priority over others; for example, critical nodes (points of minimum pressure) and nodes that supply essential services (e.g., hospitals) could be monitored as a priority. It may also occur that some nodes already have a sensor installed and that previous partial placement must be held, or that the conditions in a node are physically adverse and instrumentation is avoided. These circumstances warrant adjustments to the proposed sensor placement algorithm that may lead to future work. Another possible working line is the combination of heterogeneous sensors where different sensing specifications are included (e.g., different precision) or where the sensors measure different physical magnitudes (e.g., sensor placements combining pressure and flow sensors).

Author Contributions: Conceptualization, I.S.-R., F.-R.L.-E. and V.P.; methodology, I.S.-R.; software, I.S.-R.; validation, F.-R.L.-E., V.P. and G.V.-P.; formal analysis, G.V.-P. and H.-R.H.; data curation, I.S.-R. and H.-R.H.; writing—original draft preparation, I.S.-R.; writing—review and editing, F.-R.L.-E., V.P. and G.V.-P.; supervision, F.-R.L.-E. and V.P.; project administration, F.-R.L.-E. All authors have read and agreed to the published version of the manuscript.

Funding: This work was cofinanced by the European Regional Development Fund of the European Union in the framework of the ERDF Operational Program of Catalonia 2014–2020, under the research project 001-P-001643 Agrupació Looming Factory. Tecnológico Nacional de México (TecNM) also cofinanced this work by granting the research project 11080.21-P.

Conflicts of Interest: The authors declare no conflict of interest. The funders had no role in the design of the study; in the collection, analyses, or interpretation of data; in the writing of the manuscript, or in the decision to publish the results.

References

1. Sela, L.; Amin, S. Robust sensor placement for pipeline monitoring: Mixed integer and greedy optimization. *Adv. Eng. Inform.* **2018**, *36*, 55–63. [CrossRef]
2. Puig, V.; Ocampo-Martínez, C.; Pérez, R.; Cembrano, G.; Quevedo, J.; Escobet, T. (Eds.) *Real-Time Monitoring and Operational Control of Drinking-Water Systems*; Springer International Publishing: Berlin/Heidelberg, Germany, 2017. [CrossRef]
3. Pérez, R.; Sanz, G.; Puig, V.; Quevedo, J.; Cugueró, M.À.; Nejjari, F.; Meseguer, J.; Cembrano, G.; Mirats, J.M.; Sarrate, R. Leak localization in water networks: A model-based methodology using pressure sensors applied to a real network in Barcelona [applications of control]. *IEEE Control Syst. Mag.* **2014**, *34*, 24–36.
4. Kang, D.; Lansey, K. Optimal Meter Placement for Water Distribution System State Estimation. *J. Water Resour. Plan. Manag.* **2010**, *136*, 337–347. [CrossRef]
5. Santos-Ruiz, I.; López-Estrada, F.R.; Puig, V.; Blesa, J. Estimation of Node Pressures in Water Distribution Networks by Gaussian Process Regression. In Proceedings of the 4th Conference on Control and Fault Tolerant Systems (SysTol'19), Casablanca, Morocco, 18–20 September 2019; pp. 50–55. [CrossRef]
6. Mankad, J.; Natarajan, B.; Srinivasan, B. Integrated approach for optimal sensor placement and state estimation: A case study on water distribution networks. *ISA Trans.* 2021, *in press*. [CrossRef]
7. Simone, A.; Giustolisi, O.; Laucelli, D.B. A proposal of optimal sampling design using a modularity strategy. *Water Resour. Res.* **2016**, *52*, 6171–6185. [CrossRef]
8. Wéber, R.; Hős, C. Efficient Technique for Pipe Roughness Calibration and Sensor Placement for Water Distribution Systems. *J. Water Resour. Plan. Manag.* **2020**, *146*, 04019070. [CrossRef]
9. Kessler, A.; Ostfeld, A.; Sinai, G. Detecting Accidental Contaminations in Municipal Water Networks. *J. Water Resour. Plan. Manag.* **1998**, *124*, 192–198. [CrossRef]
10. Eliades, D.G.; Polycarpou, M.M.; Charalambous, B. A Security-Oriented Manual Quality Sampling Methodology for Water Systems. *Water Resour. Manag.* **2011**, *25*, 1219–1228. [CrossRef]

11. Vaidya, U.; Fardad, M. On optimal sensor placement for mitigation of vulnerabilities to cyber attacks in large-scale networks. In Proceedings of the 2013 European Control Conference (ECC), Zurich, Switzerland, 17–19 July 2013; pp. 3548–3553. [CrossRef]
12. Rathi, S.; Gupta, R.; Kamble, S.; Sargaonkar, A. Risk Based Analysis for Contamination Event Selection and Optimal Sensor Placement for Intermittent Water Distribution Network Security. *Water Resour. Manag.* **2016**, *30*, 2671–2685. [CrossRef]
13. Zeng, D.; Gu, L.; Lian, L.; Guo, S.; Yao, H.; Hu, J. On cost-efficient sensor placement for contaminant detection in water distribution systems. *IEEE Trans. Ind. Inform.* **2016**, *12*, 2177–2185. [CrossRef]
14. Hu, C.; Dai, L.; Yan, X.; Gong, W.; Liu, X.; Wang, L. Modified NSGA-III for sensor placement in water distribution system. *Inf. Sci.* **2020**, *509*, 488–500. [CrossRef]
15. Hooshmand, F.; Amerehi, F.; MirHassani, S.A. Risk-Based Models for Optimal Sensor Location Problems in Water Networks. *J. Water Resour. Plan. Manag.* **2020**, *146*, 04020086. [CrossRef]
16. Cugueró-Escofet, M.À.; Puig, V.; Quevedo, J. Optimal pressure sensor placement and assessment for leak location using a relaxed isolation index: Application to the Barcelona water network. *Control Eng. Pract.* **2017**, *63*, 1–12. [CrossRef]
17. Soldevila, A.; Blesa, J.; Fernandez-Canti, R.M.; Tornil-Sin, S.; Puig, V. Data-Driven Approach for Leak Localization in Water Distribution Networks Using Pressure Sensors and Spatial Interpolation. *Water* **2019**, *11*, 1500. [CrossRef]
18. Romano, M. Review of Techniques for Optimal Placement of Pressure and Flow Sensors for Leak/Burst Detection and Localisation in Water Distribution Systems. In *ICT for Smart Water Systems: Measurements and Data Science*; Scozzari, A., Mounce, S., Han, D., Soldovieri, F., Solomatine, D., Eds.; Springer International Publishing: Cham, Switzerland, 2021; pp. 27–63. [CrossRef]
19. Forconi, E.; Kapelan, Z.; Ferrante, M.; Mahmoud, H.; Capponi, C. Risk-based sensor placement methods for burst/leak detection in water distribution systems. *Water Sci. Technol. Water Supply* **2017**, *17*, 1663–1672. [CrossRef]
20. Venkateswaran, P.; Han, Q.; Eguchi, R.T.; Venkatasubramanian, N. Impact driven sensor placement for leak detection in community water networks. In Proceedings of the 2018 ACM/IEEE 9th International Conference on Cyber-Physical Systems (ICCPS), Porto, Portugal, 11–13 April 2018; pp. 77–87.
21. Casillas, M.; Puig, V.; Garza-Castañón, L.; Rosich, A. Optimal sensor placement for leak location in water distribution networks using genetic algorithms. *Sensors* **2013**, *13*, 14984–15005. [CrossRef]
22. Pérez, R.; Puig, V.; Pascual, J.; Quevedo, J.; Landeros, E.; Peralta, A. Methodology for leakage isolation using pressure sensitivity analysis in water distribution networks. *Control Eng. Pract.* **2011**, *19*, 1157–1167. [CrossRef]
23. Quiñones-Grueiro, M.; Verde, C.; Llanes-Santiago, O. Multi-objective sensor placement for leakage detection and localization in water distribution networks. In Proceedings of the 4th Conference on Control and Fault Tolerant Systems (SysToL'19), Casablanca, Morocco, 18–20 September 2019; pp. 129–134. [CrossRef]
24. Sarrate, R.; Blesa, J.; Nejjari, F.; Quevedo, J. Sensor placement for leak detection and location in water distribution networks. *Water Sci. Technol. Water Supply* **2014**, *14*, 795–803. [CrossRef]
25. Zhao, M.; Zhang, C.; Liu, H.; Fu, G.; Wang, Y. Optimal sensor placement for pipe burst detection in water distribution systems using cost–benefit analysis. *J. Hydroinform.* **2020**, *22*, 606–618. [CrossRef]
26. Cheng, W.; Chen, Y.; Xu, G. Optimizing Sensor Placement and Quantity for Pipe Burst Detection in a Water Distribution Network. *J. Water Resour. Plan. Manag.* **2020**, *146*, 04020088. [CrossRef]
27. Ponti, A.; Candelieri, A.; Archetti, F. A New Evolutionary Approach to Optimal Sensor Placement in Water Distribution Networks. *Water* **2021**, *13*, 1625. [CrossRef]
28. Casillas, M.V.; Garza-Castañón, L.E.; Puig, V. Optimal sensor placement for leak location in water distribution networks using evolutionary algorithms. *Water* **2015**, *7*, 6496–6515. [CrossRef]
29. Blesa, J.; Nejjari, F.; Sarrate, R. Robust sensor placement for leak location: Analysis and design. *J. Hydroinform.* **2016**, *18*, 136–148. [CrossRef]
30. Thomas M.; Cover, J.A.T. *Elements of Information Theory*, 2nd ed.; Wiley Series in Telecommunications and Signal Processing; Wiley-Interscience: Hoboken, NJ, USA, 2006.
31. Cooper, J.P.; Robinson, L. *Computer Modeling of Water Distribution Systems*, 4th ed.; Manual of Water Supply Practices—M32; American Water Works Association, AWWA: Denver, CO, USA, 2017.
32. Ding, C.; Peng, H. Minimum redundancy feature selection from microarray gene expression data. *J. Bioinform. Comput. Biol.* **2005**, *3*, 185–205. [CrossRef] [PubMed]
33. Fujiwara, O.; Khang, D.B. A two-phase decomposition method for optimal design of looped water distribution networks. *Water Resour. Res.* **1990**, *26*, 539–549. [CrossRef]
34. Alves, D.; Blesa, J.; Duviella, E.; Rajaoarisoa, L. Robust Data-Driven Leak Localization in Water Distribution Networks Using Pressure Measurements and Topological Information. *Sensors* **2021**, *21*, 7551. [CrossRef]
35. Rossman, L.A.; Woo, H.; Tryby, M.; Shang, F.; Janke, R.; Haxton, T. *EPANET 2.2 User Manual*; Technical Report EPA/600/R-20/133; U.S. Environmental Protection Agency: Washington, DC, USA, 2020.
36. Eliades, D.G.; Kyriakou, M.; Vrachimis, S.; Polycarpou, M.M. EPANET-MATLAB Toolkit: An Open-Source Software for Interfacing EPANET with MATLAB. In Proceedings of the 14th International Conference on Computing and Control for the Water Industry (CCWI), Amsterdam, The Netherlands, 7–9 November 2016; p. 8. [CrossRef]
37. Ferrandez-Gamot, L.; Busson, P.; Blesa, J.; Tornil-Sin, S.; Puig, V.; Duviella, E.; Soldevila, A. Leak localization in water distribution networks using pressure residuals and classifiers. *IFAC-PapersOnLine* **2015**, *48*, 220–225. [CrossRef]

38. Santos-Ruiz, I.; Blesa, J.; Puig, V.; López-Estrada, F. Leak localization in water distribution networks using classifiers with cosenoidal features. *IFAC-PapersOnLine* **2020**, *53*, 16697–16702. [CrossRef]
39. Soldevila, A.; Blesa, J.; Tornil-Sin, S.; Fernandez-Canti, R.M.; Puig, V. Sensor placement for classifier-based leak localization in water distribution networks using hybrid feature selection. *Comput. Chem. Eng.* **2018**, *108*, 152–162. [CrossRef]

sensors

MDPI

Article

Robust Data-Driven Leak Localization in Water Distribution Networks Using Pressure Measurements and Topological Information

Débora Alves [1,2,*], Joaquim Blesa [1,3,4,*], Eric Duviella [2] and Lala Rajaoarisoa [2]

1 Supervision, Safety and Automatic Control Research Center (CS2AC), Universitat Politècnica de Catalunya, Gaia Building, Rambla Sant Nebridi, 22, 08222 Terrassa , Spain
2 IMT Nord Europe, Université de Lille, CERI Digital Systems, F-59000 Lille, France; eric.duviella@imt-lille-douai.fr (E.D.); lala.rajaoarisoa@imt-lille-douai.fr (L.R.)
3 Institut de Robòtica i Informàtica Industrial (CSIC-UPC), Carrer Llorens Artigas, 4-6, 08028 Barcelona, Spain
4 Serra Húnter Fellow Automatic Control Department (ESAII), Universitat Politècnica de Catalunya (UPC), Pau Gargallo, 5, 08028 Barcelona, Spain
* Correspondence: debora.costa.da.silva@upc.edu (D.A.); joaquim.blesa@upc.edu (J.B.)

Abstract: This article presents a new data-driven method for locating leaks in water distribution networks (WDNs). It is triggered after a leak has been detected in the WDN. The proposed approach is based on the use of inlet pressure and flow measurements, other pressure measurements available at some selected inner nodes of the WDN, and the topological information of the network. A reduced-order model structure is used to calculate non-leak pressure estimations at sensed inner nodes. Residuals are generated using the comparison between these estimations and leak pressure measurements. In a leak scenario, it is possible to determine the relative incidence of a leak in a node by using the network topology and what it means to correlate the probable leaking nodes with the available residual information. Topological information and residual information can be integrated into a likelihood index used to determine the most probable leak node in the WDN at a given instant k or, through applying the Bayes' rule, in a time horizon. The likelihood index is based on a new incidence factor that considers the most probable path of water from reservoirs to pressure sensors and potential leak nodes. In addition, a pressure sensor validation method based on pressure residuals that allows the detection of sensor faults is proposed.

Keywords: water distribution networks; leak localization; data-driven

Citation: Alves, D.; Blesa, J.; Duviella, E.; Rajaoarisoa, L. Robust Data-Driven Leak Localization in Water Distribution Networks Using Pressure Measurements and Topological Information. *Sensors* **2021**, *21*, 7551. https://doi.org/10.3390/s21227551

Academic Editor: Jason K. Levy

Received: 8 September 2021
Accepted: 6 November 2021
Published: 13 November 2021

1. Introduction

Water distribution networks are complex systems that are difficult to manage and monitor with extreme importance nowadays. The detection and location of leaks have become crucial for water distribution because when there are bursts or leaks, this can generate not only economic losses but also an environmental issue and represents a potential risk to public health with contaminated water [1]. Another concern is the scarcity of water that can occur in 2025, which may affect half the world's population that will not have access to safe and accessible water for their basic needs [2]. However, with all these risks, currently, this infrastructure does not perform satisfactorily in practice. According to [3], a global volume of water loss called Non-Revenue Water (NRW) has been calculated at 346 million cubic meters per day or 126 billion cubic meters per year.

The infrastructure in a medium-sized city can have pipes that span hundreds of kilometers connected to hundreds of nodes (pipe junctions or customers that connect to the network). Therefore, several factors can generate water loss during transport between the treatment plants and the reservoir for consumers, usually attributed to several causes, including leaks, measurement errors, and theft. Water loss can be divided into two terms, "real losses" and "apparent losses". Apparent losses are constituted by badly read measurements, data handling errors, and illegal water tapping. In contrast, the real losses comprise leakage from all system parts and overflow at storage tanks. Real losses are divided into "background leakage" made up of small undetectable and into detectable leaks relevant for detection as they represent significant losses for the water distribution company.

Effective leak management is vital for all of the factors mentioned above to save financial resources and water. The methods of leak localization can be classified into two categories: Hardware-based system and Software-based.

The Hardware-based utilizes hardware sensors to detect a leak directly and help the localization of the leak. As there are various types of sensors and instruments available, they can be further subclassified as: acoustic [4,5] and non-acoustic detection methods [6].

Software-based methods generally rely on an algorithm or model for detecting leaks. Unlike hardware-based methods, these methods do not seek to locate the leak point accurately but minimize possible leakage areas. Since these methods are based on information, such as the pressure of the pipe network, flow data, and so forth, they work well on any type of pipe. These methods can be divided into physical modeling methods and data-driven methods. The physical modeling methods or model-based methods identify the leak using a numerical model and compare the results with the field data, for example, Ref. [7] which uses pressure sensitivity analysis, Ref. [8] uses leak signature space, Ref. [9] analyzes the sensitivity matrix and residuals, and [10] uses pressure and flow measurements to perform leakage detection through model-invalidation. On the other hand, data-driven methods analyze the monitoring data, combining tools such as artificial intelligence (e.g., classifiers [11–14] or artificial neural networks [15,16]). Thus, it is possible to identify potential areas of the leak based on certain rules or principles without resorting to the simulation of the physical model results [17]. However, these methods need, in general, an important number of non-leak and leak data scenarios in the training process to obtain reasonable results. As an exhaustive amount of leak scenarios are not available in general, a hydraulic simulator can be used to generate leak data. This work deals with the problem of leak localization and it is assumed that it is available a leak detection method that determines if a leak is present or not in the WDN. In particular, a non-numerical localization method, focused on a data-driven approach, is proposed.

Like other recent works [18,19], it requires only topological information of the network and historical data without leakage of the available measurements. In this work, the topological information provides the most probable paths for extra flows produced by leaks. A new incidence factor from every combination of nodes and sensors is computed with this information. Every incidence factor determines how a leak in a particular node affects a specific pressure sensor. On the other hand, historical data are used to calculate non-leak pressure estimations at sensed inner nodes. Residuals are generated using the comparison between these estimations and leak pressure measurements. Incidence factors are integrated with residuals in likelihood indexes to give the most probable leak node in a leak scenario. In addition, pressure residuals are used to detect sensor faults by means of a novel sensor validation algorithm.

The remainder of this paper is organized as follows: Section 2 presents the theory of graphs applied to WDN and explains the structure of the reduced-order model used in this work. The developed leak localization has been elaborated in Section 3. In Section 4 a sensor validation method that allows the detection of pressure sensor faults is presented. Section 5 introduces the case studies of Hanoi and Modena's WDNs. Section 6 presents the conclusions and future scope of the research work.

2. Water Distribution Networks

2.1. Preliminaries

A water distribution network is composed of m pipes, n internal consumer nodes and can be described by a directed graph $\mathcal{G} = \{\mathcal{V}, \mathcal{E}\}$, [20], with $\mathcal{V} = \{v_1, \ldots, v_n\}$ is the set of vertices that represent connections between the components of the network, additionally the last $\{v_{n-n_I+1}, \ldots, v_n\}$, represent the vertices of the system's input, n_I being the number of the inlets, with $n_I \geq 1$. The elements of the set $\mathcal{E} = \{e_1, \ldots, e_m\}$ are the edges, which represent the m pipes in the network.

The graph $\mathcal{G}$ can be represented by the incidence matrix $\mathbf{H} = [h_{ij}]$, in which the elements h_{ij} are defined as:

$$h_{ij} = \begin{cases} -1 & \text{if the } j\text{th edge is entering } i\text{th vertex.} \\ 0 & \text{if the } j\text{th edge is not connected to} \\ & \text{the } i\text{th vertex.} \\ 1 & \text{if the } j\text{th edge is leaving } i\text{th vertex.} \end{cases}$$

The direction of the edge represents a reference direction for the flow in the corresponding pipe. The incidence matrix is composed of $H \in \{-1, 0, 1\}^{n \times m}$ with each row corresponding to a node and column corresponding to a pipe.

The WDN must fulfill mass conservation law, which expresses the conservation of mass in each vertex, described by:

$$\mathbf{H} \cdot \mathbf{q} = \mathbf{d}, \tag{1}$$

where $\mathbf{d} \in \mathbb{R}^n$ is the vector of nodal demands, with $\mathbf{d}_i > 0$ when the flow is into the node i, and $\mathbf{q} \in \mathbb{R}^m$ is the vector of flows in the edges. By virtue of the mass conservation, it is possible to have only $n-1$ independent nodal demand, $\sum_{i=1}^{n} d_i = 0$, therefore the supply flow must equal the end-user demands as there is no storage in the network.

Let $\mathbf{p}$ be the vector of absolute pressures at the nodes and $\Delta\mathbf{p}$ be the vector of differential pressures across the pipes, both in meters of water column [mwc], then the energy law for water networks gives:

$$\Delta\mathbf{p} = \mathbf{H}^T\mathbf{p} = f(\mathbf{q}) - \mathbf{H}^T\mathbf{h}, \tag{2}$$

where $\mathbf{p} \in \mathbb{R}^n$, and $f : \mathbb{R}^m \to \mathbb{R}^m$, $f(\mathbf{q}) = (f_1(q_1), \ldots, f_m(q_m))$. The function $f_j(\cdot)$ describes the flow dependent pressure drop due to the hydraulic resistance in the jth edge. The relationship between pipe flow and energy loss caused by friction in individual pipes can be computed using the Hazen–Williams formula [21] for expression $f_j(\cdot)$:

$$f_j(q_j) = \frac{10.7 \cdot L_j}{\rho_j^{1.852} \cdot D_j^{4.87}} \cdot q_j^{1.852}, \tag{3}$$

where L_j is the length of the pipe and D_j is the diameter of the pipe, both in meters [m], q_j is the pipe flow in m^3/s and ρ_j is the pipe roughness coefficient.

The term $\mathbf{H}^T\mathbf{h}$ is the pressure drop across the pipes due to the difference in geodesic level (i.e., elevation) in meters [m] between the ends of the pipes with $\mathbf{h} \in \mathbb{R}^n$ the vector of geodesic levels at each vertex.

2.2. Structure of the Reduced Order Model

The reduced-order network model is used in this paper to calculate the nominal pressure at the measured internal nodes. The model uses the pressure dependence of the network's internal nodes with the pressure and flow measurements of the inlets. The details of the model derivation can be found in [22,23].

A network can be divided into nodes connected with reservoirs (the inlets nodes) and internal nodes that compose the system. To facilitate the explanation in this work, the information regarding inlet nodes will be represented by (r) superscript and those of the internal nodes, which will be expressed by the (i_n) superscript. In particular, vector $\mathbf{p}^{(in)}$ will contain pressure node values $p_1, \ldots, p_{n-n_I}$ and $\mathbf{p}^{(r)}$ inlet pressure values $p_{n+1-n_I}, \ldots, p_n$.

The network needs to fulfill some conditions for using the reduced model proposed:

Condition 1: corresponds to the demands of the internal nodes of the system, where Equation (1) can be defined as:

$$\mathbf{d}(k) = -\mathbf{v}(k)\sigma(k), \tag{4}$$

where $\sigma(k)$ denotes the total inlet flow into the network at time instant k, the vector $\mathbf{v}(k)$ defines the distribution of the total demand in the internal nodes at every time k, with the property $\sum_i^n v_i(k) = 1$. Notice that if all consumers are residential, all nodes demand have the same consumption profile, in consequence, the $\mathbf{v}(k)$ will be constant $\mathbf{v}(k) = \mathbf{v}$.

Condition 2: is a particularly case when the vector $\mathbf{p}^{(r)}$ of control inputs fulfill the following case,

$$\mathbf{p}^{(r)}(k) + \mathbf{h}^{(r)} = \kappa(k)\mathbb{1}, \tag{5}$$

for some $\kappa \in \mathbb{R}$, which is the total head at the inlets in $[\text{mwc}]$ and where $\mathbb{1}$ denote the vector consisting of ones. In [23], there is a discussion on this definition's feasibility where the controllers should satisfy this premise at least in networks with the low total consumption.

If these two conditions are fulfilled, the pressure at the ith internal node can be expressed by:

$$p_i^{(in)}(k) = \alpha_i\sigma^2(k) + \sum_{j=1}^{n_I} \beta_{ij}(k)p_j^{(r)}(k), \tag{6}$$

where α_i is parameter dependent on the network topology and the distribution of demands in the network, and β_{ij} is dependent on the network topology with $j = 1, \ldots, n_I$. The total inlet flow σ is typically well-known since inlet flows are measured. α_i is a parameter dependent on the network topology and the distribution of demands in the network, and β_{ij} is dependent on the network topology with $j = 1, \ldots, n_I$. The total inlet flow σ is typically well-known since inlet flows are measured.

Some methods of identifying parameters can be used to identify parameters α_i and β_{ij} since model (6) of $p_i^{(in)}$ is linear [24], using the measures of σ, $\mathbf{p}^{(r)}$ and $p_i^{(in)}$ with nodes that contain pressure sensors that will be denoted as p_{s_i} $\forall i = 1, \ldots, n_s$ in the following, where n_s is the number of sensors installed in the inner nodes.

Once inner pressure model (6) has been calibrated, the accuracy of the model can be assessed by applying the computation of the model error or pressure residual defined by:

$$r_{s_i} = \hat{p}_{s_i}(c) - p_{s_i}(c), \quad \forall i = 1, \ldots, n_s, \tag{7}$$

where c denotes the boundary conditions (heads and inflows in inlets) necessary to compute pressure estimation by means of (6). For example, minimum and maximum residual bounds $\underline{\sigma_i}$ and $\bar{\sigma}_i$, considering the available data, can be computed for every sensor $i = 1, \ldots, n_s$ to obtain an idea of the accuracy of model (6). Sensor noises and error models can produce residual errors. If big values of residual bounds are obtained, improvements in model (6) should be considered. For example, the assumption that all the nodes have the same consumption profile can lead to a big error in some networks. In this case, the error could be decreased if model (6) is calibrated only using data from the same hour but on different days. It would be assumed that different users can have different profiles at a given hour, but a particular user will have the same profile at a particular hour for all the different days. This possible improvement will imply the calibration of 24 different models (6) (one for each hour) and will require more historical data to obtain good accuracy. Another method to obtain an estimation of the pressure in inner nodes $\hat{p}_{s_i}(c)$ is to use historical data directly as a lookup table, as was proposed in [18]. That is, given particular operating conditions, c provides the inner pressures from historical data that had the closest operating conditions $\hat{c}$ to c. Residuals (7) considering leak pressure measurements will be used in the leak localization, as will be explained in detail in the next section.

3. Leak Localization

The location of the leak on the WDN is typically divided into two steps: leak detection and leak localization [25]. The focus in this work is the leak localization assuming that the detection has already been effectuated. In addition, it is assumed that the leaks can only happen in the nodes of the network (as considered in [7,26], or [8]), making the number of nodes equal to the number of potential leaks. The nodes correspond to water users, pipe junctions, and other structures such as hydrants. However, if the number of nodes will not provide a representative discretization of the network, some artificial nodes could be considered.

In this Section, two leak localization methods will be proposed. The first one will only use available measurements, and its diagnosis will point to one of the inner pressure sensors installed in the WDN. Therefore, the detected leak should be in an area around this sensor (cluster). The second method will combine the information of the first method with the topological information: characteristics of the pipes and connections between the nodes of the WDN, in a likelihood index that will allow the leak localization at the node level.

3.1. Leak Localization at Cluster Level

As stated before, the proposed leak localization is applied after the leak detection. In addition to inlet pressure and flow sensors, it is assumed that n_s pressure sensors are installed at different inner nodes. Consider a leak l_j acting on the node j of the network, and the used measurements are assumed to be captured under a leaky situation. Additionally, admitting leak-free historical data of all the sensors are available. The residual pressure in internal nodes that contain a sensor, defined in (7), can be computed as:

$$r_{s_i} = \hat{p}_{s_i}(c) - p_{s_i}(c^{l_j}), \quad \forall i = 1, \ldots, n_s, \tag{8}$$

where $\hat{p}_{s_i}(c)$ is the pressure estimation considering boundary conditions c in a leak-free scenario. On the other hand, $p_{s_i}(c^{l_j})$ is the pressure value measured by the inner pressure sensor i under boundary conditions c^{l_j} (the same heads and inflows in inlets as in c but with a leak in node j).

Following the ideas in [18], positive residuals can be obtained from the following transformation:

$$\bar{r}_{s_i} = r_{s_i} - \min(r_{s_1}, \ldots, r_{s_{ns}}) \quad \forall i = 1, \ldots, n_s. \tag{9}$$

Then, as the leak localization can be achieved by determining the residual pressure component with maximum size (see [22,27]), leak localization can be formulated as:

$$\hat{\jmath} = \underset{i \in \{1,\ldots,n_s\}}{\arg\max} \{\bar{r}_{s_i}\}. \tag{10}$$

Notice that the result of the leak localization method (10) is one of the n_s pressure sensor locations.

Then, the leak localization results in $\hat{\jmath}$ point not only to sensor location s_j but also to the nodes that produce a higher incidence for this sensor than the other sensors (cluster j).

3.2. Leak Localization at Node Level

Considering the Hazen–Willians Equation (3) for every pipe (edge e_z) a resistance R_z can be defined:

$$R_z = \frac{10.7 \cdot L_z}{\rho_z^{1.852} \cdot D_z^{4.87}}. \tag{11}$$

Among the multiple pipe paths that can connect every pair of nodes ij, a path $\mathcal{P}_{ij}^{min}$ with a minimum total resistance R_{ij} can be computed by means of :

$$\mathcal{P}_{ij}^{min} = \underset{\mathcal{P}_{ij}^{(k)} \in \mathcal{P}_{ij}}{\arg\min} \sum_{e_z \in \mathcal{P}_{ij}^{(k)}} R_z, \tag{12}$$

where $\mathcal{P}_{ij} = \{\mathcal{P}_{ij}^{(r)}, \ldots, \mathcal{P}_{ij}^{(e)}\}$ denotes the set of paths connecting nodes i and j.

On the other hand, the minimum path from the n_I inlets to a node j , $\mathcal{I}_j^{min}$, can be obtained by applying the computation of the minimum paths from the n_I inlets to node j by means of (12) and determine which is the one with the minimum resistance among the n_I paths.

When a leak is produced in node j, $\mathcal{I}_j^{min}$ is the most probable path for the extra flow produced by the leak. So the effect of a leak in node j to sensor s_i depends on the intersection of the paths from inlets to node j and the node where the sensor is located s_i: $\mathcal{I}_j^{min}$ and $\mathcal{I}_{s_i}^{min}$. To quantify the degree of incidence of the leak to the sensor, an incidence factor g_{j,s_i} is defined as:

$$g_{j,s_i} = R_{j,s_i}^c \bar{g}_{j,s_i}, \tag{13}$$

where R_{j,s_i}^c is the resistance of the path defined by $\mathcal{I}_j^{min} \cap \mathcal{I}_{s_i}^{min}$, the superscript c refers to the common path between node and sensors, and $\bar{g}_{j,s_i}$ is a normalization factor that takes into account the inverse of the resistance from the node j to the different sensors:

$$\bar{g}_{j,s_i} = \begin{cases} \dfrac{\frac{1}{R_{j,s_i}}}{\sum_{l=1}^{n_s} \frac{1}{R_{j,s_i}}} & \text{if } j \neq s_i \\ 1 & \text{if } j = s_i. \end{cases}$$

The n_s incidence factors associated to a leak in node j, g_{j,s_i} $i = 1, \ldots, n_s$ can be normalized:

$$\lambda_{j,s_i} = \frac{g_{j,s_i}}{\sum_{l=1}^{n_s} g_{j,s_l}}, \tag{14}$$

where coefficient λ_{j,s_i} determines the relative incidence of a leak in node j to sensor s_i regarding all the n_s sensors and the need to fulfill:

$$\sum_{i=1}^{n_s} \lambda_{j,s_i} = 1 \quad \forall j = 1, \ldots, n - n_I. \tag{15}$$

For every node $j = 1, \ldots, n - n_I$, the most sensitive sensor to a leak in this node can be computed as:

$$\hat{\jmath} = \underset{i \in \{1, \ldots, n_s\}}{\arg\max} \{\lambda_{j,s_i}\}. \tag{16}$$

The n_s clusters used in the leak localization defined in (10) can be computed using the set of nodes that provide the same value of $\hat{\jmath}$. The following equation is the definition of the cluster associated with the sensed node l:

$$C_l = \{v_j \in V | \underset{i \in 1, \ldots, n_s}{\arg\max} \{\lambda_{j,s_i}\} = l\}, \tag{17}$$

where $l = 1, \ldots, n_s$. The topological information of λ_{j,s_i} and the measurement information of residuals $\bar{r}_{s_i}$ can be integrated in a parameter θ_j defined as:

$$\theta_j = \frac{1}{\bar{\theta}} \sum_{i=1}^{n_s} \lambda_{j,s_i} \bar{r}_{s_i} \tag{18}$$

where $\bar{\theta}$ is a normalization factor. Then, θ_j can be interpreted as a likelihood index, and the leak localization at cluster level defined in (10) can be formulated at node level as:

$$\hat{\jmath} = \underset{j \in \{1,\dots,n-n_I\}}{\arg\max} \{\theta_j\}. \tag{19}$$

In order to improve the performance of the leak localization method, the information of the residuals at different time instants k can be taken into account applying the Bayes' rule as:

$$P_j(k) = \frac{P_j(k-1)\theta_j(k)}{\sum_{l=1}^{n-n_I} P_l(k-1)\theta_l(k)}, \tag{20}$$

where $P_j(k-1)$ is the prior probability whose initial value $P_j(k-1)$ has to be determined (for example $P_j(0) = 1/(n-n_I)$). Then, the leak node localization can be estimated by using posterior leak probabilities by:

$$\hat{\jmath}(k) = \underset{j \in \{1,\dots,n-n_I\}}{\arg\max} \{P_j(k)\}. \tag{21}$$

4. Sensor Validation

When a leak is not detected by the leak detection method, anomalous values of pressure residuals $r_{s_i}(k)$ $i = 1,\dots,n_s$ defined in (7) can be used to detect sensor faults. In the same operating conditions, the historical data of inner pressure sensors (leak-free data or data for a particular leak scenario) can be used first to calibrate a pressure estimation model as described in Section 2.2. Secondly, to determine residual bounds $\underline{\sigma_i}$ and $\bar{\sigma}_i$ that allows the implementation of pressure sensor fault detection through checking:

$$\begin{cases} r_{s_i}(k) \in [\underline{\sigma_i}, \bar{\sigma}_i] \Rightarrow \text{No Fault } (\phi_i(k) = 0) \\ r_{s_i}(k) \notin [\underline{\sigma_i}, \bar{\sigma}_i] \Rightarrow \text{Fault in sensor } s_i \ (\phi_i(k) = 1). \end{cases} \tag{22}$$

The accuracy of this fault detection method depends on the length of residual bounds $\underline{\sigma_i}$ and $\bar{\sigma}_i$ and, therefore, on the accuracy of pressure estimation (6). In order to increase the accuracy of the fault detection method, spatial residuals [28] between pressure residuals (7) can be computed

$$Sr_{s_i,s_j}(k) = r_{s_i}(k) - r_{s_j}(k) \quad \forall i = 1,\dots,n_s - 1 \quad \text{and} \quad j = i+1,\dots,n_s. \tag{23}$$

In the same way as the pressure residuals, spatial residual bounds $\underline{\varepsilon}_{i,j}$ and $\bar{\varepsilon}_{i,j}$ can be computed using leak-free data, and the fault detection can be implemented as follows:

$$\begin{cases} Sr_{s_i,s_j}(k) \in \left[\underline{\varepsilon}_{i,j}, \bar{\varepsilon}_{i,j}\right] \Rightarrow \text{No Fault}(\Phi_{i,j}(k) = 0) \\ Sr_{s_i,s_j}(k) \notin \left[\underline{\varepsilon}_{i,j}, \bar{\varepsilon}_{i,j}\right] \Rightarrow \text{Fault}(\Phi_{i,j}(k) = 1). \end{cases} \tag{24}$$

As model errors will affect in a similar way as close pressure sensors, it is expected that some spatial residual bounds will be smaller than pressure residual bounds. Therefore, fault detection defined by (24) will be more sensitive to pressure sensor faults than the one defined by (22). The accuracy of the sensor fault detection can be increased by means of average computing residuals in a time window leading to smaller residual bounds.

Once a residual has been violated, that is, at least one of the sensor faulty signals $\phi_i(k)$ $i = 1,\dots,n_s$ or spatial faulty signals $\Phi_{i,j}(k)$ $\quad i = 1,\dots,n_s - 1$ and $j = i+1,\dots,n_s$ is equal to one, the sensor fault isolation can be implemented in two stages as described in Algorithm 1:

Algorithm 1 Sensor validation search for sensor fault

Stage 1: In the case of the activation of one or more sensor faulty signals $\phi_i(k)\ i = 1,\ldots,n_s$, as these signals are uniquely related to sensors $s_i\ i = 1,\ldots,n_s$, the isolation is trivial and faulty sensors must be discarded for future leak localization, and the number of available healthy sensors n_s should be updated.

Stage 2: Only considers Spatial faulty signals $\Phi_{i,j}(k)$ of the n_s non-faulty sensors from *Stage*1. As these fault signals are potentially affected by two possible sensor faults s_i and s_j, the fault isolation can be implemented iteratively by the following steps:

```
for i ← 1, n_s − 1 do
  for j ← i + 1, n_s do
    if Φ_{i,j}(k) == 1 then
```

$$\hat{\imath} = \arg\max_{i\in\{1,\ldots,n_s\}} \{ \sum_{j=i+1}^{n_s-1} \Phi_{i,j}(k) + \sum_{j=1}^{i-1} \Phi_{j,i}(k) \}. \tag{25}$$

```
      Discard sensor s_î, eliminate faulty signals related to this sensors, update n_s.
    end if
  end for
end for
```

In the case that two or more sensors obtain the same cost function in (25) and less than the maximum possible value $n_s - 1$, the computation of (25) should be performed in a time window until new Spatial faulty signals are activated.

5. Case Study

5.1. Hanoi WDN

The Network used for the case study is a reduced city's network model from Hanoi's WDN (Vietnam). It is composed of one inlet (reservoir), 34 pipes, and 31 nodes, represented by Figure 1.

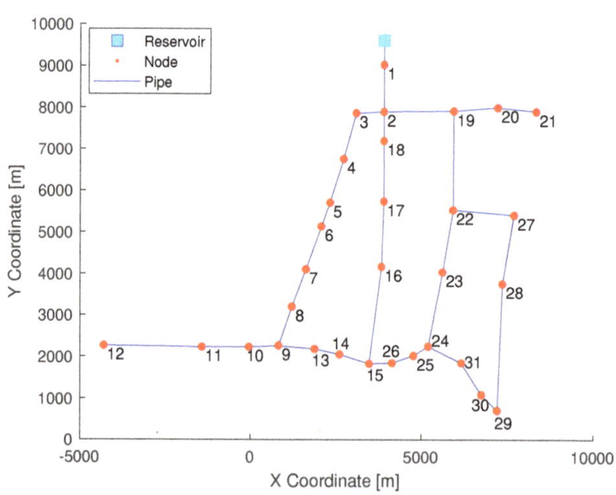

Figure 1. Simplified Hanoi topological WDN.

To analyze the performance of the proposed approach, data with different conditions have been generated artificially using the EPANET hydraulic simulator [29]. In order to consider realistic scenarios, some uncertainty has been added to the data [30]: the magnitude of the leak is random with a range of 25 to 75 [l/s], that is, between 1% and 2.5% of the average inlet flow of the WDN. Furthermore, white noise has been combined to emulate the noise present in real measurements, and uncertainty of 10% (uniform distribution) was added in the nominal demand value.

The daily water consumption pattern used for the calibration of Equation (6) is shown in Figure 2, having four days of operation.

The sample rate is 10 min, but average hourly measurements are calculated to reduce uncertainties on the diagnostic stage.

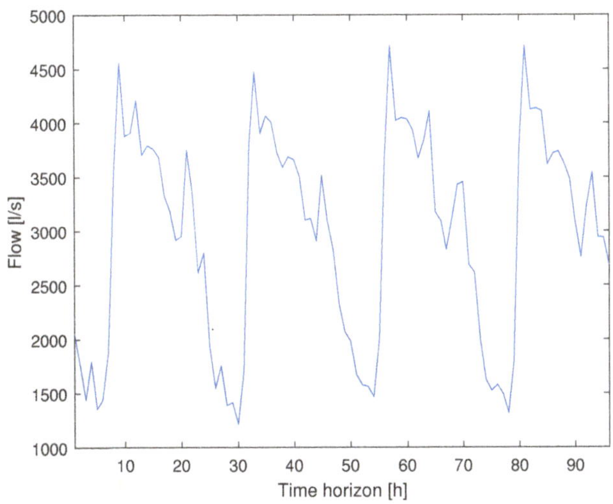

Figure 2. Flow consumption.

Results

The evaluation of the performance of the proposed leak localization method at node level defined in Equation (21) will be analyzed utilizing Average Topological Distance (ATD) [11]. The ATD represents the node's distance between the node predicted as leaking and the actual node with the leak. To calculate the ATD, it is first necessary to create a matrix containing the minimum topological distance (in nodes or meters), $A \in \mathbb{R}^{n-n_I \times n-n_I}$.

Finally, the confusion matrix $\Gamma_{i,j}(n - n_I \times n - n_I)$ defined in [18] and depicted in Table 1 is used to assess the performance of Equation (21). The rows of this matrix correspond to the leak scenario and the columns to where the leak is located ($\hat{l}$) by the leak localization method.

Table 1. Confusion matrix Γ.

	$\hat{l}_1$	$\cdots$	$\hat{l}_i$	$\cdots$	$\hat{l}_{n-n_I}$
l_1	$\Gamma_{1,1}$	$\cdots$	$\Gamma_{1,i}$	$\cdots$	$\Gamma_{1,n-n_I}$
$\vdots$	$\vdots$	$\vdots$	$\vdots$	$\vdots$	$\vdots$
l_i	$\Gamma_{i,1}$	$\cdots$	$\Gamma_{i,i}$	$\cdots$	$\Gamma_{i,n-n_I}$
$\vdots$	$\vdots$	$\vdots$	$\vdots$	$\vdots$	$\vdots$
l_{n-n_I}	$\Gamma_{n-n_I,1}$	$\cdots$	$\Gamma_{n-n_I,i}$	$\cdots$	$\Gamma_{n-n_I,n-n_I}$

Considering the confusion matrix Γ, the ATD can be computed as follows:

$$ATD = \frac{\sum_{i=1}^{n-n_I}\sum_{j=1}^{n-n_I}\Gamma_{i,j}A_{i,j}}{\sum_{i=1}^{n-n_I}\sum_{j=1}^{n-n_I}\Gamma_{i,j}}. \tag{26}$$

Four cases have been considered with different quantities of sensors in the network to analyze how this affects the final result. Table 2 presents the distribution of the selected nodes to contain a sensor. As seen in [31], the positioning of the sensors produces different results. As this work did not discuss the adequate sensors' arrangement, they were chosen to consider an improvement in the results regarding the number of sensors.

Table 2. Nodes with sensors.

Case	Nodes with Sensors
1	12, 17, 23, 29
2	6, 12, 17, 23, 29, 21
3	6, 12, 15, 17, 23, 21, 27, 30
4	6, 9, 12, 15, 17, 24, 21, 22, 28, 29, 31

Using the inlet flow data and non-leak historical pressure measurements of the selected sensors, the β_i and the α_i with $i = 1,\ldots,31$ in (6) have been identified (notice that $n_I = 1$). With these parameters, the pressure estimations under a non-leak condition in the network can be calculated considering inlet measurements using Equation (6) and posteriorly applied to calculate the residuals (9) with measured pressures in leak scenarios. In addition, non-leak pressure measurements and estimations are used to generate fault-free pressure residuals $r_{s_i}(k)$ and bounds $\underline{\sigma_i}$, $\bar{\sigma}_i$ $i = 1,\ldots,n_s$, as well as spatial residuals $Sr_{s_i,s_j}(k)$ and bounds $\underline{\varepsilon}_{i,j}$, $\bar{\varepsilon}_{i,j}$ $\forall i = 1,\ldots,n_s - 1$ and $j = i+1,\ldots,n_s$.

For every sensor configuration, normalized incidence factors (14) have been computed with topological information: node connections and pipe characteristics (length, diameter, and roughness). Figure 3 has the objective of comparing the information of the incidence of single leaks to pressure sensors obtained by a hydraulic model with the one obtained by means of topological information. The nodes selected to have sensors are the ones defined in the first case in Table 2. In particular, Figure 3c shows the clustering that groups the nodes that produce the highest effect in a specific pressure sensor. Nodes in yellow define the cluster of nodes where a leak produces a maximum pressure deviation from the non-leak scenario in the sensor installed in node 12, and the same is true for nodes in violet, red and green regarding pressure sensors in nodes 17, 23, and 29, respectively. Finally, nodes in black are nodes that produce a similar variation of pressure (difference of variation less than 0.1 [mwc]) in at least two different pressure sensors. In order to obtain this information, a hydraulic model to compute the difference of non-leak and leak pressures in all the nodes for the different leaks is required. On the other hand, Figure 3a shows the clustering that takes into account the shortest weighted pipe length (hydraulic distance), that is, the sum of $(L_z/D_z^{4.87})$ for all edge e_z in the path to the sensors, being the smallest one used to define the most resemblance to the sensor, used in Ref. [18]. Finally, the clustering depicted in Figure 3b is defined by Equation (17), which is based on the common resistance path explained in Section 3. These two last clusters that only require topological information could be used in the leak localization method at the cluster level defined in Equation (10). It is important to emphasize that the clustering based on the resistance common path, proposed in this paper and depicted in Figure 3b, resembles the clustering based on the actual leak effect in the network (given by the model) depicted in Figure 3c much more than the clustering based in the hydraulic distance depicted in Figure 3a. Therefore, the clustering proposed in this paper provides more accurate information for leak localization purposes than that based on the hydraulic distance. For example, as shown in Figure 3c, when a leak is present in nodes 3, 4, 5, 6, 7, 8 or 9, the sensor most affected by the leak is the sensor in node 12. This information is the same as the one provided by the clustering depicted in Figure 3b and is computed only with topological information. However, using the clustering of Figure 3a based on the hydraulic distance between nodes and sensors, the closest sensor to these nodes is the sensor in node 17.

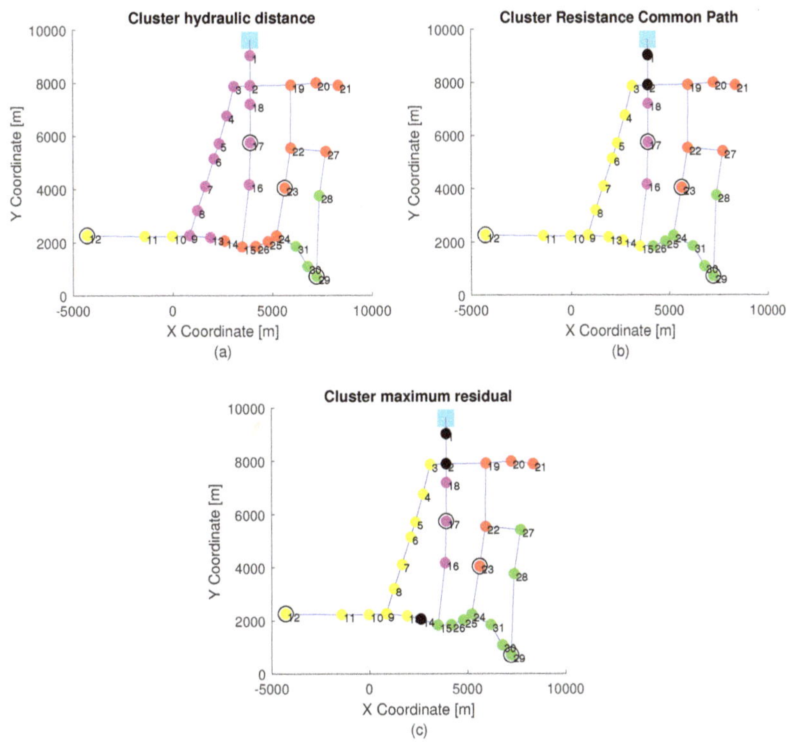

Figure 3. The n_s clustering generated with the aspects: (**a**) shortest weighted pipe length, (**b**) The resistance takes into account the common path R^c_{j,s_i}; (**c**) is the maximum residual. Nodes in yellow, violet, red and green define the cluster related to sensor installed in node 12, 17, 23, and 29, respectively.

Figure 4 shows the correlation analyses of the relative incidence index λ_{j,s_i} defined in Equation (14) for all the nodes $j = 1, \ldots, 31$ depicted in every subplot for every sensor s_i $i = 1, \ldots, 4$. As this index is normalized, its values are in the range [0,1). The nodes with the higher index (more brown color) are those that produce a higher effect in the pressure sensor s_i.

Figure 5a displays the evolution of the ATD (in nodes) obtained by the leak localization method based on the Kriging spatial interpolation methodology presented in [18] with the time horizon (in hours) used recursively by the Bayes' rule in (20). Four different sensor configurations are considered with 4, 6, 8, and 10 sensors placed optimally in order to maximize the performance of the leak localization proposed [18]. The performance can be compared with the one obtained by the new leak localization method proposed in this paper at node level defined in Equation (19) with the same dataset and the same sensor configurations as in [18], depicted in Figure 5b and with the sensor configurations shown in Table 2, depicted in Figure 5c.

Figure 5a shows that the leak detection performance of the Kriging method improves significantly from four to eight sensors and more moderately compared to ten sensors, still having a good result, even with noise data managing to reach an ATD equal to 2.5 node. When compared to the newly proposed leak localization method, as can be seen in Figure 5b,c, the new leak localization method always outperforms the Krigring method, even in the case of using the sensor configurations proposed in [18] that were computed to optimize the performance of the Kriging method. Figure 5c shows that the sensor configurations proposed in [18] are not optimal for the proposed method but the performance can be improved by changing the sensor configurations, in this case manually.

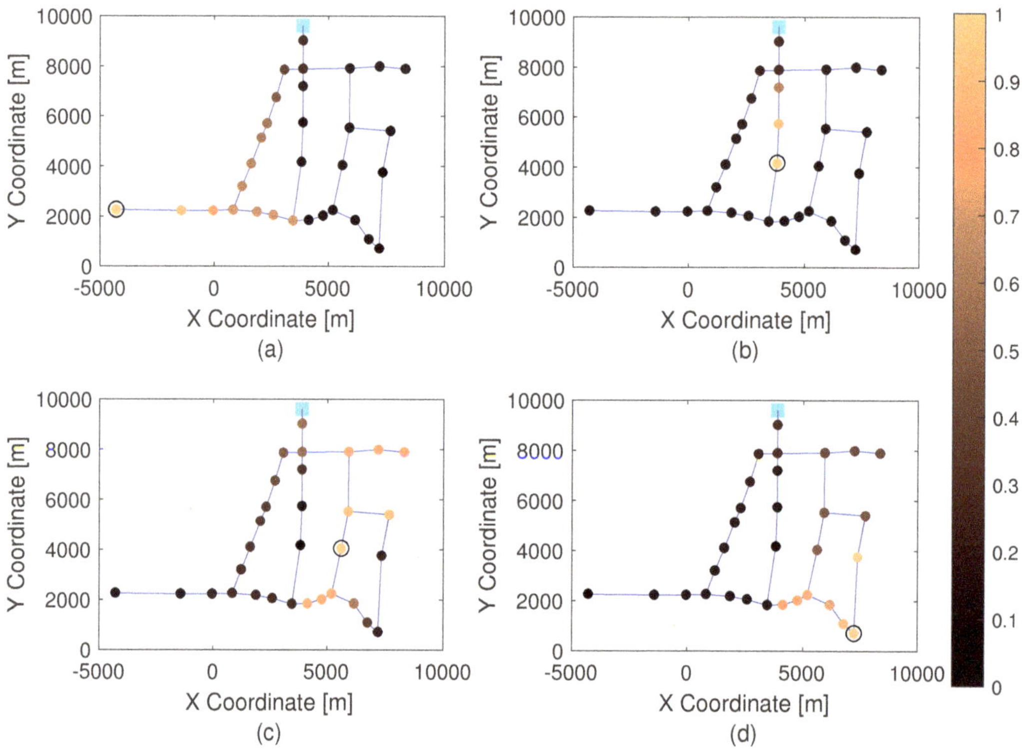

Figure 4. Relative incidence index λ_{j,s_i} for all the nodes ($j = 1, \ldots, 31$), corresponding to: (**a**) 1st sensor ($i = 1$), (**b**) 2nd senor ($i = 2$), (**c**) 3rd sensor ($i = 3$), and (**d**) 4th sensor ($i = 4$).

In order to illustrate the performance of the proposed sensor validation method, Case 1 (four sensors) will be considered. The four-sensor residuals computed by Equation (7) have been considered in a time window of 24 h using leak-free data leading to upper residual bounds equal to:

$$[\bar{\sigma}_1, \bar{\sigma}_2, \bar{\sigma}_3, \bar{\sigma}_4] = [0.11, 0.06, 0.09, 0.11]$$

and the lower residual bounds equal to:

$$[\underline{\sigma}_1, \underline{\sigma}_2, \underline{\sigma}_3, \underline{\sigma}_4] = [-0.14, -0.10, -0.10, -0.08].$$

In the same way, the six spatial residuals defined by (23) have been computed in the same conditions as sensor residuals leading to spatial residual bounds:

$$[\bar{\varepsilon}_{1,2}, \bar{\varepsilon}_{1,3}, \bar{\varepsilon}_{1,4}, \bar{\varepsilon}_{2,3}, \bar{\varepsilon}_{2,4}, \bar{\varepsilon}_{3,4}] = [0.06, 0.07, 0.08, 0.04, 0.05, 0.03]$$

and

$$[\underline{\varepsilon}_{1,2}, \underline{\varepsilon}_{1,3}, \underline{\varepsilon}_{1,4}, \underline{\varepsilon}_{2,3}, \underline{\varepsilon}_{2,4}, \underline{\varepsilon}_{3,4}] = [-0.04, -0.06, -0.08, -0.06, -0.09, -0.03].$$

Figures 6 and 7 depict the evolution of sensor and spatial residuals with their respective residual bounds in a fault scenario of sensor 1 that corresponds to the pressure sensor in node 12. The fault is a drift of 0.1 $[mcw]$ that starts on the 5th day. As shown in Figure 6, by applying (22) to sensor residuals it is impossible to detect the fault until the end of the day 9 (i.e., 4 days later) when residual sensor 1 violates the bounds. However, by applying (24) to spatial residuals it is possible to detect the fault in 10 h: Sr_{s_1,s_2} violates its bounds in 10 h, and Sr_{s_1,s_3}, Sr_{s_1,s_4} violate their bounds in 16 h and 22 h, respectively.

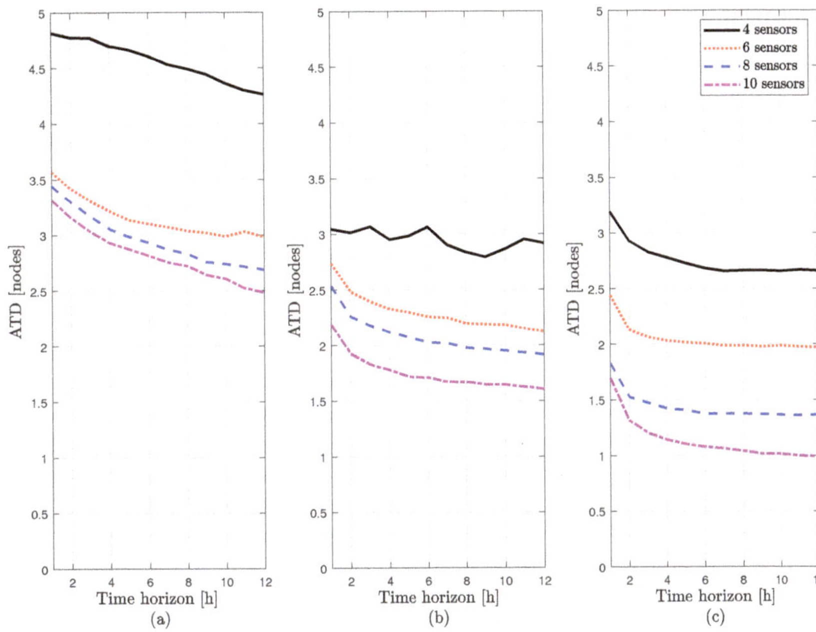

Figure 5. Evolution of the ATD between the methods: (**a**) using the Kriging interpolation method presented in [18], (**b**) using the new leak localization method with the same sensor configurations as in [18] and (**c**) using the new localization method with sensor configurations of Table 2.

5.2. Modena WDN

The second case study selected to test the performance is the reduced model of the real water distribution network of the Italian city Modena. This large-scale network is comprised of 268 junctions (nodes) connected through 317 pipes and served by four reservoirs. There are no pumps in the network since it is entirely gravity-fed [32,33].

The EPANET hydraulic simulator was used to generate artificial data to analyze the performance of the proposed method. The following simulation conditions were considered:

- The leak scenario consists of data samples collected every 10 min and filtered to hourly values to reduce the uncertainty in the data;
- The uncertainty of demand is considered by introducing the uncertainty of 10 [%] (normal distribution) of the nominal demand value. In addition, white noise is deemed to emulate the noise in the measurements;
- The leak size is randomly selected with a range of 3 to 6 [l/s], representing 1% to 2.5% of the network consumption.

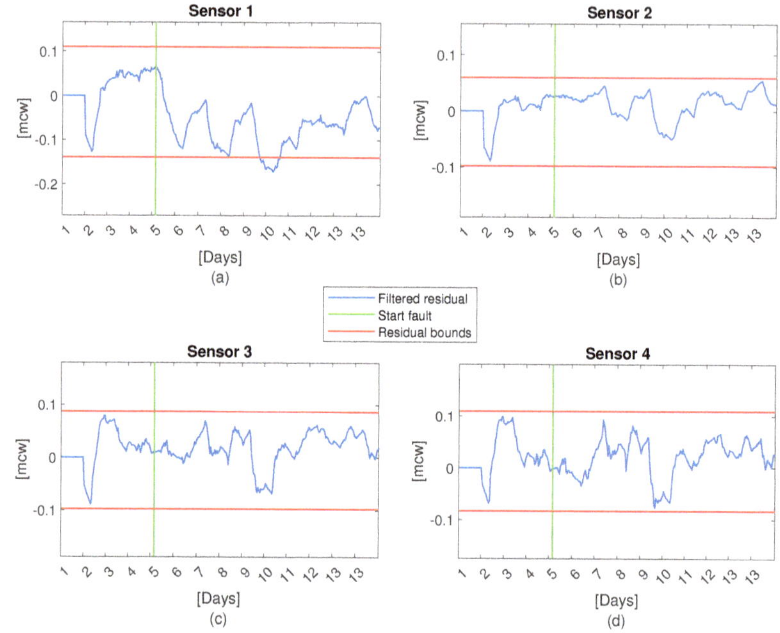

Figure 6. Graph of the filtered residual with a fault in sensor number 1 (**a**) 1st sensor, (**b**) 2nd senor, (**c**) 3rd sensor, and (**d**) 4th sensor.

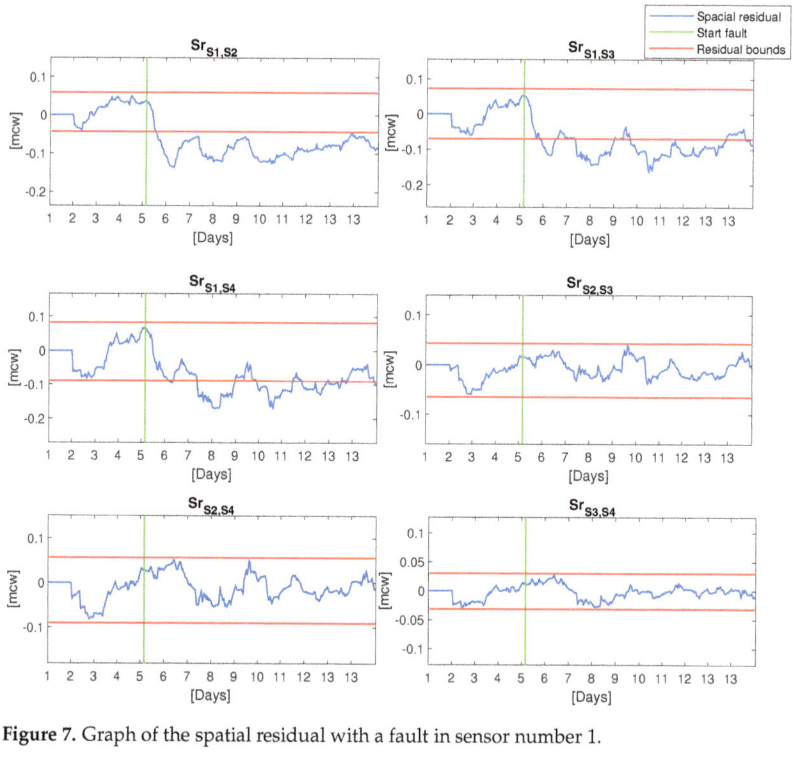

Figure 7. Graph of the spatial residual with a fault in sensor number 1.

The sensor bias, sensor drift, and abrupt sensor failure of sensor faults were proposed to analyze the sensor validation method. The sensor bias fault was simulated as a step change, and the drift fault was given as a time-varying ramp signal. In both cases, the fault magnitude was randomly chosen with a range of 0.1 to 0.2 [mwc]. The last fault was simulated by turning the sensor output to zero.

Results

As applied in the previous case study, the Average Topological Distance (ATD) was used to assess the performance of the proposed leak localization method at the node level defined in (19). Two scenarios have been considered with five and ten pressure sensors that are presented in Figure 8a,b respectively. As emphasized in the last section, performance in the leak localization task is highly dependent on the number of sensors installed in the network [34–36].

Figure 9 shows the result of ATD evolution as defined in (26) applied with Bayes' posterior time reasoning (20) to represent the leak location performance of the proposed method. This figure shows that the leak localization performance reached an ATD of 8 and 5.5 nodes with 5 and 10 inner pressure sensors installed in the network respectively. Considering that the proposed leak localization method only requires topological information and non-leak historical data in available measurements, the obtained performance is reasonably good.

On the other hand, a total of 6000 scenarios were simulated with 10 days each to evaluate the sensor validation method for the five sensor configurations depicted in Figure 8a. Thus, 1000 scenarios were generated for each sensor with sensor bias, sensor drift, and abrupt sensor failure applied randomly, and the remainder 1000 without faults.

To calculate the residual and spatial residuals bounds, a 6-month leak-free scenario was generated. The five sensor residuals computed by Equation (8) considering the time window of 24 h and increasing 24% observed bounds. Leading to upper residual bounds equal to:

$$[\bar{\sigma}_1, \bar{\sigma}_2, \bar{\sigma}_3, \bar{\sigma}_4, \bar{\sigma}_5] = [0.10, 0.06, 0.04, 0.01, 0.04]$$

and to lower residual bounds equal to:

$$[\underline{\sigma}_1, \underline{\sigma}_2, \underline{\sigma}_3, \underline{\sigma}_4, \underline{\sigma}_5] = [-0.08, -0.05, -0.03, -0.01, -0.06].$$

Following, the ten spatial residuals defined by (23) were computed in the same conditions as sensor residuals leading to spatial residual bounds:

$$\begin{aligned}[\bar{\varepsilon}_{1,2}, \quad & \bar{\varepsilon}_{1,3}, \bar{\varepsilon}_{1,4}, \bar{\varepsilon}_{1,5}, \bar{\varepsilon}_{2,3}, \bar{\varepsilon}_{2,4}, \bar{\varepsilon}_{2,5}, \bar{\varepsilon}_{3,4}, \bar{\varepsilon}_{3,5}, \bar{\varepsilon}_{4,5}] = \\ & [0.07, 0.08, 0.08, 0.10, 0.06, 0.05, 0.06, 0.03, 0.06, 0.05]\end{aligned}$$

and

$$\begin{aligned}[\underline{\varepsilon}_{1,2}, \quad & \underline{\varepsilon}_{1,3}, \underline{\varepsilon}_{1,4}, \underline{\varepsilon}_{1,5}, \underline{\varepsilon}_{2,3}, \underline{\varepsilon}_{2,4}, \underline{\varepsilon}_{2,5}, \underline{\varepsilon}_{3,4}, \underline{\varepsilon}_{3,5}, \underline{\varepsilon}_{4,5}] = \\ & [-0.07, -0.09, -0.08, -0.09, -0.05, -0.04, -0.06, -0.03, -0.04, -0.04].\end{aligned}$$

For this study, the evaluation metric applied was classification accuracy. To this purpose, the confusion matrix was used, which presents the classification accuracy and the misclassification error, and the horizontal axis of the confusion matrix describes the predicted labels of samples, while the longitudinal axis depicts the true labels of samples. The right side shows the percentages of correctly and incorrectly classified observations for each true class.

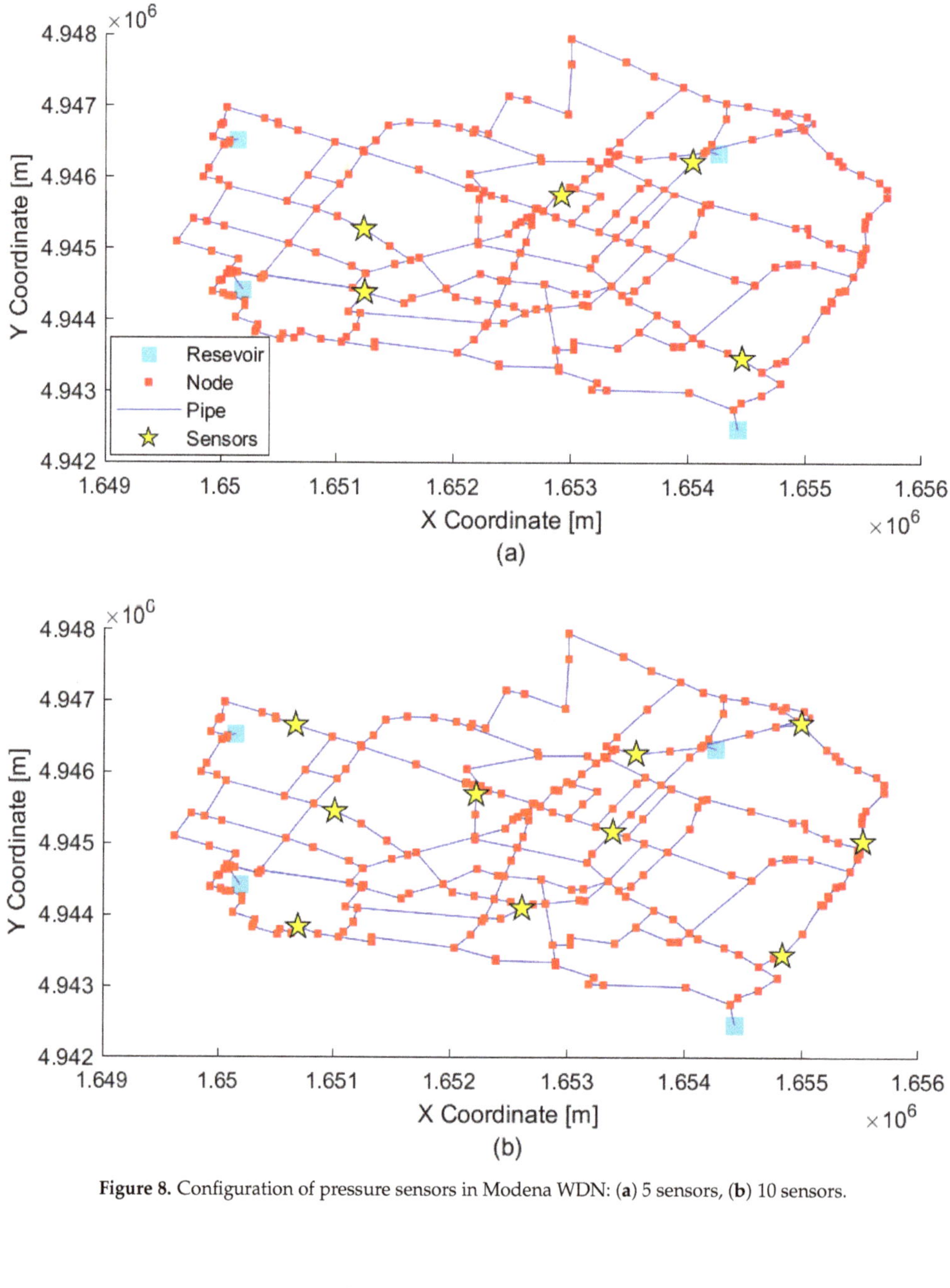

Figure 8. Configuration of pressure sensors in Modena WDN: (**a**) 5 sensors, (**b**) 10 sensors.

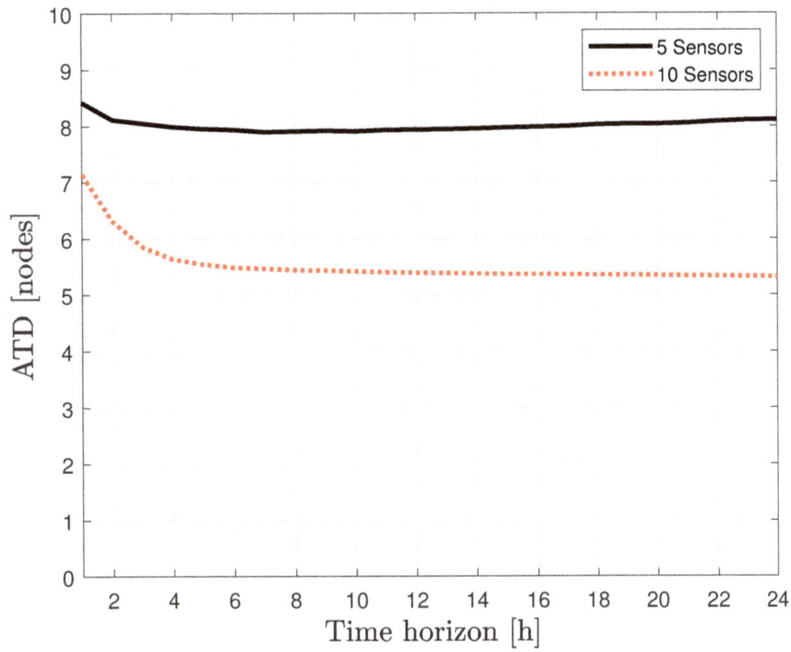

Figure 9. Evolution of the ATD.

Figure 10 illustrates the result for the confusion matrix for all scenarios generated, and depicts that the accuracy of detecting faults in the sensor is very high, where the lowest accuracy is presented in fault sensor number five with an accuracy of 95.4% and the highest in fault sensor number three with 100% accuracy. Regarding the accuracy of the scenario with no-fault, eight of the 1000 fault free scenarios presented one false alarm among the 240 samples of the scenario; therefore providing an average interval between false detections of 240,000/8 = 30,000 h.

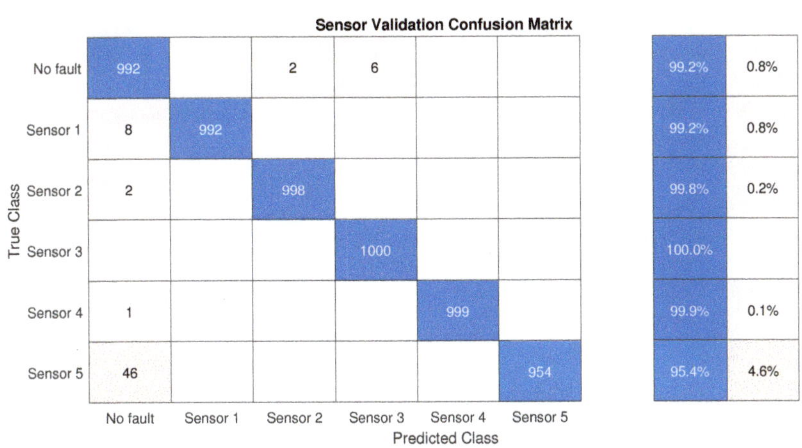

True Class \ Predicted Class	No fault	Sensor 1	Sensor 2	Sensor 3	Sensor 4	Sensor 5		
No fault	992		2	6			99.2%	0.8%
Sensor 1	8	992					99.2%	0.8%
Sensor 2	2		998				99.8%	0.2%
Sensor 3				1000			100.0%	
Sensor 4	1				999		99.9%	0.1%
Sensor 5	46					954	95.4%	4.6%

Figure 10. Confusion matrix for sensor validation method.

6. Conclusions

A new data-driven method for leak localization in WDN based on historical non-leak data and the topological information of the network is proposed. The proposed method is triggered when a leak is detected, and it is based on the evaluation of residuals generated by leak pressure measurements in some inner nodes and the estimation of leak-free pressures in these nodes utilizing a reduced-order model and historical data. Topological information is used to compute a new incidence factor that considers the most probable path of water from reservoirs to pressure sensors and potential leak nodes. The proposed incidence factor combined with residual information generates a likelihood index that allows leak localization at the node level. In addition, a sensor validation method based on the sensor pressure residuals, which is able to detect and isolate pressure sensor faults, is proposed.

The proposed method's general performance for leak location and sensor validation is evaluated in reduced models of the Hanoi and Modena water distribution networks. The results of the leak localization are compared to another technique published with satisfactory results. Future works can be developed to improve the leak localization and sensor validation performances, with a study of an algorithm to automatically determine the optimal sensors required to maximize the leak localization performance.

Author Contributions: Conceptualization, D.A., J.B., E.D. and L.R.; methodology, D.A., J.B., E.D. and L.R.; software, D.A.; validation, D.A. and J.B.; formal analysis, D.A., J.B., E.D. and L.R.; investigation, D.A., J.B., E.D. and L.R.; data curation, D.A.; writing—original draft preparation, D.A.; writing—review and editing, D.A., J.B., E.D. and L.R.; visualization, D.A. and J.B.; supervision, J.B., E.D. and L.R.; project administration, J.B., E.D. and L.R. All authors have read and agreed to the published version of the manuscript.

Funding: This work has been partially funded by SMART Project (ref.num. EFA153/16 Interreg Cooperation Program POCTEFA 2014-2020), L-BEST Project (PID2020-115905RB-C21) funded by MCIN/ AEI /10.13039/501100011033 and AGAUR ACCIO RIS3CAT UTILITIES 4.0–P1 ACTIV 4.0. ref.COMRDI-16-1-0054-03.

Institutional Review Board Statement: Not applicable.

Informed Consent Statement: Not applicable.

Data Availability Statement: The data can be found in https://github.com/adeboracris/Epanet_Leak_localization (accessed on 10 November 2021).

Conflicts of Interest: The authors declare no conflict of interest.

References

1. Puust, R.; Kapelan, Z.; Savic, D.; Koppel, T. A review of methods for leakage management in pipe networks. *Urban Water J.* **2010**, *7*, 25–45. [CrossRef]
2. Gosling, S.N.; Arnell, N.W. A global assessment of the impact of climate change on water scarcity. *Clim. Chang.* **2016**, *134*, 371–385. [CrossRef]
3. Liemberger, R.; Wyatt, A. Quantifying the global non-revenue water problem. *Water Supply* **2019**,*19*, 831–837. [CrossRef]
4. Muggleton, J.; Brennan, M.; Pinnington, R.; Gao, Y. A novel sensor for measuring the acoustic pressure in buried plastic water pipes. *J. Sound Vib.* **2006**, *295*, 1085–1098. [CrossRef]
5. Shimanskiy, S.; Iijima, T.; Naoi, Y. Development of microphone leak detection technology on Fugen NPP. *Prog. Nucl. Energy* **2003**, *43*, 357–364. [CrossRef]
6. Hunaidi, O.; Chu, W.; Wang, A.; Guan, W. Detecting leaks in plastic pipes. *J.-Am. Water Work. Assoc.* **2000**, *92*, 82–94. [CrossRef]
7. Pérez, R.; Puig, V.; Pascual, J.; Quevedo, J.; Landeros, E.; Peralta, A. Methodology for leakage isolation using pressure sensitivity analysis in water distribution networks. *Control Eng. Pract.* **2011**, *19*, 1157–1167. [CrossRef]
8. Casillas, M.V.; Garza-Castañón, L.E.; Puig, V.; Vargas-Martinez, A. Leak signature space: An original representation for robust leak location in water distribution networks. *Water* **2015**, *7*, 1129–1148. [CrossRef]
9. Pérez, R.; Cugueró, J.; Sanz, G.; Cugueró, M.A.; Blesa, J. Leak Monitoring. In *Real-Time Monitoring and Operational Control of Drinking-Water Systems*; Puig, V., Ocampo-Martínez, C., Pérez, R., Cembrano, G., Quevedo, J., Escobet, T., Eds.; Springer International Publishing: Cham, Switzerland, 2017; pp. 115–130.
10. Vrachimis, S.G.; Timotheou, S.; Eliades, D.G.; Polycarpou, M.M. Leakage detection and localization in water distribution systems: A model invalidation approach. *Control Eng. Pract.* **2021**, *110*, 104755. [CrossRef]

11. Soldevila, A.; Blesa, J.; Tornil-Sin, S.; Duviella, E.; Fernandez-Canti, R.M.; Puig, V. Leak localization in water distribution networks using a mixed model-based/data-driven approach. *Control Eng. Pract.* **2016**, *55*, 162 – 173. [CrossRef]
12. Quiñones-Grueiro, M.; Bernal-de Lázaro, J.M.; Verde, C.; Prieto-Moreno, A.; Llanes-Santiago, O. Comparison of Classifiers for Leak Location in Water Distribution Networks. *IFAC-PapersOnLine* **2018**, *51*, 407–413. [CrossRef]
13. Sun, C.; Parellada, B.; Puig, V.; Cembrano, G. Leak localization in water distribution networks using pressure and data-driven classifier approach. *Water* **2020**, *12*, 54. [CrossRef]
14. Santos-Ruiz, I.; Blesa, J.; Puig, V.; López-Estrada, F. Leak localization in water distribution networks using classifiers with cosenoidal features. *IFAC-PapersOnLine* **2020**, *53*, 16697–16702. [CrossRef]
15. Javadiha, M.; Blesa, J.; Soldevila, A.; Puig, V. Leak Localization in Water Distribution Networks using Deep Learning. In Proceedings of the 2019 6th International Conference on Control, Decision and Information Technologies (CoDIT), Paris, France, 23–26 April 2019; pp. 1426–1431. [CrossRef]
16. Pérez-Pérez, E.; López-Estrada, F.; Valencia-Palomo, G.; Torres, L.; Puig, V.; Mina-Antonio, J. Leak diagnosis in pipelines using a combined artificial neural network approach. *Control Eng. Pract.* **2021**, *107*, 104677. [CrossRef]
17. Wu, Y.; Liu, S. A review of data-driven approaches for burst detection in water distribution systems. *Urban Water J.* **2017**, *14*, 972–983. [CrossRef]
18. Soldevila, A.; Blesa, J.; Fernandez-Canti, R.M.; Tornil-Sin, S.; Puig, V. Data-driven approach for leak localization in water distribution networks using pressure sensors and spatial interpolation. *Water* **2019**, *11*, 1500. [CrossRef]
19. Soldevila, A.; Blesa, J.; Jensen, T.N.; Tornil-Sin, S.; Fernández-Cantí, R.M.; Puig, V. Leak Localization Method for Water-Distribution Networks Using a Data-Driven Model and Dempster–Shafer Reasoning. *IEEE Trans. Control Syst. Technol.* **2021**, *29*, 937–948. [CrossRef]
20. Deo, N. *Graph Theory with Applications to Engineering and Computer Science*, 1st ed.; Dover Books on Mathematics: Mineola, NY, USA, 2016; ISBN 13 978-0486807935.
21. Pérez, R.; Sanz, G. Modelling and simulation of drinking-water networks. In *Real-time Monitoring and Operational Control of Drinking-Water Systems*; Springer: Berlin/Heidelberg, Germany, 2017; pp. 37–52.
22. Jensen, T.N.; Kallesøe, C.S. Application of a novel leakage detection framework for municipal water supply on aau water supply lab. In Proceedings of the IEEE 2016 3rd Conference on Control and Fault-Tolerant Systems (SysTol), Barcelona, Spain, 7–9 September 2016; pp. 428–433.
23. Jense, T.N.; Kallesøe, C.S.; Bendtse, J.D.; Wisniewsk, R. Plug-and-play Commissionable Models for Water Networks with Multiple Inlets. In Proceedings of the IEEE 2018 European Control Conference (ECC), Limassol, Cyprus, 12–15 June 2018; pp. 1–6.
24. Simpkins, A. System identification: Theory for the user, (ljung, l.; 1999)[on the shelf]. *IEEE Robot. Autom. Mag.* **2012**, *19*, 95–96. [CrossRef]
25. Bort, C.G.; Righetti, M.; Bertola, P. Methodology for leakage isolation using pressure sensitivity and correlation analysis in water distribution systems. *Procedia Eng.* **2014**, *89*, 1561–1568. [CrossRef]
26. Soldevila, A.; Fernandez-Canti, R.M.; Blesa, J.; Tornil-Sin, S.; Puig, V. Leak localization in water distribution networks using Bayesian classifiers. *J. Process Control* **2017**, *55*, 1–9. [CrossRef]
27. Romano, M.; Woodward, K.; Kapelan, Z. Statistical Process Control Based System for Approximate Location of Pipe Bursts and Leaks in Water Distribution Systems. *Procedia Eng.* **2017**, *186*, 236–243. [CrossRef]
28. Blesa, J.; Jiménez, P.; Rotondo, D.; Nejjari, F.; Puig, V. An Interval NLPV Parity Equations Approach for Fault Detection and Isolation of a Wind Farm. *IEEE Trans. Ind. Electron.* **2015**, *62*, 3794–3805. [CrossRef]
29. Rossman, L.A. *EPANET 2 User's Manual*; U.S. EPA: Cincinnati, OH, USA, 2000.
30. Blesa, J.; Pérez, R. Modelling uncertainty for leak localization in Water Networks. *IFAC-PapersOnLine* **2018**, *51*, 730–735. [CrossRef]
31. Soldevila, A.; Blesa, J.; Tornil-Sin, S.; Fernandez-Canti, R.M.; Puig, V. Sensor placement for classifier-based leak localization in water distribution networks using hybrid feature selection. *Comput. Chem. Eng.* **2018**, *108*, 152–162. [CrossRef]
32. Wang, Q.; Guidolin, M.; Savic, D.; Kapelan, Z. Two-objective design of benchmark problems of a water distribution system via MOEAs: Towards the best-known approximation of the true Pareto front. *J. Water Resour. Plan. Manag.* **2015**, *141*, 04014060. [CrossRef]
33. Bragalli, C.; D'Ambrosio, C.; Lee, J.; Lodi, A.; Toth, P. *Water Network Design by MINLP*; Rep. No. RC24495; IBM Research: Yorktown Heights, NY, USA, 2008.
34. Quiñones-Grueiro, M.; Verde, C.; Llanes-Santiago, O. Multi-objective sensor placement for leakage detection and localization in water distribution networks. In Proceedings of the IEEE 2019 4th Conference on Control and Fault Tolerant Systems (SysTol), Casablanca, Morocco, 18–20 September 2019; pp. 129–134.
35. Ares-Milián, M.J.; Quiñones-Grueiro, M.; Verde, C.; Llanes-Santiago, O. A Leak Zone Location Approach in Water Distribution Networks Combining Data-Driven and Model-Based Methods. *Water* **2021**, *13*, 2924. [CrossRef]
36. Quiñones-Grueiro, M.; Milián, M.A.; Rivero, M.S.; Neto, A.J.S.; Llanes-Santiago, O. Robust leak localization in water distribution networks using computational intelligence. *Neurocomputing* **2021**, *438*, 195–208. [CrossRef]

sensors

MDPI

Article

Two Simultaneous Leak Diagnosis in Pipelines Based on Input–Output Numerical Differentiation

Adrián Navarro-Díaz [1], Jorge-Alejandro Delgado-Aguiñaga [2,*], Ofelia Begovich [3] and Gildas Besançon [4]

1 Tecnologico de Monterrey, School of Engineering and Sciences, Av. General Ramón Corona 2514, Zapopan C.P. 45138, Mexico; adrian.navarro@tec.mx
2 Centro de Investigación, Innovación y Desarrollo Tecnológico CIIDETEC-UVM, Universidad del Valle de México, Periférico Sur Manuel Gómez Morín 8077, Tlaquepaque C.P. 45601, Mexico
3 CINVESTAV Guadalajara, Av. del Bosque 1145, Col. El Bajío, Zapopan C.P. 45019, Mexico; ofelia.begovich@cinvestav.mx
4 GIPSA-Lab, Institute of Engineering, Université Grenoble Alpes, CNRS, Grenoble IPN, 38000 Grenoble, France; gildas.besancon@gipsa-lab.grenoble-inp.fr
* Correspondence: jorge.delgado@uvmnet.edu

Abstract: This paper addresses the two simultaneous leak diagnosis problem in pipelines based on a state vector reconstruction as a strategy to improve water shortages in large cities by only considering the availability of the flow rate and pressure head measurements at both ends of the pipeline. The proposed algorithm considers the parameters of both leaks as new state variables with constant dynamics, which results in an extended state representation. By applying a suitable persistent input, an invertible mapping in x can be obtained as a function of the input and output, including their time derivatives of the third-order. The state vector can then be reconstructed by means of an algebraic-like observer through the computation of time derivatives using a Numerical Differentiation with Annihilatorsconsidering its inherent noise rejection properties. Experimental results showed that leak parameters were reconstructed with accuracy using a test bed plant built at Cinvestav Guadalajara.

Keywords: fault diagnosis; pipelines; multiple leaks; numerical differentiation; experimental results

Citation: Navarro-Díaz, A.; Delgado-Aguiñaga, J.-A.; Begovich, O.; Besançon, G. Two Simultaneous Leak Diagnosis in Pipelines Based on Input–Output Numerical Differentiation. *Sensors* **2021**, *21*, 8035. https://doi.org/10.3390/s21238035

Academic Editors: Miquel À. Cugueró-Escofet and Vicenç Puig

Received: 3 November 2021
Accepted: 26 November 2021
Published: 1 December 2021

1. Introduction

In recent decades, climate change and the overuse of natural water resources have caused water scarcity in big cities. Furthermore, water distribution systems operators (WDSOs) are facing major water losses as high as 65% due to pipeline leaks caused by lack of maintenance, illegal intrusion, or accidents.

According to a study performed by the Organisation for Economic Co-operation (OECD), entitled Water Governance in Cities [1], aging water networks have a negative impact in terms of efficiency. One of the consequences is water loss from pipeline leaks. On average, water loss in the surveyed cities (in the referred report) was 21% in 2016. However, for Mexican cities, water loss was more than 40% (Chihuahua, Mexico City, San Luis Potosi) or even up to 65% (Tuxtla), see Figure 1.

On the other hand, to satisfy the current demand, government policies are focused on bringing more water from far away places instead of solving water losses due to leaks. This means that the amount of water lost is currently considered in the water budget. Interestingly, the amount of water needed to meet the demand in deficit is very similar to what is lost through leaks. In other words, it could be possible to satisfy the current water demand by minimizing the water losses due to leaks without the need for bringing more water from far away places.

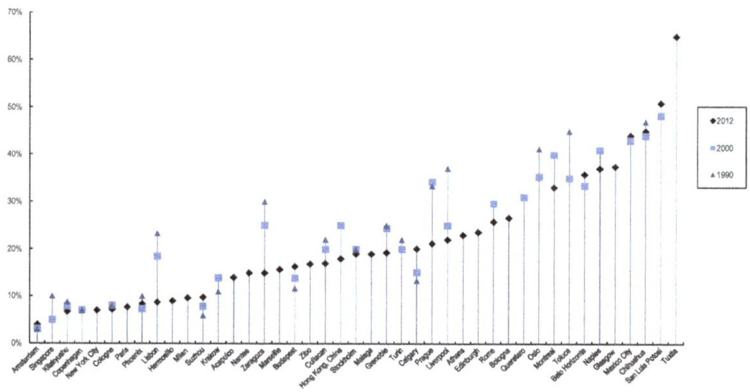

Figure 1. Proportion of water loss in surveyed cities (leakage rate).

The implementation of leak detection and isolation (*LDI*) systems has demonstrated a reduction in water losses in pipelines. *LDI* systems are algorithms that perform the following tasks: *detection* and *localization* of one or several leaks in a pipeline system. Thus, once a leak is identified, the repair technicians can fix the leak and avoid the water loss. Many works that address the *LDI* problem for one leak have been reported [2–6]. For instance, in [2], a leak isolation methodology using a fitting loss coefficient calibration is presented on the basis of two stages: in the first stage, the equivalent straight length (*ESL*) is fixed by a model-based observer designed as an extended Kalman filter (*EKF*); in the second stage, an algebraic observer is started with the *ESL* value fixed by the previous observer. Finally, the estimated leak position is recovered in original coordinates since the observer deals with *ESL* coordinates. Authors in [3] presented a methodology for leak detection and isolation in pipelines based on data fusion using two approaches: a steady-state estimation and an *EKF*. Authors concluded that the solution of the *LDI* problem improves significantly when a steady-state estimation is incorporated to the estimation provided by the *EKF*. In other words, the solution provided by the *EKF* is less accurate by itself. In [4], authors propose a bank of observers together with a Genetic Algorithm (*GA*), which is exploited to minimize the integration of the square observation error. The minimum integral observation error is reached in the observer where the estimated leak parameters match the real values. Experimental results evidence an accurate leak position estimation in a test bed pilot plant. In [5], a combined artificial neural network (*ANN*) for leak diagnosis in pipelines is presented. The *ANN* scheme estimates the location and friction factor based on measurement data. An average error of 0.629% was obtained for leak location in the experiments. More recently, in [6], a new approach for solving the *LDI* problem in pipelines is introduced on the basis of a Kalman filter for linear parameter varying (*LPV*) systems. The off-line computation of the filter gain allows the computational effort to be reduced and the authors claim that the *LPV* design outperforms the classical *EKF* design in terms of parameter-estimation accuracy.

On the other hand, the multi-leak case study has also been considered from two different perspectives: (i) for sequential leaks (non-concurrent case) [7–9] and (ii) for the simultaneous leak problem (concurrent case) [10].

In [7], a model adaptation strategy is proposed to isolate non-concurrent multiple leaks based on extended Kalman filters. Experimental results show the potential of this approach by allowing to monitor each new leak, no matter where it appears. Following this direction, a scheme is proposed in [8] for detecting and locating multiple sequential leaks based on a combination of an adaptive observer to identify the hydraulic gradient in real time and a leak location observer to estimate the leak position and its outflow. Experimental results of a pilot pipeline showed a satisfactory estimation in spite of operational changes and leaks. More recently, in [9], a pressure distribution analysis is proposed to diagnose the location of leaks via an experimental study and computational fluid dynamics (*CFD*) simulation. Multiple flow rate testing is conducted to detect the locations of two leaks.

In the same way, a more complex case of the multi-leak problem is when two or more leaks occur at the same time, also known as the concurrent case. In [10], the orthogonal collocation method (*OCM*) is used to obtain an approximate solution of the water hammer equations (*WHE*). An estimator is then designed based on the spatially-discretized model to detect multiple leaks by identifying their positions and leak coefficients by applying a persistent input [11]. The results are presented via simulation.

Regarding the one leak problem, state observer-based techniques have been proposed and successfully evaluated since the observer convergence is guaranteed, even by applying constant inputs. This is because the structure of underlying state-space representation fulfills a uniform observability condition (which is independent of the input) [12]. Conversely, when two or more leaks occur simultaneously, such an observability condition is no longer satisfied and the observability depends on the input. Particularly in steady state, the output obtained by two or more leaks is equivalent to the one obtained by a single *virtual* leak, which is known as indistinguishability [11].

1.1. Problem Statement

- In a real pipeline system, several leaks can occur and usually they are not fixed as they appear; this means that the leak problem becomes a challenging multi-leak problem known as the simultaneous leak case. In addition, this situation worsens when water management companies frequently lack flow rate and pressure head records of the leak events.
- Therefore, a methodology to address the simultaneous leak case is proposed on the basis of an input–output numerical differentiation-based strategy by applying a persistent input in the sense of [11].

1.2. Methods

- By considering a state-space representation of a pipeline with two leaks in which the leak parameters are considered as new state variables with constant dynamics, the extended state can be reconstructed via its expression in terms of input, output, and the corresponding time derivatives.
- A persistent input is experimentally generated via a frequency variation of the pump driver that produces a sine-like pressure signal. This persistent input allows the parameters of each leak to be reconstructed. This approach could be extended to a more general case of simultaneous leaks if the applied input is regularly persistent, such that the observability condition is guaranteed [11]. However, this approach could also be limited to physical constraints since a persistent input might cause additional leaks due to the flow transient effect that it produces.

Hereinafter, the paper is organized as follows: In Section 2, a mathematical model is derived from the well-known water hammer equations and the two simultaneous leak problem is stated. State vector reconstruction based on injection of input–output time derivatives is presented in Section 3. Experimental results are presented in Section 4 by using databases from a pilot plant built at Cinvestav Guadalajara. Finally, several conclusions and future perspectives are given in Section 5.

2. Preliminaries

2.1. Pipeline Mathematical Model

2.1.1. Governing Equations

The transient fluid through a pipeline can be described by conservation mass and momentum equations known as water hammer equations, which are a couple of quasi-linear hyperbolic partial differential equations (*PDE*). Generally, *PDE* are derived considering the following assumptions: the fluid is slightly compressible, the duct wall is slightly

deformable, and the convective velocity changes are negligible. The cross section and the fluid density are assumed to be constant [13]:
Momentum Equation

$$\frac{\partial Q(z,t)}{\partial t} + gA\frac{\partial H(z,t)}{\partial z} + \mu Q(z,t)|Q(z,t)| = 0 \tag{1}$$

Continuity Equation

$$\frac{\partial H(z,t)}{\partial t} + \frac{b^2}{gA}\frac{\partial Q(z,t)}{\partial z} = 0 \tag{2}$$

Here, Q stands for the flow rate [m^3/s]; H is the pressure head [m]; z is the length coordinate [m]; t is the time coordinate [s]; g is the gravity acceleration [m/s^2]; A is the cross-section area [m^2]; b is the pressure wave speed in the fluid [m/s]; $\mu = \tau/2\phi A$, where ϕ is the inner diameter [m] and τ is the friction factor. The dynamics in Equations (1) and (2) are fully defined by related pairs of initial and boundary conditions.

2.1.2. Finite Difference Approximation

For the purpose of obtaining a finite dimensional model from (1) and (2), the partial differential equations are discretized with respect to the spatial variable z, as in [14,15], by using the following approximations:

$$\text{Section } i: \begin{cases} \frac{\partial H(z_i,t)}{\partial z} \backsimeq \frac{H_{i+1}-H_i}{\Delta z_i} & \forall i = 1, \cdots, n \\ \frac{\partial Q(z_{i-1},t)}{\partial z} \backsimeq \frac{Q_i - Q_{i-1}}{\Delta z_{i-1}} & \forall i = 2, \cdots, n \end{cases} \tag{3}$$

where index i stands for the variable discretized in (1) and (2) at section i. To solve the *LDI* problem for two simultaneous leaks, Equations (1) and (2) admit a simple spatial discretization as shown in Figure 2, where sections are defined according to the two leakage positions:

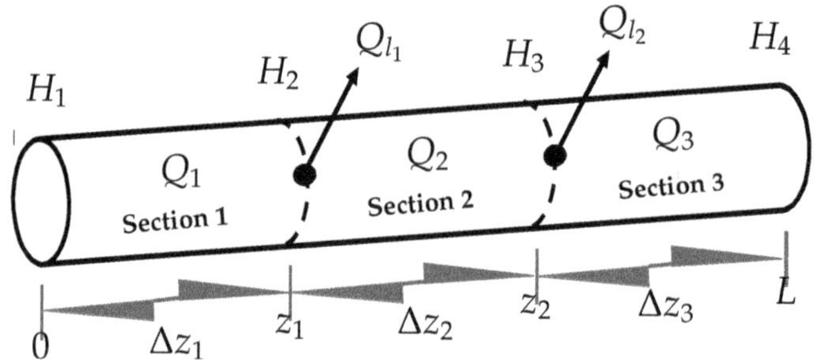

Figure 2. Discretization of the pipeline with two arbitrarily located leaks.

Here, $Q_{l_{1,2}}$ represent the leaking flows that can be modeled as:

$$Q_{l_{1,2}} = \lambda_{1,2}\sqrt{H_{2,3}} \tag{4}$$

where, $\lambda_{1,2}$ is a constant that depends on the orifice size and the discharge coefficient, and $H_{2,3}$ is the head pressure at the leak point.

Thus, assuming a lumped-parameter model for the flow equations and considering that the pressure head and flow rate measurements are available at both ends of the pipeline via sensors, a low order dynamical representation of the system with two leaks can be written using approximation (3) in Equations (1) and (2), as follows:

$$\begin{bmatrix} \dot{Q}_1 \\ \dot{H}_2 \\ \dot{Q}_2 \\ \dot{H}_3 \\ \dot{Q}_3 \end{bmatrix} = \begin{bmatrix} \frac{-gA}{\Delta z_1}(H_2 - H_1) - \frac{\tau}{2\phi A}Q_1|Q_1| \\ \frac{-b_2^2}{gA\Delta z_1}(Q_2 - Q_1 + \lambda_1\sqrt{H_2}) \\ \frac{-gA}{\Delta z_2}(H_3 - H_2) - \frac{\tau}{2\phi A}Q_2|Q_2| \\ \frac{-b_2^2}{gA\Delta z_2}(Q_3 - Q_2 + \lambda_2\sqrt{H_3}) \\ \frac{-gA}{\Delta z_3}(H_4 - H_3) - \frac{\tau}{2\phi A}Q_3|Q_3| \end{bmatrix} \tag{5}$$

Notice that, due to mass conservation, $Q_{l_{1,2}}$ must satisfy the next relation:

$$Q_{l_{1,2}} = Q_{b_{1,2}} - Q_{a_{1,2}} \tag{6}$$

where $Q_{b_{1,2}}$ and $Q_{a_{1,2}}$ are the flows in an infinitesimal length before and after the leak, respectively.

2.1.3. Pipeline Equivalent Straight Length

It is worth pointing out that the mathematical model (5) assumes a straight pipe. This is not a loss of generality because even if the pipe is not straight, it is possible to obtain an equivalent straight length (*ESL*) of the pipe. The *ESL* is the straight length of a virtual pipe (with the same parameters as the original duct) that would give rise to the same pressure drop as the real pipeline. Such an equivalence is calculated by considering losses due to each "non-straight element" (i.e., fitting) in accordance with the Darcy–Weisbach formula [2,16,17]:

$$l_e = \frac{\phi}{\tau}K \tag{7}$$

where, l_e means the equivalent straight length of a specific fitting, K is the so-called loss coefficient parameter, which is normally provided by the pipe manufacturer, and τ is the friction coefficient. Thus, the total ESL of the pipe, L_e, can be calculated as follows:

$$L_e = L_r + \frac{\phi}{\tau}\sum_{i=0}^{n_f} K_i \tag{8}$$

where L_r stands for the sum of all pipeline straight length elements, K_i is the fitting loss coefficient for the i-th fitting, and n_f the number of the pipeline fittings.

2.1.4. Friction Model

The friction factor (τ in Equation (1)) represents the loss of pressure of a fluid due to the interactions between the fluid and the internal surface roughness of the pipe. Thus, such a friction factor is a function of the Reynolds number, Re, and the pipe's roughness, ϵ [18,19].

In many practical cases, τ is deemed to be a constant value, which is commonly taken from the Moody chart; nevertheless, in pipes with a relative roughness usually less than 1×10^{-3} m, the zone where the friction factor is almost constant (i.e., the complete turbulence zone), is difficult to reach. Consequently, when the *LDI* scheme is applied to a plastic pipe, it is preferable to obtain a more accurate friction value by using a formula or an algorithm to estimate such value. In this work, the authors propose the use of the well know Swamee–Jain equation to directly calculate the coefficient of friction:

$$\tau_i(Q_i) = \frac{0.25}{\left[\log_{10}\left(\frac{\epsilon}{3.7\phi} + \frac{5.74}{\mathrm{Re}_i^{0.9}}\right)\right]^2} \tag{9}$$

where subscript i denotes the section number of the pipeline (see Figure 2). The Reynolds number is, in turn, function of the flow rate, Q_i, as follows:

$$\mathrm{Re}_i = \frac{Q_i \phi}{\nu A} \tag{10}$$

where, ν is the kinematic viscosity of the water. Notice that, due to leak occurrence, the flow rate is different in each pipeline section (see Figure 2), causing a significant deviation of the friction factor value (since the working plastic pipe area is commonly in the transition zone). Therefore, it is important to introduce Equations (9) and (10) in the mathematical model to calculate, at any sampling time, the friction coefficient due to the flow variations in each i-th section.

2.2. Two Simultaneous Leak Problem Statement

In this work, the two simultaneous leak case is considered, i.e., a couple of leaks can appear in a pipeline at locations: $z_1 \in (0, L)$ and $z_2 \in (0, L)$, with $z_2 > z_1$. Thus, the problem is reduced to the size estimation of the pipe sections: Δz_1 ($\Delta z_1 = z_1$) and Δz_2 ($\Delta z_2 = z_2 - z_1$), see Figure 2.

Now, Equation (5) can be written in compact form as follows:

$$\begin{aligned} \dot{\zeta} &= \xi(\zeta) + \rho(\zeta)\gamma \\ \Psi &= \vartheta(\zeta) \end{aligned} \tag{11}$$

where $\zeta = [\zeta_1\, \zeta_2\, \zeta_3\, \zeta_4\, \zeta_5]^T = [Q_1\, H_2\, Q_2\, H_3\, Q_3]^T \in \mathbb{R}^5$ is the state vector, $\gamma = [H_{in}\, H_{out}]^T \in \mathbb{R}^2$ is the input vector, and $\Psi = [Q_{in}\, Q_{out}]^T \in \mathbb{R}^2$ is the output vector for some functions ξ, ρ, and ϑ.

The leak diagnosis problem for two simultaneous leaks appearing in a pipeline can then amount to the estimation of parameters Δz_1, Δz_2, λ_1, and λ_2 in (5). Let us consider those parameters as new state variables with constant dynamics [11], that is: if $\theta = [z_1\ z_2\ \lambda_1\ \lambda_2]^T$ then $\dot{\theta} = 0$. This results in an extended state: $x = [\zeta\ \theta]^T = [Q_1\ H_2\ Q_2\ H_3\ Q_3\ \Delta z_1\ \Delta z_2\ \lambda_1\ \lambda_2]^T =: [x_1\ x_2\ x_3\ x_4\ x_5\ x_6\ x_7\ x_8\ x_9]^T \in \mathbb{R}^9$. Then, considering an unidirectional flow given by $Q_i|Q_i| = Q_i^2$, the extended state representation of (11) can take a form as follows:

$$\begin{aligned} \dot{x} &= f(x, u) \\ y &= h(x) \end{aligned} \tag{12}$$

where $u \doteq [H_1\ H_4]^T \doteq [u_1\ u_2]^T$, and f, g are differentiable vector fields with the following structure:

$$f(x,u) = \begin{bmatrix} -\frac{gA}{x_6}(x_2 - u_1) - \mu_1(x_1)x_1^2 \\ -\frac{b^2}{Agx_6}(x_3 - x_1 + x_7\sqrt{x_2}) \\ -\frac{gA}{x_8}(x_4 - x_2) - \mu_2(x_3)x_3^2 \\ -\frac{b^2}{Agx_8}(x_5 - x_3 + x_9\sqrt{x_4}) \\ -\frac{gA}{L - x_6 - x_8}(u_2 - x_4) - \mu_3(x_5)x_5^2 \\ \mathbb{O}^{4\times 1} \end{bmatrix} \tag{13}$$

$$h(x) = \begin{bmatrix} y_1(x) \\ y_2(x) \end{bmatrix} = \begin{bmatrix} Q_1 \\ Q_3 \end{bmatrix} = \begin{bmatrix} x_1 \\ x_5 \end{bmatrix}$$

where $\mathbb{O}^{4\times 1}$ is the 4×1 zero matrix, $\mu_i = \frac{\tau_i}{2\phi A}$ is computed as in Equation (9), and $\tau_{1,2,3}$ are functions of x_1, x_3, and x_5, respectively.

3. State Vector Reconstruction Based on Input–Output Numerical Differentiation

3.1. Observability Discussion

For the one leak case, the so-called observability rank condition is satisfied, and such a property does not depend on the inputs [12]. However, in the case of two simultaneous leaks (or even more), this is no longer true, since the states are not distinguishable by applying constant inputs, and thus a persistent input is required [11].

In fact, to reconstruct the actual state vector x of (12), one can compute time derivatives of the output as functions of state variables as well as input and its time derivatives (using Equation (13)), such that an invertible map with respect to the full state vector x is obtained by applying an appropriate input. More precisely, by considering the two simultaneous leak case described by (12), we have two input variables and two output variables, from which input–output time derivative vectors can be generally defined up to orders p, p' for the input and q, q' for the output, as:

$$U_{(p,p')}(t) := \left[u_1 \, \dot{u}_1 \, \cdots \, u_1^{(p)} \, u_2 \, \dot{u}_2 \, \cdots \, u_2^{(p')}\right]^T \tag{14}$$

and

$$Y_{(q,q')}(t) := \left[y_1 \, \dot{y}_1 \, \cdots \, y_1^{(q)} \, y_2 \, \dot{y}_2 \, \cdots \, y_2^{(q')}\right]^T \tag{15}$$

Clearly, from (13), the output time derivatives depend on the state and input time derivatives, so that we can get:

$$Y_{(q,q')}(t) = \Gamma\left(x, U_{(p,p')}\right) \tag{16}$$

for some p, p', given q, q'.

On this basis, observability somehow means that this relationship is invertible and that it is possible to find elements among the components of Γ defining an invertible map with respect to x [20,21]. Specifically, if such an inverted map exists, and the input–output and the corresponding time derivatives are known, then, it is possible to compute each independent state in (16) as follows:

$$x = \Gamma^{-1}\left(U_{(p,p')}, Y_{(q,q')}\right) \tag{17}$$

Notice that, in general, it can be of interest to avoid or limit time derivatives of the input, but assuming that they are available or can be estimated in the same way as time derivatives of y is enough for our present application.

It should be noted that for this LDI problem, it is even enough to obtain an expression by only relating the input, output, and their time derivatives with leak parameters. We will propose this in Section 3.3, with a similar procedure as *elimination* (which, conversely to realization, consists of deriving an externally equivalent representation not containing the state [22]). We will even see how to recover the full state vector. However, let us first discuss the way to obtain input–output time derivatives.

3.2. Numerical Differentiation with Annihilators

The *LDI* scheme proposed in this work is based on Equation (17), that is, the injection of the inputs, outputs, and their time derivatives. Although there are multiple algorithms to numerically compute a time derivative, here we use a numerical differentiation with annihilators [23] due to its inherent noise rejection properties. Hereinafter, a brief explanation of such an algorithm is given, which is proposed for the first time in this paper, and it is described by Equation (25).

Let $\gamma^m(t)$ denote the m-th order derivative of a smooth signal $\gamma(t)$ defined on an interval $\mathcal{I} \subset \mathbb{R}^+$. The signal $\gamma(t)$ could represent a measurement variable corrupted by some noise, whose derivatives are not directly available.

Ignoring the noise for a moment, let $\gamma(t)$ be an analytical function on $\mathcal{I}$. So, without any loss of generality, it is possible to consider the truncated Taylor expansion at $t = 0$:

$$\gamma(t) = \sum_{i=0}^{m} a_i \frac{t^i}{i!} + \mathcal{O}(t^m) \tag{18}$$

where $a_i = \left.\frac{d^i\gamma(t)}{dt^i}\right|_{t=0}$. This implies that $\gamma(t)$ can be approximated by the polynomial $p_m(t) = \sum_{i=0}^{m} a_i \frac{t^i}{i!}$.

In this way, the i-th order time derivative estimation of $\gamma(t)$ can be tackled as a parameter estimation problem for $p_m(t)$. Using the method described in [23], it is possible to calculate each a_i ($i = 0, \ldots, m$) independently, reducing in this manner the sensitivity to noise and numerical computation errors which often appear in simultaneous estimation methods. Moreover, the independent calculation allows the use of higher order polynomials without the calculation of all their coefficients. The algorithm is described as follows (a complete explanation is given in [23]):

- Let $p_m(t)$ be the m-th order polynomial approximation of $\gamma(t)$,

$$p_m = \frac{a_0}{0!} + \frac{a_1}{1!}t + \frac{a_2}{2!}t^2 + \ldots + \frac{a_m}{m!}t^m \tag{19}$$

- Transforming Equation (19) into Laplace domain yields:

$$P_m = \frac{a_0}{s} + \frac{a_1}{s^2} + \frac{a_2}{s^3}t^2 + \ldots + \frac{a_m}{s^{m+1}} \tag{20}$$

- In order to calculate the i-th time derivative approximation, a_i, it is necessary to first annihilate every a_x ($x > i$) in (20), using the next operator:

$$\left[\prod_{l=0}^{m-i} \frac{d}{ds}\right] \cdot s^{m+1} \tag{21}$$

- Then, to annihilate every a_x ($x < i$), the following operator is subsequently applied to Equation (20):

$$\left[\prod_{h=0}^{i} \frac{d}{ds}\right] \cdot s^{-1} \tag{22}$$

- Finally, the resulting equation (after applying Equations (20)–(22)) is:

$$\begin{aligned}&(-1)^i(m-i)!i!a_i s^{-(i+1)} = \\ &\sum_{h=0}^{i}\sum_{l=0}^{m-1+h}(-1)^{i-h}\binom{i}{h}\binom{m-i+h}{l}\frac{(i-h)!(m+1)!}{(i+1-h+l)!}s^l\frac{d^l P_m(s)}{ds^l}\end{aligned} \tag{23}$$

- Now, multiplying both sides of the above equation by $s^{-(m+1)}$ yields a polynomial taking the following form:

$$\frac{c_m}{s^{i+m+2}}a_i = \frac{1}{s}\frac{d^m P_m(s)}{ds^m} + \frac{c_{m-1}}{s^2}\frac{d^{m-1}P_m(s)}{ds^{m-1}} + \ldots + \frac{c_0}{s^{m+1}}P_m(s) \tag{24}$$

- Using the Cauchy rule for iterated integrals, the time domain expression for a_i in Equation (24) yields:

$$a_i = \frac{(m+i+1)!}{c_m T^{m+i+1}}\int_0^T \left[t^m + \frac{c_{m-1}}{1!}(T-t)t^{m-1} + \ldots + \frac{c_0}{m!}(T-t)^m\right] p(t) \cdot dt \tag{25}$$

It should be noted that each constant $c_j, j = \{0, 1, \ldots, m\}$, is obtained from Equation (23), and T represents a moving window of length T for the integrals. A short time window is sufficient to obtain accurate estimations. In addition, the iterated integrals work as low pass filters that provide a smoother form of highly fluctuating noises. Therefore, no previous knowledge on the statistical properties of the noise is required to filter it out.

3.3. Extended State Vector Reconstruction

The purpose of this section is to provide a synthesis of the proposed *LDI* scheme. In this sense, the method is based on the study of the structure of the input–output differential equation; thus, the problem is solved by exploiting the observability property of the system (13). First, a set of equations describing each state just as a function of the inputs, outputs, and the corresponding time derivatives is derived. Then, taking advantage of the filtering characteristic of the numerical differentiation exposed in Section 3.2, the input–output time derivatives are computed. Thus, the leak positions and magnitudes can be calculated by a pair of algebraic equations. In the next step, the algorithm used to derive this set of equations is described.

It is easy to check that the observability rank condition for the system (12) is satisfied (for the finite number of input–output time derivatives), only by applying a persistent input. More precisely, it is also easy to see that for an input of the form $A_p \sin(\omega t)$. This wave form is easy to achieve by a variable frequency drive, a commonly installed device in a pumping station.

The rank condition of (12) is fulfilled with $p = 3$, $p' = 2$, $q = 4$, and $q' = 3$ for Equations (14) and (15). Thus, the output derivatives as well as the inverse mapping can be computed as follows (perhaps the major difficulty of the algorithm is to obtain the inverse mapping due to the complexity of the resulting equation):

First, by construction of Equation (13), the state variables x_1 and x_5 are taken directly from the measurements:

$$x_1 = y_1 \tag{26}$$

$$x_5 = y_2 \tag{27}$$

Now, the following step is to take the time derivatives of Equations (26) and (27), by using (13) and replacing x_1, x_5 by outputs according to (26) and (27) solving the resulting equation for x_2 and x_4, respectively, yields expressions without x_1 and x_5:

$$x_2 = \Phi_1(x_6, u_1, y_1, \dot{y}_1) \tag{28}$$

$$x_4 = \Phi_2(x_6, x_8, u_2, y_2, \dot{y}_2) \tag{29}$$

In the same way, the third step is to compute the derivative of Equation (28), substitute x_1, x_5, x_2, and x_4 according to (26), (27), (28), and (29), respectively. Solving this equation for x_3, yields an expression without x_1, x_5, x_2, x_4:

$$x_3 = \Phi_3(x_6, x_7, u_1, \dot{u}_1, y_1, \ldots, y_1^{(2)}) \tag{30}$$

Following the same steps, now for (29) and (30), it is possible to find an expression for x_7 and x_9 free of x_1, x_2, x_3, x_4, and x_5:

$$x_7 = \Phi_4(x_6, x_8, u_1, \ldots, u_1^{(2)}, u_2, y_1, \ldots, y_1^{(3)}, y_2, \dot{y}_2,) \tag{31}$$

$$x_9 = \Phi_5(x_6, x_8, u_1, \ldots, u_1^{(2)}, u_2, \dot{u}_2, y_1, \ldots, y_1^{(3)}, y_2, \ldots, y_2^{(2)}) \tag{32}$$

The final step is to apply the same methodology for Equations (31) and (32), but now for solving for the states x_6 and x_8. In this way, it is feasible to obtain an expression of x_6 and x_8 just as a function depending on the input and output, and their time derivatives:

$$x_6 = \Phi_6(u_1, \ldots, u_1^{(3)}, u_2, \dot{u}_2, y_1, \ldots, y_1^{(4)}, y_2, \ldots, y_2^{(3)}) \tag{33}$$

$$x_8 = \Phi_7(u_1, \ldots, u_1^{(3)}, u_2, .., u_2^{(2)}, y_1, \ldots, y_1^{(4)}, y_2 \ldots, y_2^{(3)}) \tag{34}$$

At this point, we are able to recover the whole state with the acknowledgement of the input and output time derivatives calculated through Equation (25); this is achieved by first obtaining x_6 and $\dot{x}_8$ (leak positions) from Equations (33) and (34). Once x_6 and x_8 are obtained, leak magnitudes, x_7 and x_9, can be computed by using (31) and (32). The rest of the state can then be recovered going backwards through Equations (30)–(28) .

4. Experimental Results

In this section, experimental results are presented to evaluate the proposed *LDI* methodology. The experiments are performed using several databases from the pilot plant located at Cinvestav Guadalajara. A couple of different two simultaneous leak scenarios were emulated by opening different electrovalves located along the pilot plant. A general description of the pipeline prototype is presented below with a detailed description of each experiment.

4.1. Pilot Pipeline Description

The layout of the pilot pressurized water pipe of Cinvestav Guadalajara, which is 68.2 [m] long (between sensors) with an internal diameter of 6.271×10^{-2} [m], thickness 1.27×10^{-2} [m], friction coefficient 1.66×10^{-2}, pressure wave speed 358 [m/s], and gravity acceleration 9.81 [m/s^2], is shown in Figure 3. The line is instrumented with two water-flow (FT) and pressure-head (PT) sensors at the inlet and outlet of the pipe. To emulate the leak, three control valves at position 16.8, 33.3, and 49.8 [m] are installed together with a electronic-based actuator to practically set any opening of the valve.

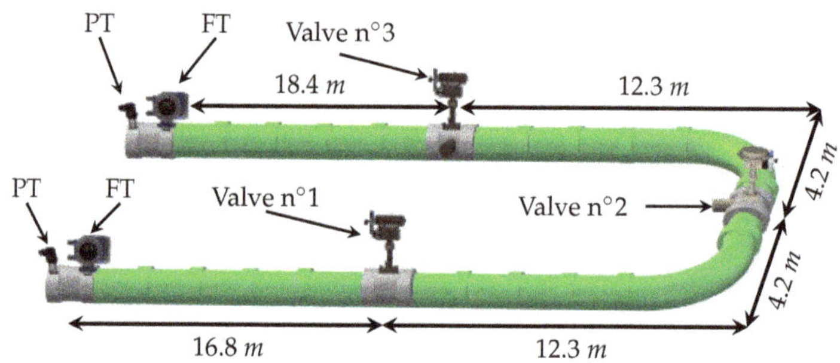

Figure 3. Schematic diagram of the pipeline's prototype.

The prototype is manufactured with a plastic material known as polypropylene copolymer random, for which technical characteristics can be found in [24]. It is integrated with a store tank of 7.5×10^{-1} [m^3], a hydraulic pump of 5 HP, and a variable-frequency driver (*VFD*) which controls the pressure in the system through the rotational speed of the pump motor (more details about the pipeline prototype can be found in [25]).

To ensure the application of a persistent input, experiments with operation point variations are carried out by the pump variation via the *VFD* (specifically, a change in the form of $A_p \sin(\omega t)$, where ω stands for the angular frequency induced via the *VFD*). Table 1 summarizes the pipeline's main parameters.

Table 1. Pipeline's main parameters.

Parameter	Symbol	Value	Units
Pipeline length	L_r	68.2	[m]
Upstream to valve $n°1$	z_1	16.8	[m]
Upstream to valve $n°2$	z_2	33.3	[m]
Upstream to valve $n°3$	z_3	49.8	[m]
Internal diameter	ϕ	6.271×10^{-2}	[m]
Pipe roughness	ϵ	7×10^{-6}	[m]
Friction factor	τ	1.66×10^{-2}	[dimensionless]
Pressure wave speed	b	358	[m/s]
Gravity acceleration	g	9.8	[m/s^2]

4.2. *LDI* Results

Hereinafter, two off-line examples of the *LDI* scheme are displayed. Both results were carried out by taking data from the pipeline prototype previously described. The experiment was performed as follows: Pump 1 is started in a steady state operation during the first 65 [s] approximately. After that, it begins to operate in some unsteady state, namely a sine-like pressure signal is introduced, just exactly like a persistent input in the sense of [11]. This sine signal was experimentally obtained by setting up the pump controller as follows:

$$\begin{aligned} VFD(t) &= 60[Hz], \forall t \leq 65[s] \\ VFD(t) &= 50 + 5sin(2.7313t)[Hz], \forall t > 65[s] \end{aligned} \tag{35}$$

At $t \approx 60$ [s], two leaks were induced simultaneously at the opening of the control valves in the pilot plant.

The leak position estimations were undertaken by the injection of the inputs and outputs for which a low pass filter was previously applied (u_1, u_2, y_1, and y_2 in (13)), and the corresponding derivatives together with Equations (33) and (34), respectively.

Then, the leak magnitudes were also estimated by using Equations (31) and (32). It is possible to recover the whole state going backwards using Equations (30)–(28). As stated earlier, the time derivatives were computed using the methodology discussed in Section 3.2.

4.2.1. Experiment 1: Leaks Induced in Valves $n°1$ and $n°3$

The first experiment consists of simultaneously inducing two leaks in valve $n°1$ and valve $n°3$ (see Figure 3). The algorithm starts once the leak is detected. This initial detection is obtained when a deviation between the upstream and downstream flow is detected $|Q_{in} - Q_{out}| > \delta$, where δ is a constant threshold defined by the designer (normally related to the signal-noise ratio): here $\delta = 1 \times 10^{-4}$ [m^3/s]. Immediately afterwards, a sinusoidal wave form as (35) is applied to ensure enough observability of the model (13) via persistent inputs (see Section 3.1). Once the sinusoidal steady state has been reached, the *LDI* scheme starts. Figure 4 shows the time evolution of the pressure head at the inlet and outlet of the pipe (system input, $H_{in} = u_1$ and $H_{out} = u_2$). As it can be seen, the *LDI* algorithm begins close to 65 [s].

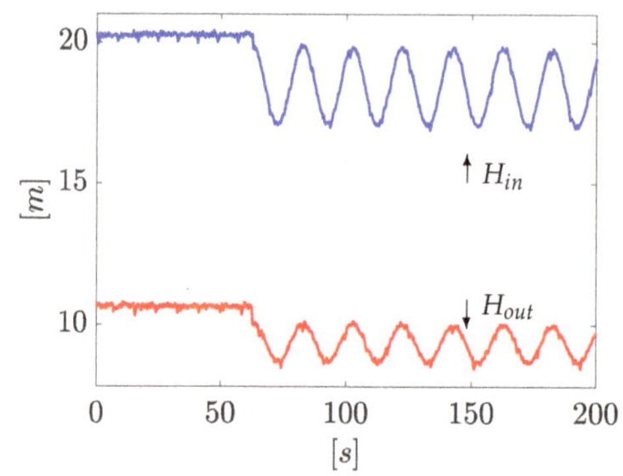

Figure 4. Pressure heads H_{in} and H_{out}.

Following with the same idea, Figure 5 shows the inlet and outlet flows (system output $Q_{in} = y_1$ and $Q_{out} = y_2$). It is clear that, due to the physical nature of the leak phenomena, the inlet flow is separated from the outlet flow:

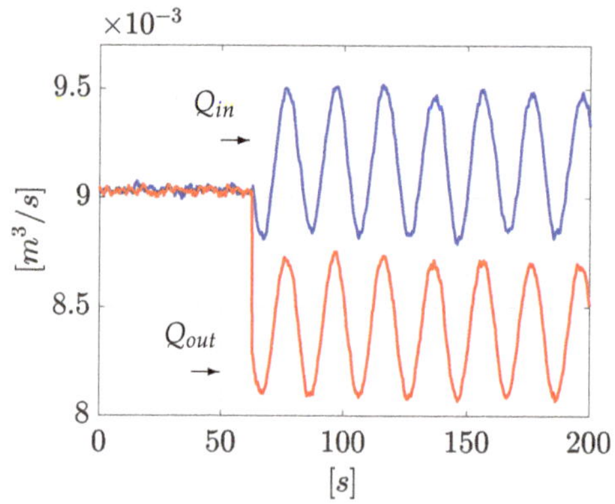

Figure 5. Flow rates Q_{in} and Q_{out}.

Now, once the input and output time derivatives have been computed with Equations (25), (33), and (34), then they are used to reconstruct the leak positions. For clarity, just one signal cycle ($T_{in} = 20$ s), as seen in Figure 6, is enough to correctly locate the two leaks, despite signal noise. To quantify the accuracy of the leak position estimation, a Mean Absolute Error index is applied. For the first leak, the estimation accuracy is 96.62% (with respect to the whole pipeline length), whereas the second leak localization accuracy is 96.33%. As Figure 6 shows, one cycle has demonstrated to be enough to isolate the leak well despite signal noise.

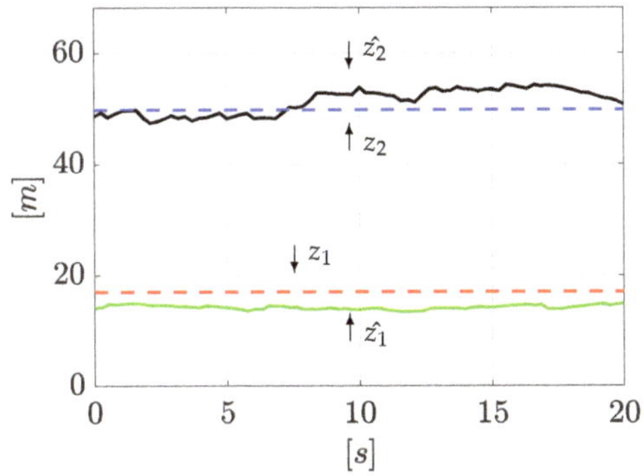

Figure 6. Estimation of leak positions.

Similarly, once the leak positions are obtained, it is possible to reconstruct the leak magnitudes using Equations (31) and (32). The corresponding results are shown in Figure 7, where estimated magnitudes are set around 1.3×10^{-4} [$m^{5/2}/s$] and 0.95×10^{-4} [$m^{5/2}/s$], respectively. The outflow computed by using Equation (4) for each leak is consistent with the total outflow.

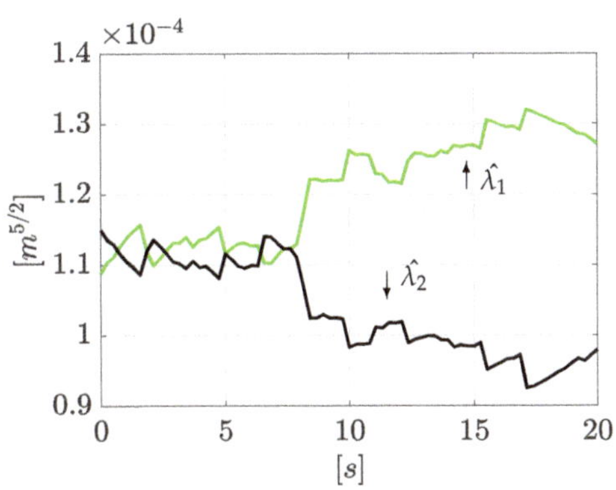

Figure 7. Estimation of leak magnitudes.

Notice that for the *LDI* problem, the estimation of the leak parameters of both leaks is enough, but, as mentioned before, it is also possible to recover the remaining states of (13), going backwards from Equations (30) and (28).

4.2.2. Experiment 2: Leaks Induced in Valve $n°1$ and $n°2$

To better illustrate the effectiveness of the method, a second experiment is exposed. The experiment setup was exactly the same as in Section 4.2.1 except that two leaks are now induced in valve $n°1$ and valve $n°2$ (see Figure 3). Figure 8 shows the time evolution of the pressure head at upstream and downstream, respectively. In the same way as before, the *LDI* algorithm starts when a flow deviation exceeds a predefined threshold, δ, ($|Q_{in} - Q_{out}| > \delta$).

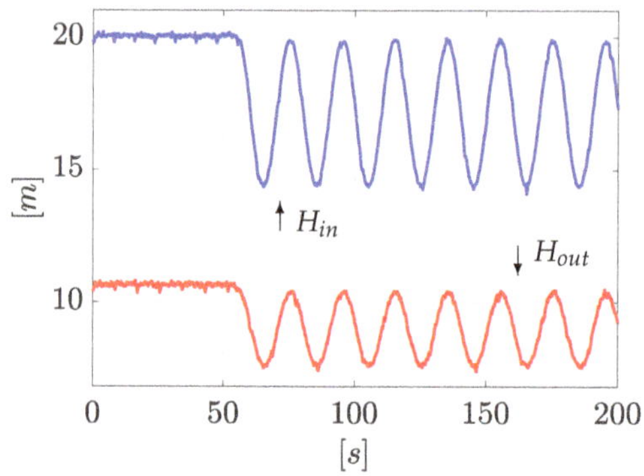

Figure 8. Pressure heads H_{in} and H_{out} (2nd experiment).

Figure 9 shows the corresponding flow rate evolution at upstream and downstream. As stated above, due to the physical nature of the leak phenomena, the inlet flow is separated from the outlet flow.

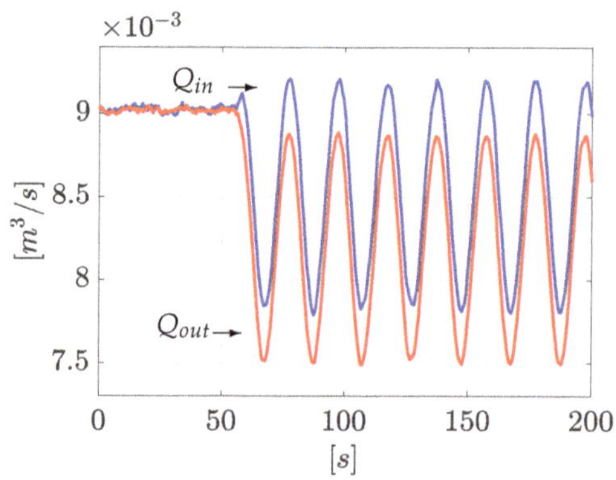

Figure 9. Flow rate Q_{in} and Q_{out} (2nd experiment).

As performed above, once the algorithm starts, the input and output time derivatives are computed following the algorithm explained in Section 3.2. Subsequently, Equations (33) and (34) are used to obtain the leak positions. The results are presented in Figure 10. As before, the leak positions are well estimated despite signal noise in just one signal cycle ($T_{in} = 20$ s). Here, the MAE yields an estimation accuracy for the first leak of 95.01%, while the second leak position presents an estimation accuracy of 97.94%.

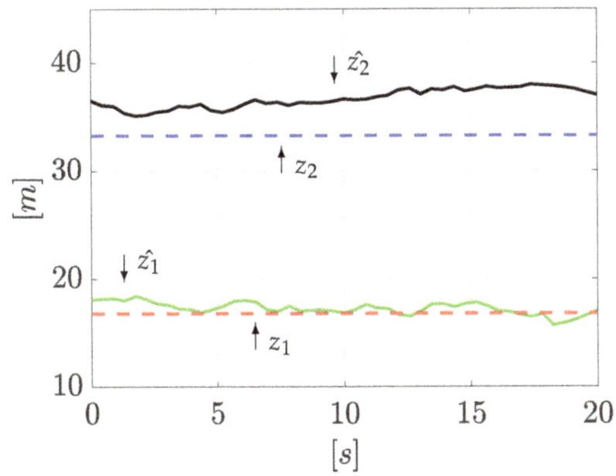

Figure 10. Estimation of leak positions (2nd experiment).

Continuing with the algorithm, the leak magnitudes are computed with Equations (31) and (32). These results are shown in Figure 11.

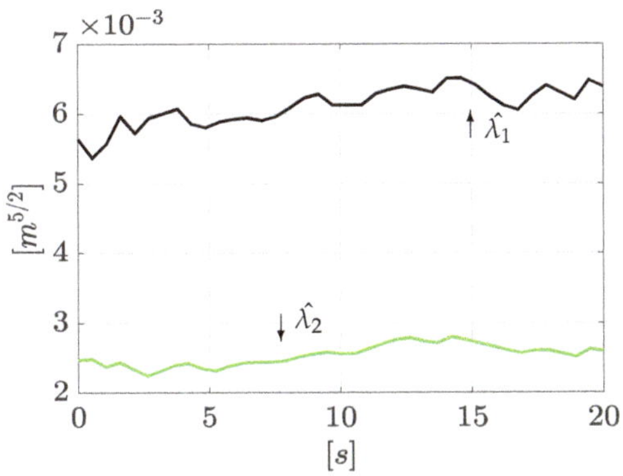

Figure 11. Estimation of leak magnitudes (2nd experiment).

5. Conclusions

The simultaneous leak detection and isolation problem is currently an open and challenging problem. Even though extensive research in the field is currently in progress, to the best of our knowledge, only simulation results have been reported until now. One of the reasons is that the distinguishability of two (or more) simultaneous leaks depends on the input.

In this work, an *LDI* methodology for the two simultaneous leak detection and isolation has been proposed based on an algebraic observer that uses the injection, the inputs and outputs of the system, and the corresponding time derivatives, by applying an appropriate input. The time derivatives are computed using *Numerical Differentiation with Annihilators* and the approach has been successfully applied to real data. This methodology could be extended to more general cases of simultaneous leaks in real operational conditions, such as the case of the *SIAPA* aqueduct in Guadalajara, Mexico.

Author Contributions: A.N.-D .: software, validation, data curation, writing—original draft preparation, methodology, formal analysis, and investigation; J.A.D.-A.: conceptualization, methodology, formal analysis, writing—review and editing, and investigation; O.B.: writing—review; G.B.: writing—review and editing. All authors have read and agreed to the published version of the manuscript.

Funding: This work was funded by the Tecnológico de Monterrey through the School of Engineering and Science.

Acknowledgments: Authors would like to thank Tecnológico de Monterrey for the financial support and the facilities granted for fulfillment of this research article. All databases were obtained by the second author during his PhD studies at Cinvestav Guadalajara.

Conflicts of Interest: The authors declare no conflict of interest.

References

1. OECD. *Water Governance in Cities*; OECD: Paris, France, 2016; p. 140. [CrossRef]
2. Navarro, A.; Begovich, O.; Sánchez, J.; Besancon, G. Real-Time Leak Isolation Based on State Estimation with Fitting Loss Coefficient Calibration in a Plastic Pipeline. *Asian J. Control* **2017**, *19*, 255–265. [CrossRef]
3. Santos-Ruíz, I.; Bermúdez, J.; López-Estrada, F.; Puig, V.; Torres, L.; Delgado-Aguiñaga, J. Online leak diagnosis in pipelines using an EKF-based and steady-state mixed approach. *Control Eng. Pract.* **2018**, *81*, 55–64. [CrossRef]
4. Navarro, A.; Delgado-Aguiñaga, J.A.; Sánchez-Torres, J.D.; Begovich, O.; Besançon, G. Evolutionary Observer Ensemble for Leak Diagnosis in Water Pipelines. *Processes* **2019**, *7*, 913. [CrossRef]
5. Pérez-Pérez, E.; López-Estrada, F.; Valencia-Palomo, G.; Torres, L.; Puig, V.; Mina-Antonio, J. Leak diagnosis in pipelines using a combined artificial neural network approach. *Control Eng. Pract.* **2021**, *107*, 104677. [CrossRef]
6. Delgado-Aguiñaga, J.; Puig, V.; Becerra-López, F. Leak diagnosis in pipelines based on a Kalman filter for Linear Parameter Varying systems. *Control Eng. Pract.* **2021**, *115*, 104888. [CrossRef]
7. Delgado-Aguiñaga, J.; Besançon, G.; Begovich, O.; Carvajal, J. Multi-leak diagnosis in pipelines based on Extended Kalman Filter. *Control Eng. Pract.* **2016**, *49*, 139–148. [CrossRef]
8. Rojas, J.; Verde, C. Adaptive estimation of the hydraulic gradient for the location of multiple leaks in pipelines. *Control Eng. Pract.* **2020**, *95*, 104226. [CrossRef]
9. Fu, H.; Wang, S.; Ling, K. Detection of two-point leakages in a pipeline based on lab investigation and numerical simulation. *J. Pet. Sci. Eng.* **2021**, *204*, 108747. [CrossRef]
10. Torres, L.; Besançon, G.; Georges, D. Multi-leak estimator for pipelines based on an orthogonal collocation model. In Proceedings of the 48h IEEE Conference on Decision and Control (CDC) held jointly with 2009 28th Chinese Control Conference, Shanghai, China, 15–18 December 2009; pp. 410–415. [CrossRef]
11. Besançon, G. *Nonlinear Observers and Applications*; Springer: Berlin/Heidelberg, Germany, 2007; Volume 363.
12. Torres, L.; Besançon, G.; Georges, D.; Navarro, A.; Begovich, O. Examples of pipeline monitoring with nonlinear observers and real-data validation. In Proceedings of the 8th International Multi-Conference on Systems, Signals and Devices, Sousse, Tunisia, 22–25 March 2011.
13. Roberson, J.A.; Cassidy, J.J.; Chaudhry, M.H. *Hydraulic Engineering*; John Wiley & Sons: Hoboken, NJ, USA, 1998.
14. Verde, C. Accommodation of multi-leak location in a pipeline. *Control Eng. Pract.* **2005**, *13*, 1071–1078. [CrossRef]
15. Besançon, G.; Georges, D.; Begovich, O.; Verde, C.; Aldana, C. Direct observer design for leak detection and estimation in pipelines. In Proceedings of the 2007 European Control Conference (ECC), Kos, Greece, 2–5 July 2007; pp. 5666–5670.
16. Mott, R.L.; Noor, F.M.; Aziz, A.A. *Applied Fluid Mechanics*; Pearson Prentice Hall: Singapore, 2006.
17. Navarro, A.; Begovich, O.; Besancon, G. Calibration of fitting loss coefficients for modelling purpose of a plastic pipeline. In Proceedings of the ETFA2011, Toulouse, France, 5–9 September 2011; pp. 1–6.
18. Takacs, G. *Sucker-Rod Pumping Handbook: Production Engineering Fundamentals and Long-Stroke Rod Pumping*; Gulf Professional Publishing: Houston, TX, USA, 2015.
19. E Shashi Menon, P. *Piping Calculations Manual*; McGraw-Hill Education: New York, NY, USA, 2005.
20. Diop, S.; Fliess, M. Nonlinear observability, identifiability, and persistent trajectories. In Proceedings of the 30th IEEE Conference on Decision and Control, Brighton, UK, 11–13 December 1991; pp. 714–719.
21. Isidori, A. *Nonlinear Control Systems*; Springer: Berlin/Heidelberg, Germany, 1995.
22. Diop, S. Elimination in control theory. *Math. Control. Signals Syst.* **1991**, *4*, 17–32. [CrossRef]
23. Mboup, M.; Join, C.; Fliess, M. Numerical differentiation with annihilators in noisy environment. *Numer. Algorithms* **2009**, *50*, 439–467. [CrossRef]
24. Rotoplas. *Tuboplus: Mejor Tubería, Mejor Agua*; Rotoplas: Mexico City, Mexico, 2018.
25. Begovich, O.; Pizano-Moreno, A.; Besançon, G. Online implementation of a leak isolation algorithm in a plastic pipeline prototype. *Lat. Am. Appl. Res.* **2012**, *42*, 131–140.

sensors

MDPI

Article

Chlorine Concentration Modelling and Supervision in Water Distribution Systems

Ramon Pérez [1,2], Albert Martínez-Torrents [2], Manuel Martínez [2,3], Sergi Grau [4,5], Laura Vinardell [2], Ricard Tomàs [4], Xavier Martínez-Lladó [2] and Irene Jubany [2,*]

1 Research Center for Supervision, Safety and Automatic Control, Universitat Politècnica de Catalunya, Rambla Sant Nebridi 10, 08222 Terrassa, Spain; ramon.perez@upc.edu
2 Sustainability Area, Eurecat, Centre Tecnològic de Catalunya, Plaça de la Ciència, 2, 08243 Manresa, Spain; albert.martinez@eurecat.org (A.M.-T.); manuel.martinez@eurecat.org (M.M.); laura.vinardell@eurecat.org (L.V.); xavier.martinez@eurecat.org (X.M.-L.)
3 Department of Mechanical Engineering, Universitat Rovira i Virgili, Av. Països Catalans 26, 43007 Tarragona, Spain
4 Aigües de Manresa S.A., Plana de l'Om 6, 3r-3a, 08240 Manresa, Spain; sgrau@aiguesmanresa.cat (S.G.); rtomas@aiguesmanresa.cat (R.T.)
5 Department of Mining, Industrial and ICT Engineering, Universitat Politècnica de Catalunya, Av. de les Bases de Manresa, 61-73, 08242 Manresa, Spain
* Correspondence: irene.jubany@eurecat.org; Tel.: +34-93-877-73-73

Abstract: The quality of the drinking water distributed through the networks has become the main concern of most operators. This work focuses on one of the most important variables of the drinking water distribution networks (WDN) that use disinfection, chlorine. This powerful disinfectant must be dosed carefully in order to reduce disinfection byproducts (DBPs). The literature demonstrates researchers' interest in modelling chlorine decay and using several different approaches. Nevertheless, the full-scale application of these models is far from being a reality in the supervision of water distribution networks. This paper combines the use of validated chlorine prediction models with an intensive study of a large amount of data and its influence on the model's parameters. These parameters are estimated and validated using data coming from the Supervisory Control and Data Acquisition (SCADA) software, a full-scale water distribution system, and using off-line analytics. The result is a powerful methodology for calibrating a chlorine decay model on-line which coherently evolves over time along with the significant variables that influence it.

Keywords: chlorine; water distribution networks; modelling; supervision; decay model

Citation: Pérez, R.; Martínez-Torrents, A.; Martínez, M.; Grau, S.; Vinardell, L.; Tomàs, R.; Martínez-Lladó, X.; Jubany, I. Chlorine Concentration Modelling and Supervision in Water Distribution Systems. *Sensors* **2022**, *22*, 5578. https://doi.org/10.3390/s22155578

Academic Editors: Eduard Llobet and Angela Maria Stortini

Received: 9 May 2022
Accepted: 19 July 2022
Published: 26 July 2022

1. Introduction

Disinfection is one of the most important steps in water treatment, as it must ensure the microbiological safety of the water generated, not only after treatment, but also throughout the transport process to the consumption point. Many countries use chlorine-based chemicals (sodium hypochlorite, chlorine dioxide, chloramines, etc.) to achieve this objective, as they guarantee the degree of residual disinfection potential that is required by their laws [1]. If required, booster disinfection stations are installed at different points in the network. Their need and best location can be optimized using models and tools based on estimates of the chlorine concentration. Chlorine concentration is precisely one of the most relevant parameters to consider for the water distribution network (WDN) quality management. Although chlorine ensures the absence of pathogens, it is the main cause of the formation of disinfection byproducts (DBPs) [2]. Most of these compounds are toxic or carcinogenic for human health and need to be controlled to ensure drinking water safety [3]. Thus, European legislation limits the concentration of some DBPs in drinking water [4].

Nowadays, given the lack of reliable and applicable models for predicting chlorine behavior, disinfection management is not optimal in most WDNs, since it is based on a point-specific control as opposed to the consideration of the whole network [5]. There are often dark points where the chlorine level may be too low, along with over-chlorination at other points (particularly in summer), with a subsequent increase in both operating costs and DBP concentrations.

The absence of robust models for predicting chlorine behavior in WDNs is fundamentally due to two aspects: (1) the complexity of modelling the hydraulics of the WDNs and (2) the need for on-line quality data. Although authors report good results in chlorine prediction in full-scale networks in some studies [6], the predictions become less accurate when the environmental conditions or the water composition change from those of the calibration. Such a situation is very common in WDNs fed with treated surface water.

Regarding the first aspect, WDNs are highly meshed and complex systems, the behavior of which is difficult to predict. The introduction of flow, level and pressure sensors, and automated metering readers (AMR) for consumption has recently increased the model accuracy [7,8]. Thus, the intense use of a large amount of hydraulic data together with hydraulic models and numerical simulators allow prediction of residence time, which is one of the main parameters needed for successful water quality prediction.

Regarding the second aspect, it is mandatory to obtain information on water quality in the effluent of the drinking water treatment plant and the relevant points in the WDNs in order to predict the behavior of chlorine. Several studies [6,9] base their decay models on parameters that are easily measured on-line, such as temperature, pH, redox potential, conductivity, turbidity, and chlorine concentration. Nevertheless, the calibration and maintenance of these models for their on-line use is seldom performed.

Common models for chlorine modelling in WDNs are first-order linear differential equations [10] such as:

$$\frac{dC}{dt} = K_b \cdot C^{\alpha} \tag{1}$$

where K_b is a constant that contains the different parameters and physicochemical phenomena that may affect chlorine decay, such as natural organic matter, inorganic compounds, or temperature [6]. More sophisticated models have also been studied, including second-order models [11].

Furthermore, knowing the effect of the parameters influencing chlorine decay is also important, as this will, in turn, allow the prediction of such decay and, in some cases, the application of corrective measures to reduce its effect.

Equations of different complexity have been used to model the effect of some of these parameters. One of the most influencing and studied parameters is temperature, which is usually based on Arrhenius's model [12] or other power models. Liu et al. [13] took pH and temperature into account in their models, thus differentiating the effect of HOCl and OCl species on pH. Similarly, Arevalo, in his doctoral dissertation [14], used a model that considered temperature and UV254 as an indication of organic matter. In this case, two decay constants were used, one related to chlorine decay on bulk water and one related to chlorine decay on water close to pipe wall.

Hassan et al. [15] studied the specific case of organic matter adsorbed onto goethite, which is the predominant iron oxide in pipe deposits, in order to see how effectively the presence of organic matter increased the decay rate. Their main conclusions were: (i) an increase in temperature causes an increase in the decay constant and therefore in the decay rate, (ii) the pH has not been seen to greatly affect decay, (iii) a higher initial chlorine concentration leads to a lower decay rate, (iv) a higher organic matter concentration, in general dissolved organic matter (DOC), causes an increase in the rate of decay, (v) an increase in the velocity of the water flow through the pipe causes an increase in the rate of decay, and (vi) the concentration of ammonium, nitrites, iron, and manganese seems to affect the rate of decay, causing its increase.

Chlorine decay first-order equations can also be used in software like EPANET, which has sufficient power to simulate and predict the concentration of chlorine in the network. EPANET is a public domain software for WDN modelling developed by the United States

Environmental Protection Agency (US EPA). This software can perform transient simulations of hydraulic behavior and water quality in pressurized pipe networks. In order to properly model water quality and its time evolution at consumer points, it is mandatory to have a reliable hydraulic model of the WDN. However, like any simulation software, EPANET depends on the availability and application of continuous data into robust models. The default built-in models have fixed calibration parameters that are not easily extrapolated in most real cases. The approach some authors take to overcome these barriers is to modify and recalibrate the default models included in the EPANET database based on the real system data to be modelled [16].

Another important aspect when modelling chemical reactions in pipes is the different behavior in wall and in bulk water. In the literature, bulk reactions are usually considered first-order and wall reactions are considered zero-order. Values for the bulk reaction coefficient are usually obtained using laboratory measurements [17]. Nevertheless, changing water characteristics in the network requires updating the model. There are a few approaches for the quality model calibration. This calibration requires a validated hydraulic model and water quality data. This is often carried out in a well-monitored part of the network and then generalized to the whole network [18,19].

This paper focuses on the chlorine decay process and variables that affect it, the models for concentration prediction, and their application within WDNs in a specific case-study. First, an on-line calibration procedure, with available data from the transport network, is adapted to the full-scale system and performed over a long period so that the evolution of the decay parameters can be studied. This model is used to predict the chlorine concentration in the distribution network and validated with discrete monitoring data. The dependence of chlorine decay on the relevant variables is also studied. Finally, this dependence is compared with the evolution of the parameters estimated using the on-line calibration method. The aim is to illustrate how the intense use of models and available data can provide a better understanding of the behavior of chlorine in a WDN and, thus, be used to support decision-making to improve water quality.

2. Materials and Methods

2.1. Case Study Network

The case study in this work is a WDN in Catalonia (Spain) (see location in Figure 1) managed by Aigües de Manresa, who provided the network configuration information and hydraulic and water quality data for a period of 14 months (2017–2018). The water supplied comes from the Llobregat River and goes through a prechlorination step with sodium hypochlorite (Apliclor Water Solutions S.L., Sant Martí Sesgueioles, Spain) or chlorine dioxide generated using sodium chlorite (Apliclor Water Solutions S.L., Sant Martí Sesgueioles, Spain) and hydrochloric acid (Apliclor Water Solutions S.L., Sant Martí Sesgueioles, Spain) (depending on the season), a sand filtration process, and a final disinfection step with sodium hypochlorite.

Two parts of the network were used in this study: the transport network and the district metered area (DMA). The transport network (Figure 2) consists of two water storage tanks (T_1 and T_2) equipped with sensors for chlorine concentration (input of T_1 (Cl_1) and output of T_2 (Cl_2)), flowmeters (outflows from the tanks, Q_1 and Q_2), and water level (H_1 and H_2). Water flows from T_1 to T_2 through a 6859 m main. T_2 is a boosting station with known sodium hypochlorite (Apliclor Water Solutions S.L., Sant Martí Sesgueioles, Spain) addition. The geometry of the tanks (volume) and pipes (length and diameter) are known. Water from T_2 is distributed to the rest of the network (through Q_2) of the DMA.

The DMA corresponds to a residential area. The hydraulic model includes 572 nodes and 610 pipes with a total length of 31 km, providing water to 300 consumers. Water flows by gravity. There are two quality-sampling points where the chlorine concentration is measured weekly. Figure 3 presents the model of this DMA visualized in EPANET. The input tank (corresponding to T_2 in Figure 2) and the two sampling points are highlighted (S_1 and S_2).

Figure 1. Study site location, a network in Catalonia supplied by the Llobregat River.

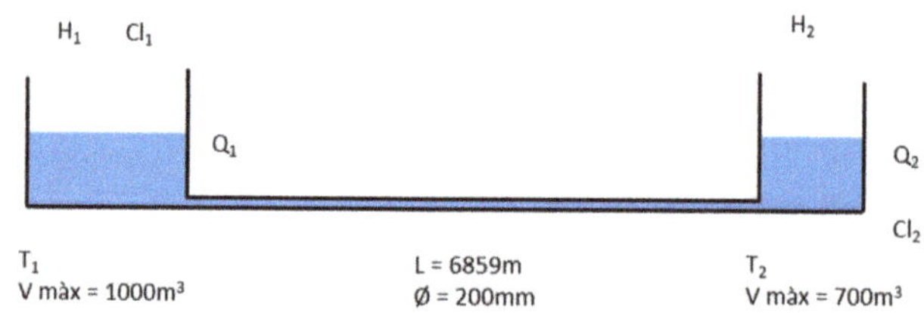

Figure 2. Outline of the network section used for the on-line calibration.

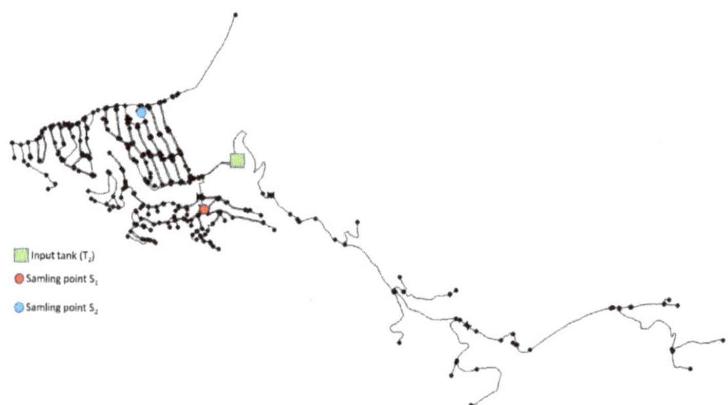

Figure 3. DMA network model in EPANET.

2.2. On-Line Calibration

A very well parametrized system in terms of hydraulics and chlorine concentration (at least at two points) is required to calibrate the chlorine decay constant of a WDN. As in this study, the transport network often fulfils this condition. Therefore, the network used for on-line calibration in this study was the transport network shown in Figure 2. The objective was to find the decay constants for the model that best explained the chlorine concentration measured at the output of T_2.

The chosen model was a first-order model. Higher-order models could be used with no fundamental changes in the methodology. Equation (2) shows that the chlorine concentration (Cl_2) at the outflow of T_2 depends on the input chlorine concentration (Cl_1) and the residence time in the system (t). The solution of Equation (1) is as follows:

$$Cl_2 = Cl_1 \cdot e^{-K_b \cdot t} \tag{2}$$

where K_b is the decay constant and α is considered as 1.

This dependence is defined by the decay constant K_b, which was calibrated on-line using the measurements available so that it was adapted throughout the year to the different water characteristics and environmental conditions. Estimations were performed on a weekly basis, since some information was only available at this frequency (chlorine dosing in T_2).

The residence time (RT) in T_1 was calculated from the hydraulic information available using (3). The weekly mean residence time at tank T_1 and the pipe was calculated using the flowmeter data (Q_1) and the volume of this subsystem.

$$\overline{RT}_1 = \frac{\overline{V}_1 + V_{pipe}}{\overline{Q}_1} \tag{3}$$

In order to estimate the mean water volume in T_1, the level data of the tank (H_1) and the geometric information was used. The residence time in T_2 was calculated using the mean values of the volume obtained from the level data (H_2) and the mean values of the tank effluent (Q_2), as shown in (4):

$$\overline{RT}_2 = \frac{\overline{V}_2}{\overline{Q}_2} \tag{4}$$

The chlorine concentration increase due to rechlorination ($\overline{Cl}_{added}$) was calculated using the added volume of chlorine divided by the mean volume of water treated:

$$\overline{Cl}_{added} = \frac{\overline{\Delta V_{Cl}} \cdot 143}{\int Q_2} \tag{5}$$

where $\overline{\Delta V_{Cl}}$ is the volume in liters of the concentrated chlorine added weekly to the network and 143 is the concentration of the added chlorine in g/L (value obtained from the conversion of the 15% NaClO to reactive chlorine, see Section S1 in the Supplementary Material).

Finally, a decay K_b constant was calculated which explained the chlorine concentration $\overline{Cl}_2$ at the outflow of T_2 given the residence time calculated using (6)

$$\overline{Cl}_2 = \overline{Cl}_1 \cdot e^{-K_b(\overline{RT}_1 + R\overline{RT}_2)} + \overline{Cl}_{added} \cdot e^{-K_b \overline{RT}_{T_2}} \tag{6}$$

where $\overline{Cl}_2$ was considered equal to 0.6 ppm, which is the set point of the chlorine control system in the boosting station. The algorithm for K_b calibration is shown in the Supplementary Material (Section S2).

The first order decay model is the most used in the water industry. Its decay constant includes all the dependencies related to environmental and water characteristics. Thus, the continuous updating of this constant is the guarantee of its reliability. The limitation of this methodology is the information required. Hydraulic information, that allows the determination of the residence time, must be available. Multiple chlorine concentration measurements and the exact volume of added chlorine between these measurements are also mandatory data.

2.3. Chlorine Decay Calibration and Validation

The chlorine decay first-order model validation for the transport network was carried out using available the on-line data of the chlorine concentration in the output of T2 considering the residence time in this tank. The data used covered the period from February 2017 to April 2018. The chlorine decay model was also validated for its use in the distribution network in the section where chlorine concentration is monitored.

Applying the calibrated decay model directly to the distribution network produced poor results. This was expected, due to the difference between the transport network and the distribution network regarding pipe size, materials, age, etc. To adjust the model, the available data period was divided into two sets: one for training the new distribution quality model and the remaining data for validation. There were 35 samples available in S_1 and 11 samples in S_2. Thus, the first 21 samples in S_1 were used for the training and the remaining ones for the validation. The algorithm used for this calibration is shown in the Supplementary Material (Section S3).

2.4. Parametrised Chlorine Decay Model

The decay constant, determined from the available on-line data, evolved clearly throughout 2017. The question arose if this could be due to the effect of the available variables such as temperature, the initial chlorine concentration, the cumulative precipitation, and the turbidity at the drinking water treatment plant or not.

This suggested the idea of analyzing the variables that influence chlorine decay in order to generate an empirical model based on the available independent variables. The availability of considerable data (temporal and spatial) implies dealing with large amounts of data, multiple variables, and experimental noise, which hinders the direct extraction of valuable information.

Principal component analysis (PCA) is a multivariate statistical technique that allows the description of the data according to the variance [20]. This method transforms data in noncorrelated new variables by linear transformation, decreasing the data dimensions. New data description is more condensed and can describe patterns that are hard to identify in multivariable datasets. PCA has already been used to determine the physical and chemical parameters influencing chlorine decay [21]. Therefore, for being a powerful, reliable, and globally accepted tool when dealing with big data, PCA was selected to extract the main trends, patterns, and correlations among the variables (dimensions) [11].

Based on the PCA results, the chlorine decay constant was modelled using the available variables and a potential multiparametric model (7).

$$K_b = K \cdot Parameter1^a \cdot Parameter2^b \cdot Parameter3^c \cdot \ldots \tag{7}$$

Specifically, a power model (8) and an Arrhenius model (9) [12] where calibrated using experimental data (temperature in 2017) and K_b obtained from the on-line calibration using the least square error fitting method implemented in the "Solver" function in Excel.

$$K_b = K_{power} \cdot T^a \tag{8}$$

$$K_b = A \cdot exp(-E_a/RT) \tag{9}$$

where, K_{power}, a, and A are constants, E_a is the activation energy (Jmol^{-1}), R is the universal gas constant, and T is the temperature. Finally, the parametrized chlorine decay model was compared with that obtained in the on-line calibration to assess its coherence throughout the year.

3. Results and Discussion

3.1. On-Line Calibration

The decay constants K_b obtained are presented in Figure 4. In the upper graphic, the weekly evolution between February 2017 and April 2018, can be observed. A different icon was used for the data of each trimester to clearly identify the season of the year. In the lower graphic obtained, K_b are grouped by month to observe how this parameter evolves throughout the year (some months include estimations of both years). It seems clear that there may be a seasonal variation related to temperature.

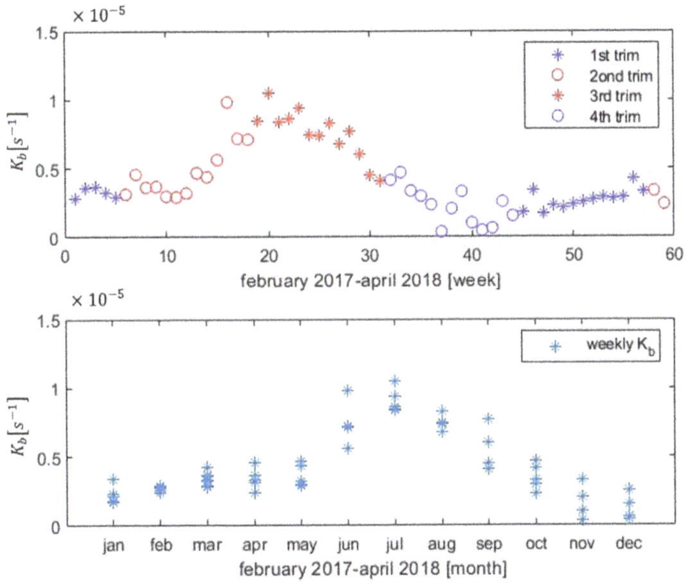

Figure 4. Up: Mean K_b calculated weekly indicating the season of the year (by trimester). Down: Data of mean K_b grouped by month.

3.2. *Chlorine Decay Validation*

The calibrated model was applied to the peak episodes observed at the output of T_2 due to the rechlorination and mixing effect. The dataset used in this validation was not used for the estimation. The dataset for calibration consisted of the mean values corresponding to the stationary state. Figure 5 shows the chlorine concentration data and the model prediction. It can be observed how this high-frequency dynamic is adjusted with the model obtained with the mean values. For this prediction, K_b evolves weekly.

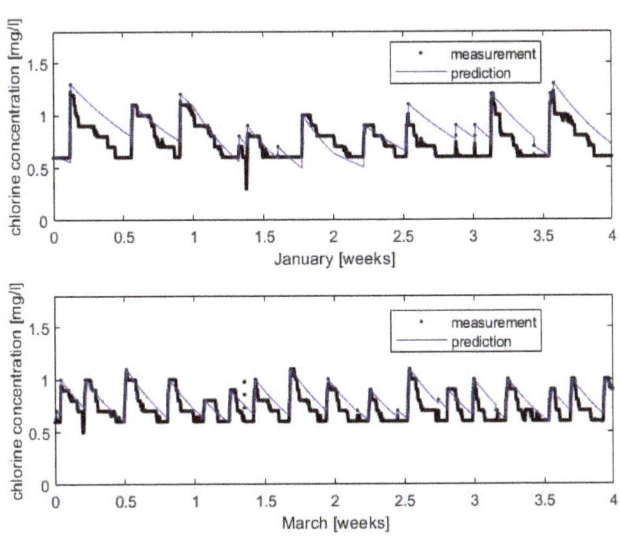

Figure 5. Chlorine decay model validation at the output of T_2. Up: data from January 2017. Down: data from March 2017. Due to high frequency of sampling measurement, data appears like a thick line.

For the distribution network simulation, a relation between the transport K_b, estimated by on-line calibration, with the distribution K_b^* was obtained by adjusting the concentration in the training set of chlorine sampling. This relation was applied to the entire period and the predicted concentration was compared with the measurements for the validation set of samples. A total of 40 days were simulated. The decay constant for both the bulk and wall were fitted using the first 21 samples of the chlorine concentration in S1. These are the first samples of upper graphic in Figure 6.

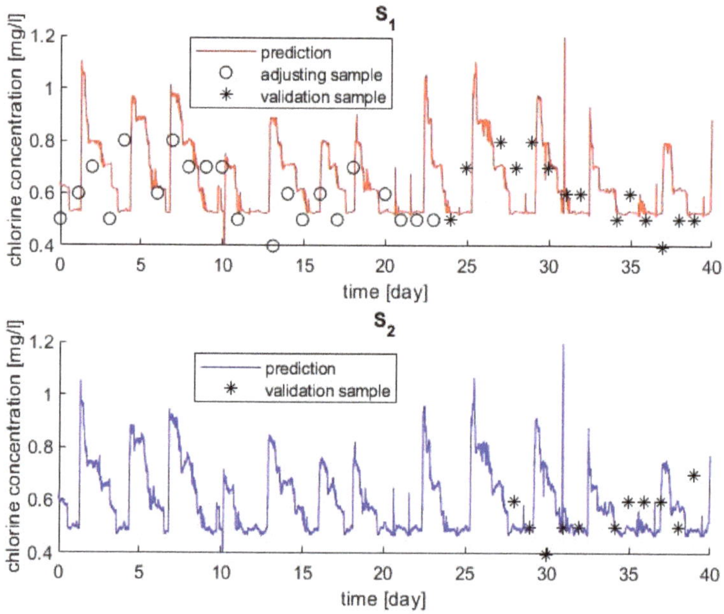

Figure 6. Simulation results for the two sampling points (**S_1** and **S_2**) compared with experimental samplings.

The result was that both decay constants minimize the error when the original K_b obtained in the transport system was divided by 2, as if the calibrated effect was distributed in the two phenomena ($K^*_{b,bulk} = K^*_{b,wall} = K_b/2$). The results obtained are compared with the available experimental data in Figure 6. The mean absolute percentage error was 16% for S_1 (including calibration and validation samples) while it was 17% using only validation samples. Therefore, not significantly different deviations were obtained for the calibration and the validation steps. Graphically, the fit may seem poor; however, the concentration is lower in S_2 than in S_1 both in prediction and measurements, and the mean values in both sampling points are coherent between the prediction and measurements. One aspect that may justify part of the mismatching is that the exact hour of the day of the manual measurements was not available and, therefore, each experimental data may not be in its exact position. This difficulty could be overcome with on-line chlorine sensors instead of manual analysis. Figure 7 presents the measured chlorine concentration at the source (T_2) and the chlorine prediction in the two sampling points (S_1 and S_2). The chlorine concentration decreases with the residence time, since the concentration in S_2 is lower than in S_1, and both are lower than in T_2.

Finally, Figure 8 shows the network nodes colored by their chlorine concentration: green, black, or red, depending on whether their concentration is too low, acceptable, or too high, respectively. In fact, no red points exist in this area and period. Such a representation is very useful in order for the network operator to make decisions. Nodes in this figure correspond to those in Figure 3 and are presented with north at the top.

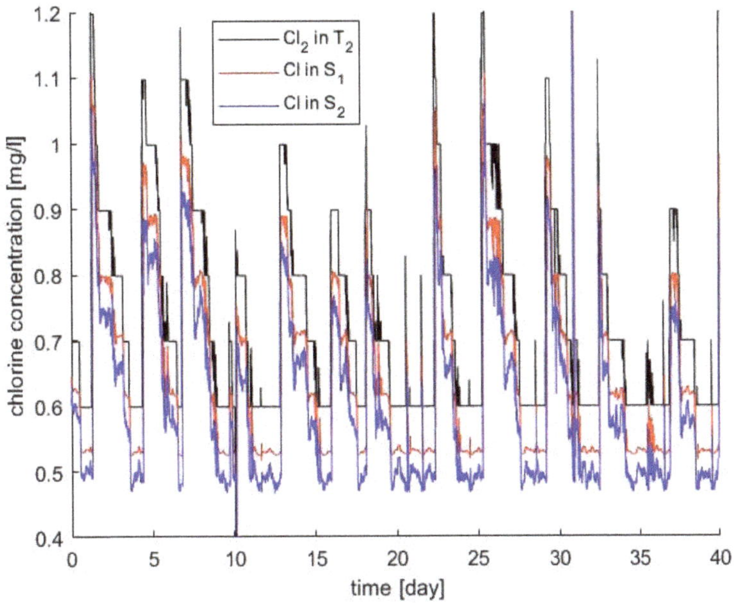

Figure 7. Measured chlorine concentration at the source (T_2) and prediction at the two sampling points (S_1 and S_2).

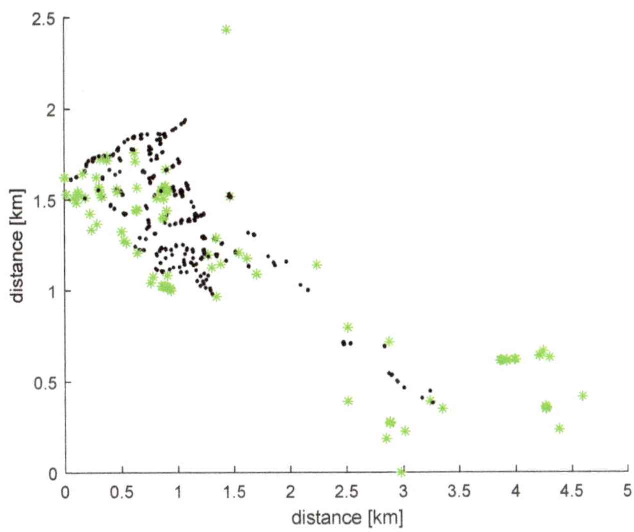

Figure 8. Distribution of network nodes with low concentration (<0.4 ppm, green), acceptable concentration (0.4 ppm > chlorine < 1 ppm, black), and excessive concentration (>1 ppm, red).

Results from the PCA applied to the decay constant (K_Cl_decay) determined in the distribution network and other data available (temperature as Tavg_C, initial chlorine concentration as Initial_Cl, cumulative precipitation as Cumulative_Prec, and turbidity at the drinking water treatment plant) are shown in Figure 9.

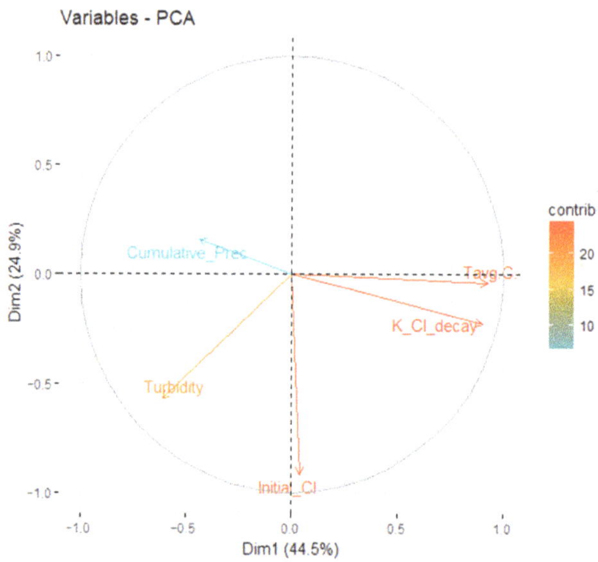

Figure 9. Variables for the two main dimensions of the PCA.

In this case, dimension one was related to the temperature, and dimension two was related to the initial chlorine concentration. Therefore, the decay constant was closely related to the temperature. Figure 10 shows the variables that had the most influence on dimension five, which were again the temperature and the chlorine decay, demonstrating the clear strong relation of the temperature on the decay constant.

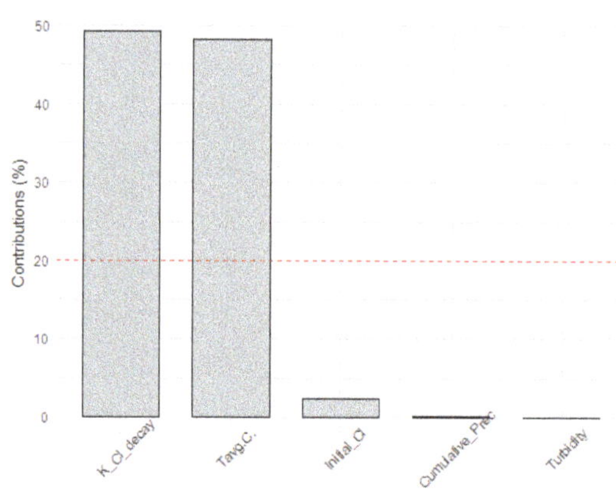

Figure 10. Contribution of variables to dimension 5. Variables with contributions below the dotted line are considered not significant for that dimension.

Thus, it was concluded that the variable that had a higher effect on the decay constant was temperature. It was observed that the variables turbidity, precipitation, and initial chlorine did not excessively improve the fit between the decay constants of the model and the decay constants obtained. Therefore, only the temperature was used, since the effort required to obtain values for the rest of the parameters did not compensate the improvement of the model adjustment.

Experimental data from 2017 and K_b obtained with the on-line calibration were used to calibrate the parameters of the two equations, an Arrhenius model (8) and a power model (9), to predict the temperature effect on the decay constant. The following Equations (10) and (11) show the results obtained.

$$K_b = 5.477 \cdot 10^{-8} \cdot T^{1.524} \left(s^{-1}\right) \quad (10)$$

$$K_b = 3950 \cdot exp(-49873/RT) \left(s^{-1}\right) \quad (11)$$

The fit obtained using the Arrhenius model and the power model were similar, although the Arrhenius one was slightly better. The Arrhenius model also determines the activation energy (J/mol), which is the minimum energy that the system needs for the reaction to take place. The ratio Ea/R obtained in this study was 5999 K, which is in accordance with other authors. For example, Courtis et al. [22] estimated 5388 K and 6701 K for two different water distribution systems, Powell et al. [23] obtained a range between 7500 and 9600 K, and Hua et al. [12] obtained a range between 8203 and 8727 K, depending on the type of water. The variability of the values for this ratio suggests that this is a water-specific parameter that might depend significantly on the natural organic matter composition [24].

Figure 11 shows the fit of the two models to data from 2017, and Figure 12 their forecast for the first days of 2018. In these figures, K Chlorine is the chlorine decay constant determined previously, K power the constant determined following the potential model, and K Arrhenius the constant obtained from the Arrhenius model. As it can be seen, the Arrhenius model provides good predictions while using only the temperature as an input parameter.

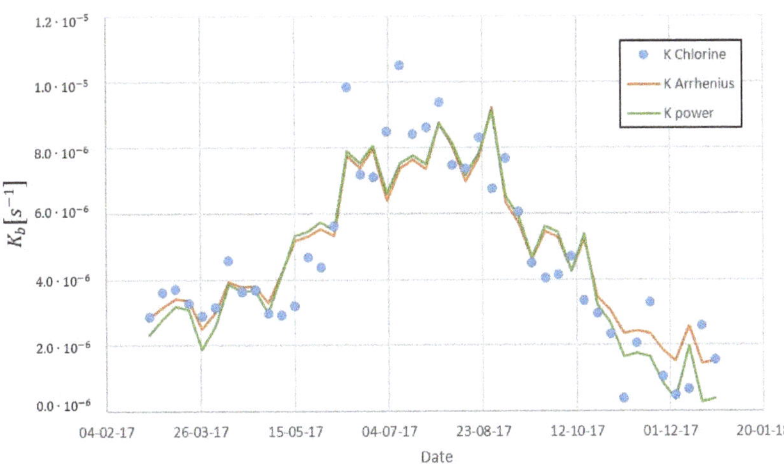

Figure 11. Chlorine decay constant fitting to temperature dependent models using data from 2017.

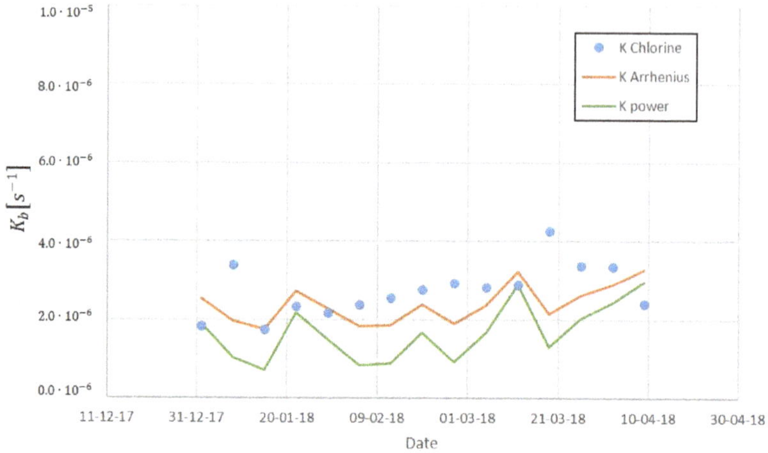

Figure 12. Chlorine decay constant forecast for the first days of 2018.

4. Conclusions

The literature review shows that only the simplest models of chlorine decay are applied to water distribution networks where the hydraulic behavior is complex enough. Even so, these models are seldom used due to the lack of proper calibration. In this paper, the performance of a decay model was evaluated when the parameters were calibrated using state-of-the-art techniques.

The calibration was first carried out in the transport network, where the on-line data allowed an on-line calibration. The relation between the decay constant in the transport and distribution networks used analytical data and, thus, it could not be done on-line.

The prediction error in the validation data was 17% and quite similar to the error obtained for the training set (16%), which meant that there was no overfitting. The decay constant obtained changed during the year following the assumed dependence of the chlorine decay on the temperature. This result suggested the possibility of using available data for predicting this decay constant.

The principal component analysis determined that the temperature was the parameter with higher effect on the decay of chlorine. The chlorine decay constant was obtained using temperature as an independent variable. The obtained constants were compared with the data-driven model obtained in the on-line calibration, showing a high correlation. While the dominant dependence on the temperature is not a novelty, it is important to ensure this unique dependency, as it guarantees that other characteristics of the water source will not be relevant. This has been studied throughout one year, and the results obtained by both models are coherent.

This procedure could also be applied to other quality parameters, such as disinfection byproduct concentrations, which are currently under investigation by the authors. The variables analyzed for chlorine decay estimation are being studied for the trihalomethanes formation prediction.

The final aim of this study is to increase knowledge within the network in order to enable decision-making processes regarding chlorine dosing (quantity and frequency) both in the disinfection process and the boosting stations, in addition to identifying whether other control systems are required to ensure the continuous good quality of supplied water to the final user at a minimum cost.

Supplementary Materials: The following supporting information can be downloaded at: https://www.mdpi.com/article/10.3390/s22155578/s1, Table S1: Pseudocode to estimate the decay chlorine constant; Table S2: Pseudocode to estimate the chlorine decay constant in the distribution network.

Author Contributions: Conceptualization, R.P., R.T. and X.M.-L.; Data curation, R.P., A.M.-T., M.M. and S.G.; Formal analysis, R.P., A.M.-T. and M.M.; Funding acquisition, R.T., X.M.-L. and I.J.; Investigation, A.M.-T., M.M. and L.V.; Methodology, R.P., A.M.-T. and M.M.; Project administration, X.M.-L. and I.J.; Resources, S.G. and R.T.; Software, R.P., A.M.-T. and M.M.; Supervision, R.P., X.M.-L. and I.J.; Validation, R.P. and X.M.-L.; Visualization, R.P., A.M.-T. and L.V.; Writing—original draft, R.P., A.M.-T. and L.V.; Writing—review and editing, R.P., L.V. and I.J. All authors have read and agreed to the published version of the manuscript.

Funding: This research received no external funding.

Data Availability Statement: The data presented in this study are available on request from the corresponding author. The data is not publicly available because the owners want to keep track of its use, as it is sensitive information from a full-scale system (hydraulic models). Nevertheless, software for data treatment and chlorine decay model estimation and validation including nonsensitive data (model_estimation.m and model_validation.m) are available at https://cs2ac.upc.edu/en (18 July 2022) in Matlab format.

Conflicts of Interest: The authors declare no conflict of interest.

References

1. Tsitsifli, S.; Kanakoudis, V. Disinfection Impacts to Drinking Water Safety–A Review. *EWaS3* **2018**, 2, 603. [CrossRef]
2. Mazhar, M.A.; Khan, N.A.; Ahmed, S.; Khan, A.H.; Hussain, A.; Rahisuddin; Changani, F.; Yousefi, M.; Ahmadi, S.; Vambol, V. Chlorination disinfection by-products in municipal drinking water–A review. *J. Clean. Prod.* **2020**, *273*, 123159. [CrossRef]
3. Richardson, S.D.; Plewa, M.J.; Wagner, E.D.; Schoeny, R.; DeMarini, D.M. Occurrence, genotoxicity, and carcinogenicity of regulated and emerging disinfection by-products in drinking water: A review and roadmap for research. *Mutat. Res. Rev. Mutat. Res.* **2007**, *636*, 178–242. [CrossRef] [PubMed]
4. European Commission. Directive (EU) 2020/2184 of the European Parlliament and of the Council of 16 December 2020 on the quality of water intended for human consumption. *Off. J. Eur. Union* **2020**, *2019*, 1–61.
5. Wang, Y. Optimization of booster chlorination for water distribution system under dual uncertainties. *Urban Water J.* **2022**, *19*, 363–373. [CrossRef]
6. Tiruneh, A.T.; Debessai, T.Y.; Bwembya, G.C.; Nkambule, S.J. A Mathematical Model for Variable Chlorine Decay Rates in Water Distribution Systems. *Model. Simul. Eng.* **2019**, *2019*, 5863905. [CrossRef]
7. Walski, M.T.; Chase, V.D.; Savic, A.D.; Grayman, W.; Beckwith, S.; Koelle, E. Methods. In *Advanced Water Distribution Modeling and Management*, 1st ed.; Bentley Institute Press: Exton, PA, USA, 2007.
8. Puig, V.; Ocampo-Martínez, C.; Pérez, R.; Cembrano, G.; Quevedo, J.; Escobet, T. *Real-Time Monitoring and Operational Control of DrinkingWater Systems*; Springer Nature: Berlin/Heidelberg, Germany, 2017.
9. Shahamat, Y.D.; Godini, K.; Khademi, S.M.R.S.; Rahmani, K.; Rahmani, A. Investigation of chlorine decay of water resource in khanbebein city, Golestan, Iran. *Int. J. Environ. Health Eng.* **2013**, *2*, 11. [CrossRef]
10. Kowalska, B.; Kowalski, D.; Musz, A. Chlorine decay in water distribution systems. *Env. Prot. Eng.* **2006**, *32*, 5–16.
11. Clark, R.M.; Sivaganesan, M. Predicting Chlorine Residuals in Drinking Water: Second Order Model. *J. Water Resour. Plan. Manag.* **2002**, *128*, 152–161. [CrossRef]
12. Hua, P.; Vasyukova, E.; Uhl, W. A variable reaction rate model for chlorine decay in drinking water due to the reaction with dissolved organic matter. *Water Res.* **2015**, *75*, 109–122. [CrossRef] [PubMed]
13. Liu, B.; Reckhow, D.A.; Li, Y. A two-site chlorine decay model for the combined effects of pH, water distribution temperature and in-home heating profiles using differential evolution. *Water Res.* **2014**, *53*, 47–57. [CrossRef] [PubMed]
14. Arevalo, J. *Modeling Free Chlorine and Chloramine Decay in a Pilot Distribution System*; University of Central Florida: Orlando, FL, USA, 2007.
15. Hassan, K.Z.A.; Bower, K.C.; Miller, C.M. Iron Oxide Enhanced Chlorine Decay and Disinfection By-Product Formation. *J. Environ. Eng.* **2006**, *132*, 1609–1616. [CrossRef]
16. Seyoum, A.G.; Tanyimboh, T.T. Integration of Hydraulic and Water Quality Modelling in Distribution Networks: EPANET-PMX. *Water Resour. Manag.* **2017**, *31*, 4485–4503. [CrossRef]
17. Fisher, I.; Kastl, G.; Sathasivan, A. Evaluation of suitable chlorine bulk-decay models for water distribution systems. *Water Res.* **2011**, *45*, 4896–4908. [CrossRef] [PubMed]
18. Xie, X.; Zeng, B.; Nachabe, M. Sampling design for water distribution network chlorine decay calibration. *Urban Water J.* **2015**, *12*, 190–199. [CrossRef]

19. Nejjari, F.; Puig, V.; Pérez, R.; Quevedo, J.; Cugueró, M.; Sanz, G.; Mirats, J. Chlorine Decay Model Calibration and Comparison: Application to a Real Water Network. *Procedia Eng.* **2014**, *70*, 1221–1230. [CrossRef]
20. Anderson, T.W. *An Introduction to Multivariate Statistical Analysis*; John Wiley & Sons: Hoboken, New Jersey, USA, 2003.
21. Chowdhury, S.; Champagne, P.; McLellan, P.J. Factors Influencing Formation of Trihalomethanes in Drinking Water: Results from Multivariate Statistical Investigation of the Ontario Drinking Water Surveillance Program Database. *Water Qual. Res. J.* **2008**, *43*, 189–199. [CrossRef]
22. Courtis, B.J.; West, J.R.; Bridgeman, J. Temporal and Spatial Variations in Bulk Chlorine Decay within a Water Supply System. *J. Env. Eng.* **2009**, *135*, 147–152. [CrossRef]
23. Powell, J.C.; Hallam, N.B.; West, J.R.; Forster, C.F.; Simms, J. Factors which control bulk chlorine decay rates. *Water Res.* **2000**, *34*, 117–126. [CrossRef]
24. Monteiro, L.; Viegas, R.; Covas, D.I.C.; Menaia, J. Modelling chlorine residual decay as influenced by temperature. *Water Environ. J.* **2015**, *29*, 331–337. [CrossRef]

sensors

MDPI

Article

Economic Linear Parameter Varying Model Predictive Control of the Aeration System of a Wastewater Treatment Plant †

Fatiha Nejjari [1], Boutrous Khoury [1], Vicenç Puig [1,2,*], Joseba Quevedo [1], Josep Pascual [1] and Sergi de Campos [3]

1 Advanced Control Systems Group, Automatic Control Department, Edifici Gaia Campus de Terrassa, Universitat Politècnica de Catalunya, Rambla Sant Nebridi, 22, 08222 Terrassa, Spain
2 Institut de Robòtica i Informàtica Industrial (CSIC-UPC), Llorens i Artigas, 4-6, 08028 Barcelona, Spain
3 ADASA Sistemas, S.A.U. C/Ignasi Iglesias, 217, 08820 El Prat de Llobregat, Spain
* Correspondence: vicenc.puig@upc.edu

† This paper is an extended version of our paper published in Proceedings of the 27th European Symposium on Computer Aided Process Engineering (ESCAPE 27), Barcelona, Spain, 1–5 October 2017.

Abstract: This work proposes an economic model predictive control (EMPC) strategy in the linear parameter varying (LPV) framework for the control of dissolved oxygen concentrations in the aerated reactors of a wastewater treatment plant (WWTP). A reduced model of the complex nonlinear plant is represented in a quasi-linear parameter varying (qLPV) form to reduce computational burden, enabling the real-time operation. To facilitate the formulation of the time-varying parameters which are functions of system states, as well as for feedback control purposes, a moving horizon estimator (MHE) that uses the qLPV WWTP model is proposed. The control strategy is investigated and evaluated based on the ASM1 simulation benchmark for performance assessment. The obtained results applying the EMPC strategy for the control of the aeration system in the WWTP of Girona (Spain) show its effectiveness.

Keywords: economic model predictive control; linear parameter varying modelling; wastewater treatment process

Citation: Nejjari, F.; Khoury, B.; Puig, V.; Quevedo, J.; Pascual, J.; de Campos, S. Economic Linear Parameter Varying Model Predictive Control of the Aeration System of a Wastewater Treatment Plant. *Sensors* **2022**, *22*, 6008. https://doi.org/10.3390/s22166008

Academic Editor: Assefa M. Melesse

Received: 10 May 2022
Accepted: 9 August 2022
Published: 11 August 2022

1. Introduction

Biological wastewater treatment plants (WWTPs) are complex nonlinear systems with large variations in their flow rates and feed concentrations. These plants have to be operated continuously taking care of strict environmental regulations. Thus, the use of advanced control strategies becomes necessary to make them more efficient.

The most widely used biological wastewater treatment is the activated sludge process (ASP). In the ASP, microorganisms are mixed with wastewater. The pollutants of the wastewater constitute the nutrient of the microorganisms. As the organisms feed on the organic pollutants in the wastewater, the pollutants are converted to more organisms, biomass, and some by-products. Following an adequate amount of treatment time, the mixture of microorganisms and wastewater, the mixed liquor flows from the aeration tank to a clarifier or settler where the sludge is separated from the treated water. Some of the settled sludge is continuously recirculated from the clarifier to the aeration tank to ensure the maintenance of adequate amounts of microorganisms in this tank. The microorganisms are again mixed with incoming wastewater where they are reactivated to consume organic nutrients. There are five major groups of microorganisms generally found in the aeration basin of the activated sludge process: (i) aerobic bacteria responsible for removing the organic nutrients, (ii) protozoa to remove and digest dispersed bacteria and suspended particles, (iii) metazoa to dominate longer age systems and clarify effluent, (iv) filamentous bacteria or bulking sludge, which are present when operating conditions change, (v) algae and fungi, which are photosynthetic organisms that are present with pH changes and older sludge.

The majority of the culture is mixed and reused with inlet wastewater to keep high reaction rates and sludge age characteristics. In particular, nitrogen is eliminated as follows: First, ammonium is oxidized producing nitrate under nitrification in the aerobic step. The nitrate that is produced is then converted into nitrogen gas by means of denitrification in the anoxic step. Thus, the control of aeration is very important because a low amount of dissolved oxygen can cause the biomass death. On the other hand, an excess of dissolved oxygen could cause the sludge to settle insufficiently. Moreover, because 60% to 80% of the global energy consumption is due to aeration and the operating costs of a WWTP [1], an excessive aeration is not desirable regarding economic efficiency.

The models that are usually considered for characterizing the WWTP processes are the ones developed by the International Association on Water Quality (IAWQ) known as Activated Sludge Models (ASMs) [2].

In this paper, optimal economic operation of the aeration system is considered to improve the efficiency and reliability of an ASP with intermittent aeration, which is used for the removal of nitrogen from domestic wastewater. The objective of the control is to design an aeration strategy (air-on and air-off periods) which minimizes the energy dissipated by the aeration system, with adherence to the limits of the effluent requirements and the operating constraints. The implementation of optimal operation strategies is therefore interesting because WWTPs face the challenge of treating water properly albeit ensuring the minimization of operational costs. This has been the driving force for the active research in the development of advanced control techniques and hierarchical control schemes to improve the operation of the WWTPs, see for example [3,4].

Model predictive control (MPC) has been the most successful advanced control approach applied to control WWTPs. This is due to the fact that MPC controllers allow in a straightforward manner the different operational requirements, the multivariate nature of the control problem (that could even include delay) and directly handling constraints on the control inputs, system outputs and/or internal states [5]. It can also include disturbance prediction, allowing to anticipate the appropriate control actions (feedforward) to achieve optimal performance according to defined criteria in the cost function, which can include different quality criteria and operational costs. Adjusting the MPC control strategy is carried out by suitable manipulating prioritization of different objectives of the performance index that could also include the use of soft constraints. In this way, MPC has become an attractive control strategy for a considerable number of WWTP applications in the last few years. Some examples of MPC control of WWTP can be found in [6,7]. In [6], a benchmarking of different hierarchical control structures for WWTPs that combines static and dynamic real-time optimization (RTO) and nonlinear model predictive control (NMPC) is presented. In [8], a procedure to find the best controlled variables in an economic sense for the activated sludge process in a wastewater treatment plant, despite the large load disturbances, is introduced.

Classical MPC formulation considers pre-established set points, and the objective functions related with error and energy effort have quadratic forms [5]. However, the determination of optimal and reachable reference set points in real time is not an easy task because of the existence of disturbances, set-point changes, time-varying parameters and model uncertainties, among others. This constitutes one of the main limitations of classical MPC. To remedy this issue, real-time optimizers (RTO) or steady-state target optimizers (SSTO) are used to pre-compute the reference set- points at a supervisory layer in the control hierarchy. Then, these pre-computed set points are sent to the lower layer, where a classical MPC behaves as a regulatory controller, forcing the process to follow the desired set points. However, in spite of the use of an RTO, not reachable trajectories might be generated because of the appearance of unexpected disturbances or set points variations, among others. Moreover, there is a delay between the different layers, because the lower layer receives the reference set points determined from the upper layer before its execution. These problems can be avoided by using economic MPC (EMPC) that optimizes process performance directly (e.g., by means of economic objective functions), eliminating the need of generating reachable reference set points [9]. The first results of the application of EMPC to DO concentration control in WWTPs have been presented in a previous work from the authors [10] considering the nonlinear model of the plant. However, this leads to a nonlinear optimization problem.

Alternatively, this paper proposes an EMPC strategy using the linear parameter-varying (LPV) framework to optimize the effluent quality and minimize the operational cost of a WWTP under operating and physical constraints. The objective is to minimize the energy used by the aeration system with the control of the dissolved oxygen (DO) concentrations in the aerated reactors and maintain the effluent concentration under the required limits. The proposed approach is based on real-time dynamic optimization methods. Optimization in MPC with nonlinear models presents a non-convex problem which is computationally demanding, especially when dealing with large-scale plants with complex dynamics such as the WWTP. Thus, the LPV framework allows the embedding of these nonlinearities in scheduling variables, which are functions of system states (i.e., qLPV). This allows obtaining a pseudo-linear model which is linear in state space but nonlinear in the parameter space and deriving a less demanding convex MPC optimization problem, since convex quadratic optimization tools can be applied. The stability and recursive feasibility of MPC with LPV models has been studied (see [11] for a review of the recent results). The application of dynamic optimization methods requires a sufficiently accurate mathematical model describing the wastewater treatment process. The present work uses the Activated Sludge Model No. 2 (ASM2) [12]. To illustrate the proposed approach a WWTP located in Girona (Spain) is considered as a case study.

In Section 2, the WWTP is described and modeled using a reduced ASM2 model, which is then represented in a qLPV form. The proposed EMPC strategy is introduced and described in Section 3, while the proposed MHE approach is presented in Section 4. The results are presented in Section 5, with simulation scenarios obtained from the application of the EMPC strategy on the Girona WWTP. Finally, some conclusions are given in Section 6.

2. WWTP Description and Modeling

2.1. WWTP Description

The Girona WWTP is a biological treatment plant designed to treat the wastewater generated by 200,000 inhabitant equivalents with a medium daily inflow of 35,000 m^3/d. The processes of the plant can be divided into two main treatment lines: water and sludge (see Figure 1). The water line is separated into three phases: pre-treatment, primary treatment and secondary treatment. The secondary treatment is designed to convert biodegradable, organic wastewater constituents and certain inorganic fractions into new cell mass and by-products. The plant uses an activated sludge system and has three lines composed of three main reactors that are divided into various compartments and three clarifiers. Each line is made of two anoxic reactors located at the beginning, three aerated tanks and an anoxic tank followed by an aerated one. With this configuration, the plant can nitrify and denitrify with great efficiency. The anoxic and aerobic tanks have volumes of 1335, 4554, 1929, and 1929 m^3 for anoxic and 1929, 1276, and 1409 m^3 for aerobic, respectively. Oxygen is supplied to aerated tanks by the aeration system, which delivers air to each of the aeration tanks. The wastewater and activated sludge are separated into three parallel secondary settlers. The volume of each secondary settler is approximately 5024 m^3. The activated sludge is internally recirculated from the last aerobic zone to the anoxic tank (210% of influent waste). Additionally, the wastewater is recirculated from the secondary settlers to the anoxic tank (45 to 100% of influent waste).

Figure 2 shows a standard WWTP technological layout. The wastewater flow enters into the biological part after the mechanical treatment. The nutrient removal takes place in the activated sludge reactor through the biological treatment. The first zone in this treatment is anaerobic, where phosphorus is released. The mixed liquor internal recirculation originates from the anoxic zone. The denitrification occurs in the second zone.

The activated sludge returned from the clarifiers bottom, and the internal recirculation from the aerobic zones end is directed toward the anoxic zone.

2.2. WWTP Modeling

The Benchmark Simulation Model (BSM1), developed within the framework of COST Actions 624 and 682 [2], has been adapted to represent the Girona WWTP (see Figure 1).

Figure 1. Girona Wastewater Treatment Plant.

The Activated Sludge Model No. 1 (ASM1) describes the biological phenomena that takes place in the biological reactors, and it is supposed that no biological reactions take place in the settlers. Due to the complexity of the nonlinear model describing the different complex processes in the plant, various reduced models have been proposed in the literature [13–15] to aid in the online implementation of certain modern control schemes (e.g., MPC), which would have otherwise presented ill-conditioned or stiff numerical problems due to slow and fast dynamic interactions. In [15,16], one can see some successful implementations using reduced WWTP models in various areas of control applied to WWTPs. The reduced model as suggested in [14], which primarily involves certain simplification criteria for a reduced order of the rigorous high dimensional WWTP model, has been adapted to conditions representing the Girona WWTP. This basically involves the derivation of the reactor model based on mass balances of the wastewater species, which are generally expressed as follows:

$$Accumulation = Inflow - Outflow + Reaction$$

Validation of the reduced model considering data from the ASM1 and the reduced model has been undertaken in [14]. In simplifying the complex model, a systematic reduction process of the high-dimensional model considers some assumptions, with the principal conditions given as follows:

- The soluble (S_S) and particulate (X_S) organic compounds are aggregated as a single variable X_{COD}, the chemical oxigen demand (COD).
- Through reduction by time scale from the theory of singular pertubation, the slow dynamics of the variables X_I, X_{BH} and X_{BA} together with the soluble inert organic compounds (S_I) are excluded.
- Finally, simplification of complicated kinetic process, assumption of no alkalinity and separation of aerobic and anoxic conditions are considered.

Under these conditions, the resultant state variables of the reduced model are therefore the chemical oxygen demand (X_{COD}), the dissolved oxygen concentration, (S_O), heterotrophic biomass, X_{BH}, ammonia concentration (S_{NH}), nitrate concentration (S_{NO}) and autotrophic biomass (X_{BA}). The control of oxygen concentration (S_0) in the aerobic tanks is via the manipulation of the control input, the oxygen transfer coefficient $K_{La}(t)$.

The states and input vectors are thus given as:

$$x(t) = \Big[X_{COD}(t), S_O(t), X_{BH}(t), S_{NH}(t), S_{NO}(t), X_{BA}(t)\Big]^T$$

$$u(t) = K_{La}(t)$$

The WWTP process is therefore described by the following dynamic equations of the reduced model:

$$\dot{X}_{COD}(t) = \frac{1}{Y_h}\Big[\theta_1(t) + \theta_2(t)\Big] + \Big(1 - f_p\Big)\Big(\theta_4(t) + \theta_5(t)\Big) + \vartheta_1(t), \tag{1}$$

$$\dot{S}_O(t) = \frac{Y_h - 1}{Y_h}\theta_1(t) + \frac{Y_a - 4.57}{Y_a}\theta_3(t) + \vartheta_2(t), \tag{2}$$

$$\dot{S}_{NH}(t) = -i_{xb}\Big[\theta_1(t) + \theta_2(t)\Big] - \Big[i_{xb} + \frac{1}{Y_a}\Big]\theta_3(t) + \Big(i_{xb} - f_p i_{xp}\Big)\Big[\theta_4(t) + \theta_5(t)\Big] + \vartheta_3(t), \tag{3}$$

$$\dot{S}_{NO}(t) = \frac{Y_h - 1}{2.86Y_h}\theta_2(t) + \frac{1}{Y_a}\theta_3(t) + \vartheta_4(t), \tag{4}$$

$$\dot{X}_{BH}(t) = \theta_1(t) + \theta_2(t) - \theta_4(t) + \vartheta_5(t), \tag{5}$$

$$\dot{X}_{BA}(t) = \theta_3(t) - \theta_5(t) + \vartheta_6(t). \tag{6}$$

where

$$\theta_1(t) = \mu_h \frac{X_{COD}(t)}{K_{COD} + X_{COD}(t)} \frac{S_O(t)}{K_{OH} + S_O(t)} X_{BH}(t)$$

$$\theta_2(t) = \mu_h \eta_{NOg} \frac{X_{COD}(t)}{K_{COD} + X_{COD}(t)} \frac{S_{NO}(t)}{K_{NO} + S_{NO}(t)} \frac{K_{OH}}{K_{OH} + S_O(t)} X_{BH}(t)$$

$$\theta_3(t) = \mu_a \frac{S_{NH}(t)}{K_{NH,A} + S_{NH}(t)} \frac{S_O(t)}{K_{O,A} + S_O(t)} X_{BA}(t)$$

$$\theta_4(t) = b_H X_{BH}(t)$$

$$\theta_5(t) = b_A X_{BA}(t)$$

With the flow rate given as $Q_{in}(t)$, V_o as the volume of the aerobic tank and considering that $S_{O_{in}}(t), S_{NO_{in}}(t), X_{BA_{in}}(t)$ are equal to zero. $\vartheta_1(t), \vartheta_2(t), \cdots, \vartheta_6(t)$ are given as follows:

$$\vartheta_1(t) = \frac{Q_{in}(t)}{V_o}\Big[X_{COD_{in}}(t) - X_{COD}(t)\Big]$$

$$\vartheta_2(t) = \frac{Q_{in}(t)}{V_o}\Big[-S_O(t)\Big] + K_{La}(t)\Big[S_{O_{sat}} - S_O(t)\Big]$$

$$\vartheta_3(t) = \frac{Q_{in}(t)}{V_o}\Big[S_{NH_{in}}(t) - S_{NH}(t)\Big]$$

$$\vartheta_4(t) = \frac{Q_{in}(t)}{V_o}\Big[-S_{NO}(t)\Big]$$

$$\vartheta_5(t) = \frac{Q_{in}(t)}{V_o}\Big[X_{BH_{in}}(t) - \frac{f_w(1+f_r)}{f_r + f_w} X_{BH}(t)\Big]$$

$$\vartheta_6(t) = \frac{Q_{in}(t)}{V_o}\Big[X_{BA}(t) - \frac{f_w(1+f_r)}{f_r + f_w} X_{BA}(t)\Big]$$

where Y_H, Y_A, f_r, f_w, b_h, b_A, i_{xb}, and f_p are the stoichiometric parameters and μ_h, K_{COD}, K_{OH}, μ_a, $K_{NH,A}$, and $K_{O,A}$ are the kinetic parameters.

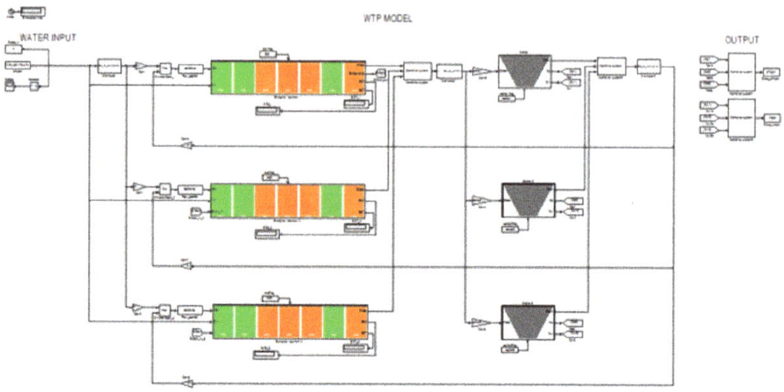

Figure 2. Layout of Girona WWTP.

2.3. LPV Representation of the WWTP

For ease of computational burden, the nonlinear reduced model is represented in a LPV form which involves the embedding of nonlinearities in varying parameters, resulting in a linear representation in state space. This procedure offers benefits when applied to MPC over its nonlinear MPC [15] and linear MPC [17] counterparts as applied on the WWTP by providing a faster run time and the avoidance of numerical problems with respect to the former and the ability to operate in a wide range of operating points with regard to the latter. The nonlinear model in this case is defined by linear systems at each time instance based on some time-varying parameters $\sigma(t) \in \mathbb{R}^{n_\sigma}$, with an assumption that the parameters $\sigma(t)$ are not known a priori but can be measured or estimated online [18]. The dynamic behavior of the LPV model is therefore described as:

$$\dot{x}(t) = A(\sigma(t))x(t) + B(\sigma(t))u(t) \tag{7}$$

$$y(t) = C(\sigma(t))x(t) + D(\sigma(t))u(t) \tag{8}$$

where $x(t) \in \mathbb{R}^{n_x}$ and $u(t) \in \mathbb{R}^{n_u}$ are the states and inputs, respectively, with $y(t) \in \mathbb{R}^{n_y}$ as the measured signals. $A(\sigma(t))$, $B(\sigma(t))$, $C(\sigma(t))$ and $D(\sigma(t))$ are time-varying matrices of appropriate dimensions that are affine in $\sigma(t) \in \mathbb{R}^{n_\sigma}$. In the quasi LPV case, the scheduling parameters are dependent on measured signals, $y_s(t) \in \mathbb{R}^k \subset y(t) \in \mathbb{R}^{n_y}$, such that

$$\sigma(t) = f(y_s(t))$$

where $f : \mathbb{R}^k \mapsto \mathbb{R}^{n_\sigma}$ is a continuous mapping [19]. With observed states and exogenous inputs ($w(t)$), nonlinearities involving the system states can be "hidden" in the varying parameters, $\sigma(t, y_s(t), w(t))$.

Therefore, from the generic nonlinear form

$$\begin{aligned} \dot{x}(t) &= f(x(t), u(t), w(t)) \\ y(t) &= g(x(t), u(t)) \end{aligned} \tag{9}$$

a linear quadruple $(A(\sigma(t, y_s(t), w(t)))$, $B(\sigma(t, y_s(t), w(t)))$, $C(\sigma(t, y_s(t), w(t))), and$ $D(\sigma(t, y_s(t), w(t)))$ estimate is formulated and incorporated into the EMPC for a convex optimization problem. In the following, the function $\sigma(t, y_s(t), w(t)))$ will simply be represented as $\sigma(t)$. The choice of scheduling parameters considering the origin of nonlinearities in the reduced model (1) are

$$\sigma_1(t) = Q_{in}(t), \qquad \sigma_2(t) = \frac{X_{COD}(t)}{K_{COD} + X_{COD}(t)} \frac{X_{BH}(t)}{K_{OH} + S_O(t)},$$
$$\sigma_3(t) = \frac{X_{COD}(t)}{K_{COD} + X_{COD}(t)} \frac{S_{NO}(t)}{K_{NO} + S_{NO}(t)} \frac{K_{OH}}{K_{OH} + S_O(t)},$$
$$\sigma_4(t) = \frac{1}{K_{OA} + S_O(t)} \frac{S_{NH}(t)}{K_{NH} + S_{NH}(t)} X_{BH}(t), \qquad \sigma_5(t) = S_O(t).$$

The dynamic LPV model is thus given as:

$$\dot{x} = A(\sigma(t))x(t) + B(\sigma(t))u(t) + Ew(t). \tag{10}$$

with the time-varying matrices, $A(\sigma(t))$, $B(\sigma(t))$, and time-invariant disturbance matrix E as:

$$A(\sigma(t)) = \begin{bmatrix} a_{11}(t) & 0 & 0 & 0 & a_{15}(t) & a_{16} \\ 0 & a_{22}(t) & 0 & 0 & a_{25}(t) & 0 \\ 0 & a_{32}(t) & 0 & 0 & a_{35}(t) & a_{36} \\ 0 & a_{42}(t) & 0 & 0 & a_{45}(t) & 0 \\ 0 & 0 & 0 & 0 & a_{55}(t) & 0 \\ 0 & a_{62}(t) & 0 & 0 & 0 & a_{66}(t) \end{bmatrix},$$
$$B(\sigma(t)) = \begin{bmatrix} 0 \\ b_{12}(t) \\ 0 \\ 0 \\ 0 \\ 0 \end{bmatrix} \text{ and } E = \frac{1}{V_O} \begin{bmatrix} 1 & 0 & 0 \\ 0 & 0 & 0 \\ 0 & 1 & 0 \\ 0 & 0 & 0 \\ 0 & 0 & 1 \\ 0 & 0 & 0 \end{bmatrix}.$$

where

$$a_{11}(t) = \frac{\sigma_1(t)}{V_o}, \qquad a_{15}(t) = -\frac{\mu_h}{Y_h}\sigma_2(t) + (1 - f_p)b_h - \frac{\mu_h \eta_{NOg}}{Y_h}\sigma_3(t), \qquad a_{16} = (1 - f_p)b_a,$$
$$a_{22}(t) = -\frac{\sigma_1(t)}{V_O} - \frac{4.57 - Y_a}{Y_a}\mu_a\sigma_4(t), \qquad a_{25}(t) = \frac{Y_h - 1}{Y_h}\mu_h\sigma_2(t)$$
$$a_{32}(t) = -\left(i_{xb} + \frac{1}{Y_a}\right)\mu_a\sigma_4(t), \qquad a_{35}(t) = (i_{xb} - f_p i_{xp})b_h - i_{xb}\mu_h\sigma_2(t) - (i_{xb}\mu_h\sigma_3(t)),$$
$$a_{36} = (i_{xb} - f_p i_{xb})b_a, \qquad a_{42}(t) = \frac{1}{Y_a}\mu_a\sigma_4(t), \qquad a_{45}(t) = \frac{Y_h - 1}{2.86Y_h}\mu_h\eta_{NOg}\sigma_3(t),$$
$$a_{55}(t) = \mu_h\sigma_2(t) - b_h - \left(\frac{\sigma_1(t)}{V_O} - \frac{f_w(1 + f_r)}{f_r + f_w}\right) - b_a, \qquad a_{62}(t) = \mu_a\sigma_4(t),$$
$$a_{66}(t) = \frac{\sigma_1(t)}{V_O}\left(-\frac{f_w(1 + f_r)}{f_r + f_w} - 1 - b_a\right), \qquad b_{12}(t) = S_{sat} - \sigma_5(t).$$

The input concentrations are

$$w(t) = \left[Q_{in}(t)X_{COD_{in}}(t) \quad Q_{in}(t)S_{NH_{in}}(t) \quad Q_{in}(t)X_{BH_{in}}(t)\right]^T.$$

Remark 1. *In this work, it is assumed that all the concentrations are measured online, but it must be noted that in practice, not all the concentrations, such as, e.g.,* $X_{COD_{in}}$ *can be measured.*

3. EMPC of a WWTP

3.1. Operational Goals

The immediate control goal of a WWTP is to meet water quality levels established by regulators while operating efficiently by reducing operational cost. As discussed in the introduction, predictive control techniques may be used to compute strategies which achieve this goal while at the same time optimizing the system performance in terms of different operational indices. To achieve this objective, the control of dissolved oxygen concentration as well as nitrates within certain limits is necessary. The MPC presents the advantage of being a non-conservative control strategy, such that in periods of low influents, with a minimal level of pollutants, the effluent quality can be achieved by regulating the levels of S_0 and S_{NO} below the stipulated reference point to avoid waste of energy. Subsequently, during periods of high influents levels, it is then important to meet the predefined set points to reduce pollutants, avoiding the violation of the standard effluent quality set by authorities [15]. In this work, a PI-EMPC control strategy is employed: PI designed by authors of the BSM1 for the regulation of S_{NO} and a designed EMPC for the control of S_O in the aeration tank. In the proposed LPV EMPC, the following objectives are then considered:

- **Economic costs.** The main economic costs associated with WWTP are primarily due to treatment and electricity costs. Water through the WWTP involves important electricity costs in pumping stations in charge of internal and external water recirculations as well as aeration in the aerobic tanks. In our case, only the aeration energy is considered with an objective of minimizing the cost associated with supply of oxygen for controlled culture growth. The performance index is described as follows

$$J_{eco}(k) = \frac{S_{O_{sat}}}{1800} V_o K_{La}(k) \left[\frac{\text{kwh}}{\text{day}}\right]. \tag{11}$$

- **DO concentration control.** In order to control the S_o within some bounds in the EMPC during the aeration process, slack variables are introduced in the optimization problem, which seek to penalize the dissolved oxygen states, such that they are maintained in a range to maintain effluent quality. Selecting slack variables, ($\lambda^+ > 0$ and $\lambda^- > 0$), additional terms of soft constraints (see (16c) and (16d)) and a quadratic objective index are introduced with x_{sp} as the selected DO concentration value. The introduction of the slack variables ensures that the DO concentration varies within a boundary around x_{sp} aided by the appropriate selection of weights in the objective function. The performance index is thus given as

$$J_\lambda(k) = \|\lambda(k)\|_2^2, \tag{12}$$

where $\lambda(k) = \left[\lambda^-(k), \lambda^+(k))\right]^T$.

- **Smooth set points for equipment conservation.** The operation of WWTP and main valves and pumps usually requires smooth flow set-point variations. To obtain such a smoothing effect, the proposed MPC controller includes a third term in the objective function to penalize the control signal variation between consecutive time intervals. This term is expressed as

$$J_{smo}(k) = \Delta u(k)^T W_u \Delta u(k). \tag{13}$$

Therefore, the performance function J considering the aforementioned control objectives has the form

$$J = w_1 \sum_{k=0}^{H_p-1} J_{eco}(k) + w_2 \sum_{k=0}^{H_p-1} J_{smo}(k) + w_3 \sum_{k=1}^{H_p} J_\lambda(k). \tag{14}$$

3.2. Control Strategy Computation

The control strategy is determined by the computation of an optimal sequence of control actions for a prediction horizon, H_p.

$$\tilde{u}_k = (u(k|j))_{j=0}^{H_{p-1}} = (u(k|0), u(k|1), \cdots, u(k|H_{p-1})). \tag{15}$$

We solve at each time instance k, the following optimal control problem with initial state obtained from measurements (or state estimation) of the dynamics WWTP model and prediction in the MPC loop with the qLPV plant model (10),

$$\min_{\tilde{u}_k} J(\tilde{u}_k, k) \tag{16a}$$

subject to

$$x(i+1|k) = A(\sigma(k))x(i|k) + B(\sigma(k))u(i|k) + Ew(i|k) \quad i = 0, \cdots, H_p - 1, \tag{16b}$$
$$x_{s_o}(i|k) >= x_{sp} + \lambda^+(i|k), \quad i = 1, \cdots, H_p, \tag{16c}$$
$$x_{s_o}(i|k) <= x_{sp} - \lambda^-(i|k), \quad i = 1, \cdots, H_p, \tag{16d}$$
$$u(i|k) \in \mathcal{U} \quad i = 0, \cdots, H_p - 1, \tag{16e}$$
$$x(i|k) \in \mathcal{X} \quad i = 1, \cdots, H_p, \tag{16f}$$
$$y(i|k) \in \mathcal{Y} \quad j = 0, \cdots, H_p, \tag{16g}$$
$$\lambda^+(i|k), \lambda^-(i|k) >= 0 \tag{16h}$$

where x_{s_o} is the dynamic state representing the soluble oxygen. (16c–f) are described by the box constraints:

$$\begin{aligned} \mathcal{U} &= \left\{u \in \mathbb{R}^{n_u} | u^{\min} \le u \le u^{\max}\right\}, \\ \mathcal{X} &= \left\{x \in \mathbb{R}^{n_x} | x^{\min} \le x \le x^{\max}\right\}, \\ \mathcal{Y} &= \left\{y \in \mathbb{R}^{n_y} | y^{\min} \le y \le y^{\max}\right\}. \end{aligned} \tag{17}$$

which are determined from the maximum residual concentrations imposed in order to cope with the European Union effluent standards on chemical oxygen demand COD, suspended solids S_S and total nitrogen T_N:

$$\begin{aligned} COD &\leqslant COD_{max} = 125\,\text{gm}^{-3}, \\ S_S &\leqslant S_{Smax} = 35\,\text{gm}^{-3}, \\ T_N &\leqslant T_{Nmax} = 10\,\text{gm}^{-3}. \end{aligned}$$

The first control action of the sequence $u(k|0)$ is applied to the WWTP plant to obtain the system measurements and/or MHE estimated states, which are then used in the succeeding optimization problem, resulting in a recursive procedure. Not all the state variables are measured as stated earlier; the moving horizon estimator (MHE), which is the dual of the MPC controller, estimates the unmeasurable states.

4. Moving Horizon Estimation

Since some states cannot be measured online in the operation of the WWTP, a design of an estimator, in our case the MHE, is necessary for the prediction of system outputs, bearing in mind that apart from purposes of feedback control, the quasi-LPV formulation relies on information of the system states for the model construction. By solving a constrained optimization problem, the MHE utilizes a limited N-prediction horizon of past measurements through an error minimization scheme aided by information of the system model in a prediction window to estimate the system states. The optimization problem is therefore set up with the discretized plant model as:

$$\begin{aligned} \min_{\{\hat{x}(i|k)\}_{i=-N}^{0}} \quad & \left(\hat{x}(-N|k) - x_o\right)^T P_o \left(\hat{x}(-N|k) - x_o\right) + \sum_{i=-N}^{k} \left(\epsilon(i|)^T Q \epsilon(i|k) + s(i|k)^T R s(i|k)\right) \\ \text{s.t.} \quad & \hat{x}(i+1|k) = A(\sigma(i|k))\hat{x}(i|k) + B(\sigma(i|k))u(i|k) + Ew(i|k) + \epsilon(i|k) \quad i = -N, \cdots, -1, \\ & y(i|k) = Cx(i|k) + s(i|k), \\ & \hat{x}_k \in \mathcal{X}. \end{aligned} \tag{18}$$

where $R = R^T \in \mathbb{R}^{n_y \times n_y} > 0$, $Q = Q^T \in \mathbb{R}^{n_x \times n_x} \geq 0$ and $P_o = P_o^T \in \mathbb{R}^{n_x \times n_x} \geq 0$ are the weighting matrices that are defined according to uncertainty levels induced respectively by the noise, disturbance and unknown initial conditions (x_o). $\mathcal{X}$ bounds the estimated states. At every iteration, N sets of control inputs, $\{u(i|k)\}_{i=-N}^{-1} \in \mathbb{R}^{n_u \times N}$, measurements $\{y(i|k)\}_{i=-N}^{-1} \in \mathbb{R}^{n_y \times N}$ and N sets of LPV matrices $\{A_i\}_{i=-N}^{-1} \in \mathbb{R}^{(n_x \times n_x)N}$, $\{B_i\}_{i=-N}^{-1} \in \mathbb{R}^{(n_x \times n_u)N}$ are taken as inputs into the optimization problem to predict the state sequence $\{\hat{x}(i|k)\}_{i=-N}^{0} \in \mathbb{R}^{n_x \times (N+1)}$ by solving the dynamical optimization problem (18). The last element of the sequence $\{\hat{x}(i|k)\}_{i=-N}^{0}$ is subsequently chosen as the estimated states, the measurements and inputs are then discarded, and the procedure is repeated. The ammonia concentration (S_{NH}), nitrate concentration (S_{NO}) and the soluble oxygen (S_o) are supposedly measurable; therefore, the MHE is designed for the estimation of $[X_{COD}, X_{BH}, X_{BA}]^T$ as shown in Figures 3–5 .

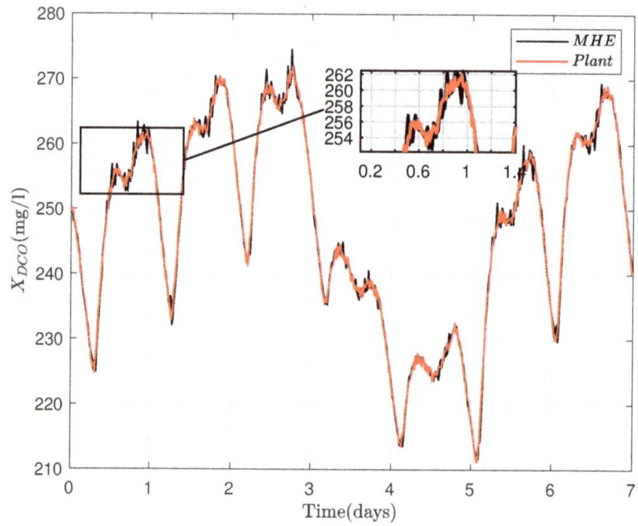

Figure 3. MHE estimate of oxygen demand concentration (X_{COD}) for 7 days.

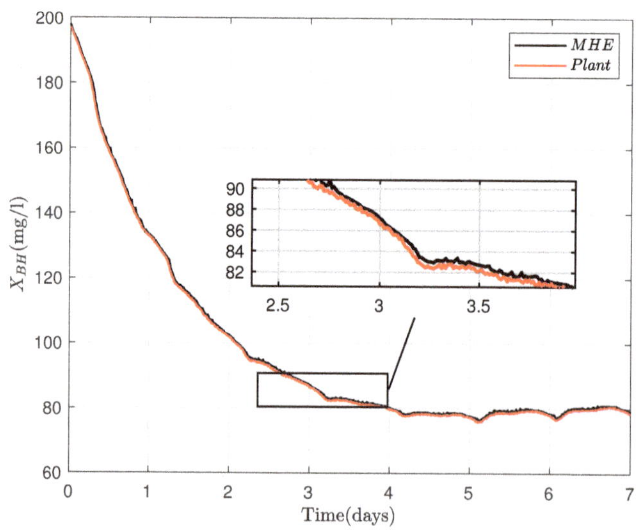

Figure 4. MHE estimate of heterotrophic biomass (X_{BH}) for 7 days.

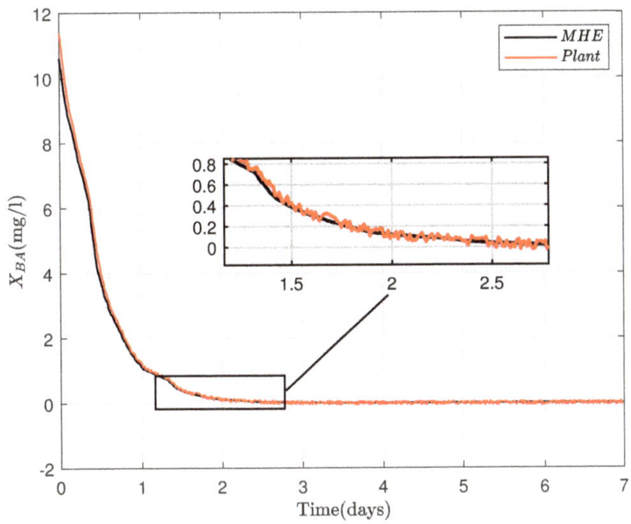

Figure 5. MHE estimate of autotrophic biomass (X_{BA}) for 7 days.

5. Simulation Results

5.1. LPV EMPC Implementation Details

To illustrate LPV EMPC approach presented in this paper, the Girona WWTP case study presented in Section 2 is used. The constituents of the influent wastewater of Girona WWTP varies during the day between the following bounds :

- Qin (between 10,000–35,000 m^3/d);
- COD (between 400–650 mg/L);
- DBO (175–225 mg/L); and
- Nitrogen (between 40–65 mg/L).

The inflow of Girona WWTP is shown in Figure 6.

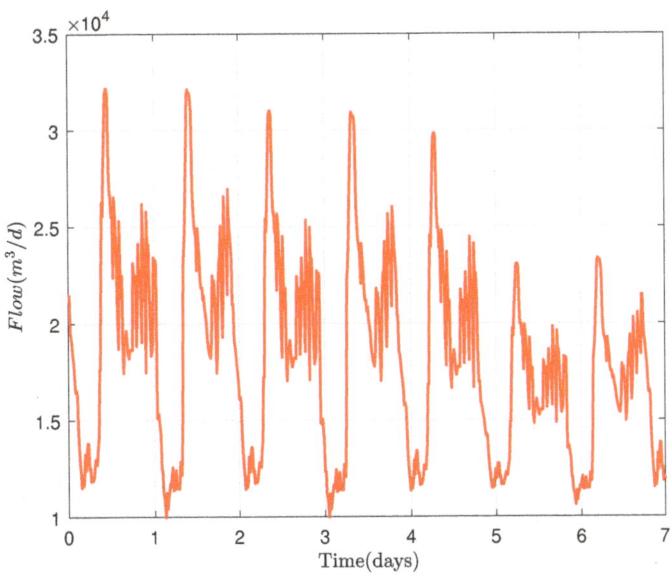

Figure 6. WWTP inflow.

With a quasi-linear approximation of the nonlinear WWTP via the LPV representation, the constrained optimization problem (16) is solved using quadratic programming formulation using the CPLEX® solver in MATLAB® on an Intel Core i7, 8 GB of RAM PC. A sampling time of 15 min and a prediction horizon of 6 h is chosen for simulation. The process is simulated for 7 days in a Simulink environment representing the dynamics of the Girona WWTP, as shown in Figure 1.

Using the weights w_i associated with the multiobjective EMPC cost function, (14) is tuned using the procedure as performed in [20,21] with the aim of maintaining the quality of the exit water at some levels within the current regulations regardless of the entry at a minimum cost.

Some control scenarios are selected to show different behaviors of the proposed scheme by altering X_{sp} and manipulating weights w_i, ideally to illustrate the different actions of aeration corresponding to different dissolved oxygen requirements for a quality effluent.

5.2. First Scenario

The first scenario consists of controlling the dissolved oxygen concentration in the exit of the biological treatment plant between the bounds $(1.5, 2.5)$. Figure 7 shows the dynamics of the DO concentration (above) and its corresponding aeration energy (below). The operation of the aeration, as stated in the preceding section, corresponds to the variation of the influents during the day; therefore, the DO concentration varies between the defined bounds in relation to the amount of pollutants at each time instance in the influents, which can be inferred from Figure 6.

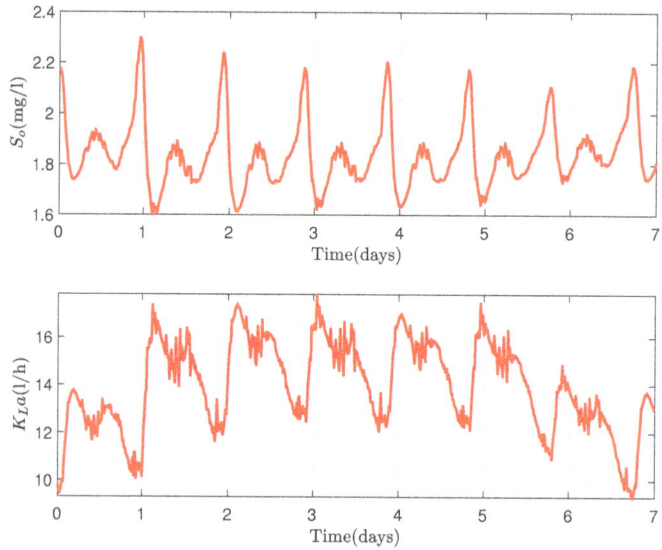

Figure 7. (**Above**): DO concentraton variation. (**Below**): Aeration flow for Scenario 1.

5.3. Second Scenario

The second scenario also consists of controlling the DO concentration between the ranges of 0.5 to 1.2 mg/L with minimum aeration energy consumption.

From Figure 8, a similar behavior of oxygen in the tanks as in the first scenario is realized with an expected less aeration energy, as less DO is required for treatment. The nitrates in the exit of the WWTP (Figure 9) range approximately between 5 and 7 mg/L.

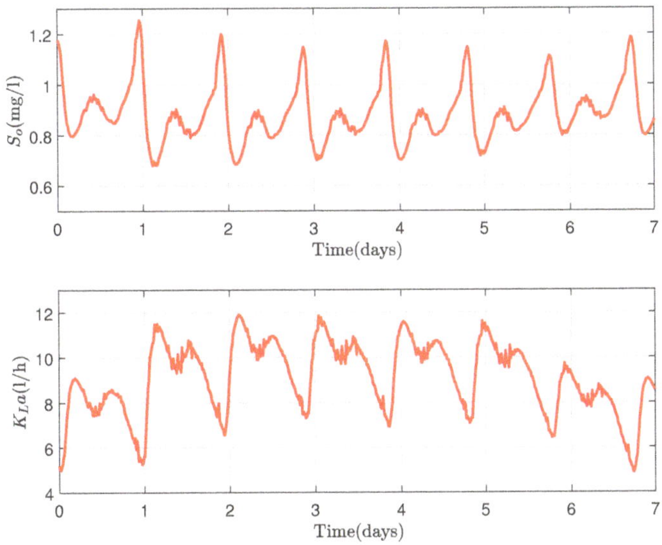

Figure 8. (**Above**): DO concentration variation and (**Below**): Aeration flow for Scenario 2.

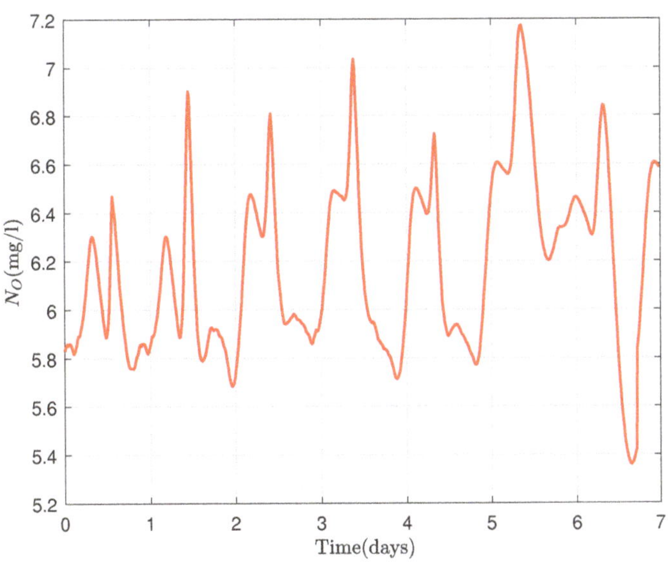

Figure 9. Nitrate concentration variation.

6. Conclusions

In this paper, an LPV EMPC strategy for the control of dissolved oxygen concentration in the aerated reactors of a WWTP is proposed and applied to the Girona (Spain) case study. The proposed approach combines two improvements with respect to the existing approaches in the literature: First, differently from standard tracking MPC, the proposed EMPC strategy optimizes the economic performance of the plant instead of following some pre-established set points. Second, a reduced model of the WWTP is represented in a quasi-LPV form allowing the real-time implementation of the controller thanks to the use of quadratic programming optimization tools. If otherwise, the nonlinear model plant was used, nonlinear programming algorithms are required that usually prevent the real-time implementation because of the large computational time. Moreover, an LPV moving horizon state estimation scheme has also been proposed that allows the implementation of the LPV EMPC with the available sensors in the WWTP. The effectiveness of the proposed scheme has been illustrated in the considered case study with two scenarios aiming at keeping the DO within some bounds.

As future work, real testing in the WWTP plant will be conducted to further validate the performance of the proposed solution. Another issue to take into consideration is the application of the proposed methodology for aerobic conditions maintenance in sewer networks [22].

Author Contributions: Conceptualization, F.N., B.K. and V.P.; methodology, F.N., B.K. and V.P.; software, B.K. and J.P.; writing—original draft preparation, F.N.; writing—review and editing, F.N., B.K, V.P., J.P., J.Q. and S.d.C.; supervision, F.N., B.K and V.P. All authors have read and agreed to the published version of the manuscript.

Funding: This work has been co-financed by the Spanish State Research Agency (AEI) and the European Regional Development Fund (ERFD) through the project SaCoAV (ref. MINECO PID2020-114244RB-I00), by the European Regional Development Fund of the European Union in the framework of the ERDF Operational Program of Catalonia 2014–2020 (ref. 001-P-001643 Looming Factory), and by the DGR of Generalitat de Catalunya (SAC group ref. 2017/SGR/482).

Institutional Review Board Statement: Not applicable.

Informed Consent Statement: Not applicable.

Data Availability Statement: All the required data are included in the paper.

Conflicts of Interest: The authors declare no conflict of interest.

Abbreviations

The following abbreviations are used in this manuscript:

ASM	Active Sludge Model
ASP	Active Sludge Process
BSM	Benchmark Simulation Model
COD	Chemical Oxygen Demand
DO	Dissolved Oxygen
EMPC	Economic Model Predictive Control
qLPV	Quasi-Linear Parameter Varying
LPV	Linear Parameter Varying
MHE	Moving Horizon Estimator
MPC	Model Predictive Control
NEMPC	Nonlinear Economic Model Predictive Control
NMPC	Nonlinear Model Predictive Control
RTO	Real-Time Optimization
SSTO	Steady-State Target Optimizator
WWTP	Wastewater Treatment Plant

References

1. Chachuat, B.; Roche, N.; Lati, M. Long-term optimal aeration strategies for small-size alternating activated sludge treatment plants. *Chem. Eng. Process.* **2005**, *44*, 593–606. [CrossRef]
2. Jeppsson, U.; Pons, M.N. The cost benchmark simulation model current state and future perspective. *Control Eng. Pract.* **2004**, *12*, 299–304. [CrossRef]
3. Holenda, B.; Domokos, E.; Redey, A.; Fazakas, J. Dissolved oxygen control of the activated sludge wastewater treatment process using model predictive control . *Comput. Chem. Eng.* **2008**, *32*, 1270–1278. [CrossRef]
4. Santin, I.; Pedret, C.; Vilanova, R. Fuzzy control and model predictive control configurations for effluent violations removal in wastewater treatment plants. *Control Eng. Pract.* **2015**, *54*, 2763–2775. [CrossRef]
5. Maciejowski, J.M. *Predictive Control: With Constraints*; Pearson Education: London, UK, 2002.

6. Vega, P.; Revollar, S.; Francisco, M.; Martin, J. Integration of set point optimization techniques into nonlinear MPC for improving the operation of WWTPs. *Comput. Chem. Eng.* **2014**, *68*, 2763–2775. [CrossRef]
7. Ostace, G.S.; Cristea, V.M.; Agachi, P. Cost reduction of the wastewater treatment plant operation by MPC based on modified ASM1 with two-step nitrification/denitrification model. *Comput. Chem. Eng.* **2011**, *35*, 2469–2479. [CrossRef]
8. Francisco, M.; Skogestad, S.; Vega, P. Model predictive control for the self-optimized operation in wastewater treatment plants: Analysis of dynamic issues. *Comput. Chem. Eng.* **2015**, *82*, 259–272. [CrossRef]
9. Rawlings, J.; Angeli, D.; Bates, C. Fundamentals of economic model predictive control. In Proceedings of the 51st IEEE Conference on Decision and Control, Maui, HI, USA, 10–13 December 2012; pp. 3851–3861.
10. Nejjari, F.; Puig, V.; Quevedo, J.; Pascual, J.; de Campos, S. Economic Model Predictive Control of Aeration Systems in a Full Scale Biological Wastewater Treatment Plant. In *Computer Aided Chemical Engineering (Proceedings of the 27th European Symposium on Computer Aided Process Engineering)*; Elsevier: Amsterdam, The Netherlands, 2017; Volume 40, pp. 1579–1584.
11. Morato, M.M.; Normey-Rico, J.E.; Sename, O. Model predictive control design for linear parameter varying systems: A survey. *Annu. Rev. Control* **2020**, *49*, 64–80. [CrossRef]
12. Olsson, G.; Newell, B. *Wastewater Treatment Systems: Modeling, Diagnosis and Control*; IWA Publishing: London, UK, 1999.
13. Jeppsson, U.; Olsson, G. Reduced Order Models for On-Line Parameter Identification of the Activated Sludge Process. *Water Sci. Technol.* **1993**, *28*, 173–183. [CrossRef]
14. Chachuat, B.; Roche, N.; Latifi, A. Réduction du modèle ASM 1 pour la commande optimale des petites stations d'épuration à boues activées. *Rev. Des Sci. de l'eau* **2003**, *16*, 5. [CrossRef]
15. Revollar, S.; Vega, P.; Vilanova, R.; Francisco, M. Optimal Control of Wastewater Treatment Plants Using Economic-Oriented Model Predictive Dynamic Strategies. *Appl. Sci.* **2017**, *7*, 813. [CrossRef]
16. Boulkroune, B.; Darouach, M.; Zasadzinski, M.; Gillé, S.; Fiorelli, D. A nonlinear observer for an activated sludge wastewater treatment process. *J. Process Control* **2009**, *19*, 1558–1565. [CrossRef]
17. Harja, G.; Vlad, G.; Nascu, I. MPC advanced control of dissolved oxygen in an activated sludge wastewater treatment plant. In Proceedings of the 2016 IEEE International Conference on Automation, Quality and Testing, Robotics (AQTR), Cluj-Napoca, Romania, 19–21 May 2016; pp. 1–6. [CrossRef]
18. Shamma, J.; Cloutiert, J. Gain-Scheduled Missile Autopilot Design using Linear Parameter Varying Transformation. *J. Guid. Control Dyn.* **1993**, *16*, 256–263. [CrossRef]
19. Kwiatkowski, A.; Werner, H. Parameter reduction for LPV systems via principle components analysis. *IFAC Proc. Vol.* **2005**, *38*, 596–601. [CrossRef]
20. Toro, R.; Ocampo-Martinez, C.; Logist, F.; Impe, J.V.; Puig, V. Tuning of Predictive Controllers for Drinking Water Networked Systems. *IFAC Proc. Vol.* **2011**, *44*, 14507–14512. [CrossRef]
21. Ocampo-Martinez, C.; Toro, R.; Puig, V.; Van Impe, J.; Logist, F. Multi-Objective-Based Tuning of Economic Model Predictive Control of Drinking Water Transport Networks. *Water* **2022**, *14*, 1222. [CrossRef]
22. Freire Diogo, A.; Oliveira, M.A. A preliminary numerical approach for the study of compressed air injection in inverted siphons. *J. Hydraul. Eng.* **2013**, *139*, 772–784. [CrossRef]

sensors

MDPI

Article

Comparison of Optimisation Algorithms for Centralised Anaerobic Co-Digestion in a Real River Basin Case Study in Catalonia

David Palma-Heredia [1,*], Marta Verdaguer [1], Vicenç Puig [2,3], Manuel Poch [1] and Miquel Àngel Cugueró-Escofet [2]

[1] Laboratory of Chemical and Environmental Engineering (LEQUIA), Institute of the Environment, Universitat de Girona, MªAurelia Capmany, 69, 17003 Girona, Spain; marta.verdaguer@udg.edu (M.V.); manuel.poch@udg.edu (M.P.)

[2] Advanced Control Systems (SAC) Research Group, Polytechnic University of Catalonia (UPC-BarcelonaTech), Terrassa Campus, Gaia Research Bldg, Rambla Sant Nebridi, 22, 08222 Terrassa, Spain; vicenc.puig@upc.edu (V.P.); miquel.angel.cuguero@upc.edu (M.À.C.-E.)

[3] Institut de Robòtica i Informàtica Industrial (CSIC-UPC), 46 Llorens i Artigas Street, 08028 Barcelona, Spain

* Correspondence: david.palma@udg.edu; Tel.: +34-972-41-98-59

Abstract: Anaerobic digestion (AnD) is a process that allows the conversion of organic waste into a source of energy such as biogas, introducing sustainability and circular economy in waste treatment. AnD is an intricate process because of multiple parameters involved, and its complexity increases when the wastes are from different types of generators. In this case, a key point to achieve good performance is optimisation methods. Currently, many tools have been developed to optimise a single AnD plant. However, the study of a network of AnD plants and multiple waste generators, all in different locations, remains unexplored. This novel approach requires the use of optimisation methodologies with the capacity to deal with a highly complex combinatorial problem. This paper proposes and compares the use of three evolutionary algorithms: ant colony optimisation (ACO), genetic algorithm (GA) and particle swarm optimisation (PSO), which are especially suited for this type of application. The algorithms successfully solve the problem, using an objective function that includes terms related to quality and logistics. Their application to a real case study in Catalonia (Spain) shows their usefulness (ACO and GA to achieve maximum biogas production and PSO for safer operation conditions) for AnD facilities.

Keywords: anaerobic co-digestion; ant colony optimisation; particle swarm optimisation; genetic algorithms; waste management; circular economy

Citation: Palma-Heredia, D.; Verdaguer, M.; Puig, V.; Poch, M.; Cugueró-Escofet, M.À. Comparison of Optimisation Algorithms for Centralised Anaerobic Co-Digestion in a Real River Basin Case Study in Catalonia. *Sensors* **2022**, *22*, 1857. https://doi.org/10.3390/s22051857

Academic Editor: Ernest W. Tollner

Received: 27 January 2022
Accepted: 24 February 2022
Published: 26 February 2022

1. Introduction

In the context of global climate change with rising and more extreme events—such as droughts and floods—which will likely provide growing uncertainty to water demand and jeopardise the availability of specific resources, there is a growing interest in the adaptation and use of technologies related to the circular economy that promote environmental sustainability. In this framework, resource recovery is a key issue for industrial and environmental processes and shows a wide spectrum of study possibilities. In water sanitation, wastewater treatment plants (WWTPs) offer a wide range of possibilities for resource recovery, mainly related to sludge treatment processes [1–7] as biogas generation via the substrate co-digestion process, which can be an alternative source for thermal and electrical energy production [8–14]. This potential for biogas generation could be translated as well to a source of renewable natural gas, which has specific composition requirements that demand high-tech sensors to assure its quality no matter its origin, as those developed in [15,16]. Due to their potential for resource recovery and the further implications in the water–food–energy nexus, WWTPs have been a research focus from different areas of expertise: from modelling and engineering design [17–24] to process dynamics, simulation and integration [25–28].

Anaerobic digestion (AnD), a complex process involved in biogas production, has a delicate balance of substrate composition. The optimal performance requires avoiding process inhibition and maximising biogas generation. The optimal balance may be achieved with the correct mixture of available substrates, but this task is challenging and difficult to achieve manually due to the high combinatorial possibilities and changing availability of substrates of heterogeneous nature. The complexity increases when the process is co-digestion, with the addition of residual substrates produced by agro-food and similar industries, each with its own dynamics of substrate generation and composition [29–32]. Additionally, not all WWTPs have an anaerobic digester. Therefore, optimisation also requires logistical challenges to process a maximum volume of the available substrates in a certain geographical area and its travel logistics (of sludge and co-substrates) from its origin to the destination digester.

Hence, dealing with such complexity is a former step to tackle optimal co-digestion in a complex network composed of many substrate sources—including WWTPs without AnD processes and industrial producers—and several co-substrate receptors. These will be located in different geographical places. As a result, the logistics of substrates will be affected by the geographical distance between actors involved and the restrictions related to the receptors.

Optimisation of the individual digester feed requires optimal blending of different co-substrates in order to fulfil the volumetric and compositional requirements of the anaerobic procedure. This problem can be understood as a multidimensional knapsack problem (MKP) [33–35]. The MKP is an NP-hard problem [36] and has been widely studied in the literature. To solve this type of problem, the use of combinatorial optimisation metaheuristics is proposed in [37,38], mainly when a high number of restrictions are presented [39].

Many tools have been developed to this end, either focused on modelling, control of the optimum co-substrate blending, or system operation, as shown in [37,39–43]. In [37,40], identification and modelling of critical parameters are performed; in [39] control schemes based on the composition qualities are developed; and in [42,43], optimised control strategies are implemented according to blend composition. In [41], logistics are also included to optimise the performance of a single anaerobic digester with co-digestion strategies.

However, in real-world installations, most of these systems are managed and supervised not in a single fashion but in a network fashion. Thus, proper system management requires simultaneous consideration of the entire AnD network to select which combination is the best for each digester to maximise the potential of the overall infrastructure. Besides, literature on this matter is relatively scarce due to its ad-hoc nature. There is literature related to optimised placement of new AnD plants, such as in [44], but it lacks optimisation of the operational part involved in the feeding of the anaerobic digesters. Very specific works can be found about optimisation of supply chain networks in the field of waste valorisation, such as in [45], where an integrated geographical information system (GIS)-based optimisation is performed, but it requires highly detailed and tailored data, so its implementation becomes time-consuming and highly dependent on data availability; furthermore, it does not tackle process optimisation regarding waste processing facilities. Regarding logistics, other works can be found for path planning optimisation such as in [46], where truck routes are traced based on GIS-oriented algorithms, or in [47], where a smart waste bin prototype is developed for sensor-based waste classification. As it can be seen, there is a gap in the literature regarding network optimisation of existing waste management facilities (such as AnD plants) that would include both logistics (i.e., minimising route impact and length) and quality (i.e., improving process performance) optimisation. This is a necessary gap to fulfil, since, as stated before, AnD networks are currently managed in an ad-hoc, manual fashion by the practitioners, which is dramatically time-consuming and needs highly qualified personnel. Although there are currently different approaches that could help overcoming specific parts of this challenge (i.e., those observed in [44–47]), none of them can currently successfully accomplish the overall task.

The approach presented in this study continues the work introduced in [48] where the optimisation problem of blending in anaerobic co-digestion (AnCD) is handled by an ant colony optimisation (ACO) algorithm in a synthetic case study with realistic conditions, simulating a centralised AnD single-stage reactor that received feed once a day. In [41], the work is extended while considering the quality, social and travel logistics of the co-substrates, analysing its importance to the overall optimisation. In addition, ACO has been implemented in real-world waste sector case studies, e.g., [48,49]. Here, the work is extended to a similar real-world case study considering multiple receptors in the geographical area of the Besòs River basin (Catalonia, Spain). The data used correspond to the real operation conditions of this area. For these conditions, the authors have also evaluated the results obtained using different optimisation approaches such as ACO, genetic algorithms (GA) and particle swarm optimisation (PSO).

ACO, GA and PSO algorithms were selected as convenient approaches to tackle a problem of the nature stated here after reviewing applications of similar nature in the literature. In [50], a review of nature-inspired algorithms is performed, including GA, ACO and PSO—among others—for AnD modelling and optimisation, showing how in the field of AnD, PSO obtains better performance in substrate feed optimisation for agricultural biogas plants than other evolutionary methods. For example, in [51] genetic algorithms are used to minimize the environmental impact caused by mine water. The main drawback for PSO pointed out in [50] is premature convergence, since particles may become trapped in local optima or suffer stagnation, but this may be solved by a partial restart of the process introducing new particles in the search space. In [52], ACO and GA are applied to optimise the route of waste collection vehicles for municipal waste collection and transportation—the highest cost of the entire waste management system—with similar performance attained by both algorithms; however, only the problem of travel logistics is considered, not the blending of municipal waste. Ref. [53] proposes a nonlinear model predictive control strategy using the MATLAB BioOptim toolbox, developed by the same authors, for optimal control of substrate feed to AnD operation of an agricultural biogas plant, with a graphical user interface (GUI) integrating a fitness function including different operating constraints and parameters such as pH, solids or methane concentration, and using evolutionary optimisers such as PSO, covariance matrix adaptation evolution strategy (CMAES). Alternatively, they propose differential evolution (DE), which achieved better performance with PSO but without considering the substrate travel logistics in the optimisation. Ref. [54] presents a prediction and optimisation method using a multi-layer perceptron artificial neural network (ANN) and PSO for the maximisation of biogas generation in a real wastewater treatment facility. A similar approach is presented in [55], where modelling and optimisation of biogas production with mixed substrates are obtained with a combination of ANN and GA methods.

Additionally, ref. [55] and references therein point out how stochastic global optimisation algorithms (SGOAs), such as PSO, the ACO, and GA, among others, are considered efficient alternatives in the design of optimal production media and optimal process operating conditions in fermentation research and can significantly reduce the process development time. Regarding the comparison of the optimisation algorithms selected here solving combinatorial optimisation (CO) problems, in [56], ACO and GA are compared, both achieving good performance but with GA exhibiting slightly better performance than ACO. In the latter reference, it is also mentioned how trimming of specific parameters for both optimisers—e.g., number of iterations, evaporation coefficient and number of ants for ACO, or chromosome population, crossover and mutation probabilities for GA—is required to achieve good performance in both cases. In [57], relationships between GA and ACO-type algorithms are detailed, presenting their similitudes and showing how they use similar principles to succeed in CO problems with globally convex structure of its solution space. Overall, SGOAs such as ACO, GA or PSO have shown good performance in the type of applications presented and hence demonstrated suitability for non-convex nonlinear multidimensional optimisation problems, as presented here. Optimal blending for AnCD is considered, e.g., in [40,55], but to the knowledge of the authors, the optimisation of such blending combined with the travel logistics of co-substrates in a centralised multi-receptor co-digestion strategy has not yet been studied. Additionally, a comparison between different suitable optimisation algorithms for such applications, i.e., ACO, GA and PSO, is presented.

The application of optimisation strategies in AnD allows a significant enhancement of co-digestion strategies [30] maximising biogas production and minimising associated risks to each AnD operation (e.g., overdosing or acidification). In this work, the performance of each optimisation approach considered is evaluated on a real case study in the area of the Besòs River basin in Catalonia, including a network of substrate generators and three anaerobic digesters. Hence, the objective of this study is to develop a tool that is able to optimise the centralised digestion process of an AnD network with multiple waste sources and waste receptors by means of three evolutionary optimisation algorithms—namely, ACO, GA and PSO. Such a tool is tested in a real case study to further analyse and compare the performance of each algorithm in the overall AnD network optimisation.

2. Material and Methods

2.1. Optimisation Algorithms Considered

The optimisation algorithms presented here fall within the set of SGOA, concretely in the subset of evolutionary algorithms (EA) for GA—which use mechanisms inspired by biological evolution, e.g., mutation or recombination to achieve the goal of optimisation—and in the subset of swarm intelligence methods for ACO and PSO based on the collective behaviour of self-organised decentralised systems, respectively. SGOA algorithms have been widely used to solve NP-hard combinatorial optimisation problems, such as that presented in this study, which deterministic optimisation methods fail to handle due to their complexity.

Regarding each proposal, ACO is a metaheuristic approach that has been shown to be effective in solving a variety of NP-hard problems [58]. The algorithm is based on simulation of the behaviour of real ants in their search for food. When ants find food, they leave a pheromone trail on their path. Then, new ants follow that trail. In this way, an increasing number of ants are concentrated in places where there is food. In a similar way, the virtual ants construct a solution moving through the graph that represents the search space of solutions. Their paths are guided by a probabilistic state transition rule, which is based on pheromone trails and specific heuristic information. The algorithmic procedure is iterative. At each iteration, the pheromone trails are updated by applying an evaporation coefficient (when the value selected is not part of a feasible solution). To avoid rapid stagnation of the solution, the ACO algorithms can use several strategies [59], such as that related to the limitation of the pheromone trails between maximum and minimum values. Max–Min Ant System [60] uses this procedure.

The GA is a metaheuristic approach also used in combinatorial optimisation problems. It is based on the mechanics of natural selection and natural genetics. GA applications cover a range of combinatorial optimisations, e.g., hydraulic model calibration [61], performance of photovoltaic systems under variable atmospheric conditions [62], and sensor placement for leak detection in water distribution networks [39,63]. The GA is based on three main parameters: selection, crossover and mutation. The population matrix is randomly generated and consists of the design variables, and the best variables are selected according to their fitness value. From these solutions, new solutions are produced via the crossover operator [64]. The mutation operator is finally employed to avoid the algorithm converging to local optima (i.e., to maintain the genetic diversity).

The GA cycle is repeated through a number of generations until a stopping criterion is met. It is worth noting that elitism is not generally considered an operator in the canonical GA. However, it is deemed a robust and effective operator because it leads the optimisation procedure towards the optimal solution. Accordingly, this operator stops the best solutions from being mutated. In this way, the best solutions of each generation would pass to the next, unaltered. Over the course of the algorithm and through a sufficient number of generations, the traits of these solutions would transfer to their offspring, increasing the chance of producing new solutions whose fitness function values might be better than their parents [64]. Some drawbacks of GAs are noted in [61], e.g., achieving a global optimum for large and complex systems is not guaranteed, which is also a drawback for ACO. In [57], the relation between GAs and ACO is noted.

PSO is a recently developed EA that includes features such as easy implementation for solving practical problems, high accuracy and fast convergence of the solution as some of its main advantages [65,66]. While similarities exist in the iterative nature of PSO and GAs, conversely, in PSO, there are no, e.g., "crossover" or "mutation" operations. Instead, PSO is based on a population of candidate solutions, defined as particles. The set of particles composes a swarm, where each individual flows through the parameter space. The flow of such particles is defined by trajectories, which are driven by the best performance of the particle and the neighbouring particles in the parameter space. The initialization of particle swarm is random. The initial solution of each particle represents an alternative solution; that is, each particle has its own initial position and speed and is randomly distributed in each position of the feasible solution space to be searched. Therefore, the initialization of the particle swarm represents the preparation of the particle swarm search. Its size is determined by its speed and position, and the particle update is based on the comparison of the fitness values between each search particle and its neighbouring search particles to determine the necessity of updating a particle. The updated particle adjusts its speed and position according to the particle's new flight path model, which is based on the best results achieved by its neighbouring particles. These conditions yield different optimal experiences for different particle subgroups, which dynamically evolve according to the current position of particles, the particle current velocities, the distance between each particle of the subgroup and its best position and the distance between each particle of the subgroup and the best position of the whole subgroup.

The PSO algorithm does not need cross-mutation or other genetically inspired operations, so the algorithm has fewer parameters and is still high efficiency [65]. These properties are suitable for both engineering applications and scientific research, and a significant number of research results have been produced in recent years [67]. For example, in [68], PSO is applied for function optimisation regarding eco-economics modelling and assessment; in [69], it is used as part of fuzzy systems developed to optimise fuel consumption of hybrid vehicles; and in [70], PSO is used to train neural network models and perform real-time optimisation.

2.2. Centralised Co-Digestion as an Optimisation Problem

Mathematical optimisation involves the selection of one solution amongst a set, according to some criterion and constraints (that is, the optimisation problem). This optimisation problem can be stated as:

$$\min_{x \in X} f(x) \text{ subject to } g(x), \tag{1}$$

where $f(x)$ is the objective or cost function, X is a feasible region and $g(x)$ are the constraints that have to hold to find a minimiser x^* of $f(x)$ such that $f(x^*) = \min_{x \in X} f(x)$. ACO, GA and PSO introduced in Section 2.1 are algorithms aimed at finding the optimal solution of the optimisation problem posed in (1)—i.e., minimise the objective function $f(\cdot)$ subject to the set of constraints $g(x)$ that apply—which here consists of the selection of the best substrates and volumes according to a set of restrictions related to the operation of the anaerobic digester. In addition, the cost function allows quantifying each alternative potential solution according to (1), involving the calculation of a value or "cost" associated with each alternative considered to find the optimal solution.

The problem statement is similar to that presented in [41], although the number of waste receptors increases from one to three, thus making necessary a reformulation of the optimisation problem involved. Specifically, it is required to increase the dimensions of all data vectors that define each generator-receptor interaction and their subsequent calculations. In addition, matrix operations are repeated per new dimension (i.e., waste receptor) added.

This optimisation problem, which can be understood as a MKP and is of combinatorial nature, can be represented as a matching problem. It is defined with a graph $G = (N, E)$ that summarises all the possible combinations. The graph consists of N vertices (or nodes) and E edges or pairs of vertices. Specifically, for the case of AnD optimisation, a bipartite graph can be used to represent the posed optimisation problem to differentiate between the set of waste generators (N_1), containing W nodes, and the set of waste receptors (N_2), containing R nodes. For the specific optimisation problem, all W nodes are connected to each R node, thus resulting in a total of $W{\cdot}R = E$ edges. Figure 1 shows a generic representation of the defined matching problem applied to AnD co-digestion optimisation.

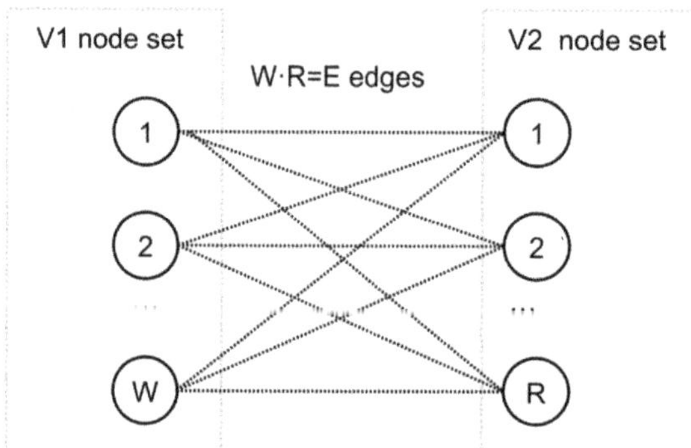

Figure 1. Generic representation of the posed matching problem.

A set of substrate generators $w \in \{1, \ldots, N\}$ is considered. The volume of each substrate V_w can be selected as a contribution to any of the AnD systems. The binary decision variable y_w^s allows generating array volumetric possibilities (V_w^s, with $s \in \{0 \ldots, l_w\}$ that are determined as a multiple of a number (e.g., 1000 by default) such that $1000 l_w = V_w$. The selection of each volumetric possibility is determined by the corresponding value of the binary decision variable, y_w^s, where $y \in \{0,1\}$, with $y_w^s = 0$ when the corresponding volumetric configuration is not selected, and $y_w^s = 1$ when it is selected. Note that for each waste generator w, there are l_w different volumetric configurations in y_w^s, but only one is selected at a time, i.e., $\sum_{s=1}^{l_w} y_w^s = 1 \; \forall w \in \{1 \ldots, N\}$.

To normalise the objective function, selected volumes V_w^s are divided by the maximum volume from their corresponding waste generator (V_w). This approach provides values of the cost function between 0 and 1, where the closest to 1 the better the solution. However, note that the ACO algorithm looks for a maximum of the objective function, while GA and PSO look for a minimum. This behaviour is considered using the constant $K \in \{-1,1\}$, which depends on the algorithm considered: for the ACO algorithm $K = 1$ and for the GA and PSO algorithms $K = -1$. Hence, cost index B would take positive values between 0 and 1 for ACO and negative values between 0 and -1 for GA and PSO. Instead, the absolute value is taken for all three algorithms.

F_w^c $(c = 1, \ldots, 3)$ and T_w are the set of dimensionless coefficients corresponding to the substrate characterisation and the quality term $\left(\sum_{c=1}^{3} F_w^c\right)\rho_q$, already used and explained previously in [41,46] and defined as shown in Figure 2.

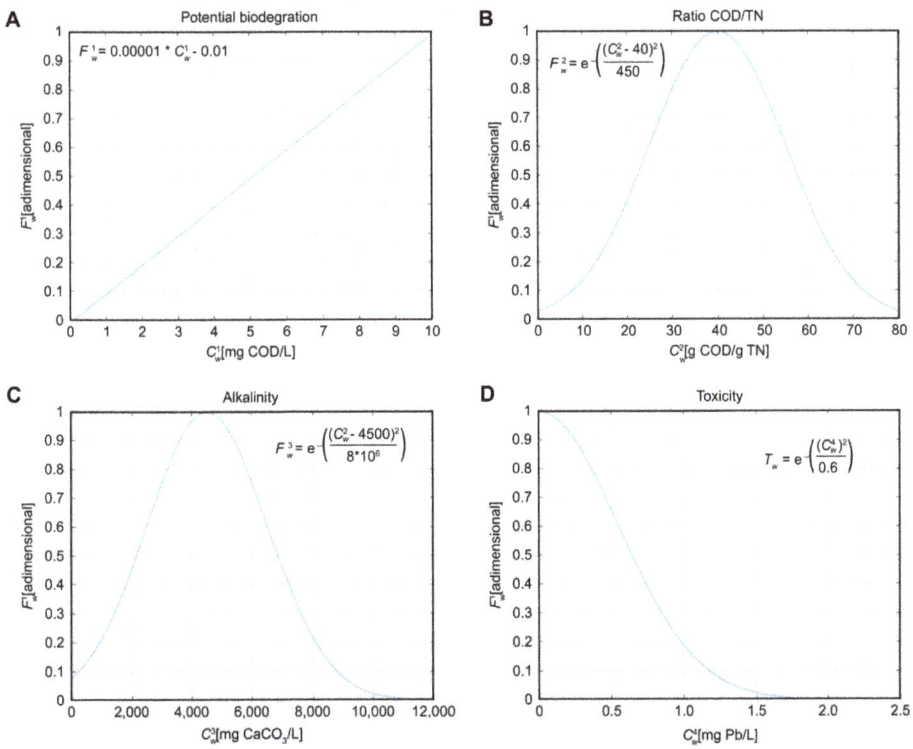

Figure 2. (**A**) F_w^1, (**B**) F_w^2, (**C**) F_w^3 and (**D**) T_w equations used for dimensionless coefficient calculation.

F_w^1 is a coefficient related to the potential biogas production, measured as a function of the Chemical Oxygen Demand (*COD*) content. F_w^2 indicates the ratio of COD/TN (where *TN* refers to Total Nitrogen), a useful measure to prevent situations of acidification and other undesired reactions of the AnD process, as long as it is maintained around the range of 20–60. F_w^3 is linked to the alkalinity (Alk) concentration, and it is associated with a restriction ranging from 2500 to 6000 mg $CaCO_3/L$ integrated within all optimisation algorithms. T_w is a coefficient of the utmost importance since it describes the toxicity level of all waste fluxes, which should be kept at the lowest level possible (specifically below 2.1 mg Pb/L).

The N different substrate generators are located at different distances (d_w) from each anaerobic digester. The conveyance of the selected volumes implies a travel distance d_w (in km) with an economic cost x_w (in €/km) and a social impact $I_w = 1, \ldots, 3$ (dimensionless). The higher the value of I_w, the higher the social impact of the related route (e.g., proximity to sensitive areas due to pollution, traffic density, or pedestrian presence). Since each route is different for each generator, different values are assigned to approach the logistic impact of the corresponding waste generator, so a value for I_w is assigned for each sludge/substrate generator depending on its route to the ST.

The coefficient weight ρ_q (dimensionless) is related to the quality term $\left(\sum_{c=1}^{3} F_w^c\right)\rho_q$, and the coefficient ρ_x (dimensionless) is the coefficient that weights the logistics term $\frac{\rho_x}{X_w d_w I_w}$. Each weight is given a value of 0.5 to provide a balance between the quality and logistics terms in the optimisation. Selected volumes of each substrate to each receptor contribute to the input to the AnD network, and the aforementioned parameters constitute the objective or cost function $f(x)$.

Additionally, the optimisation problem presented considers a set of restrictions $g(x)$ related to each of the total inputs to each of the receptor systems, based on those presented in [41,48]. The first restriction is the sum of accepted substrates $\sum_{w=1}^{N}\sum_{s=0}^{l_w} y_w^s V_w^s$ must not exceed the maximum acceptable volume V for each AnD system. Moreover, the COD/TN ratio, related to the dimensionless coefficient F_w^2, must be kept within the range $\left[C_{min}^2, C_{max}^2\right]$. The alkalinity concentration, related to the dimensionless coefficient F_w^3, must also be kept within the range $\left[C_{min}^3, C_{max}^3\right]$. The toxicity level, related to T_w, does not require restriction since the corresponding coefficient T_w is considered to be restrictive enough, as shown in Figure 2.

In addition, an estimation of the produced biogas is made assuming a conversion factor of 0.268 m^3 biogas/kg COD. Finally, note that the cost function presented in this work is adapted from [41,48], where the ACO algorithm was used for waste management optimisation in a similar fashion but limited to one AnD receptor.

The objective function $f(x)$ for the presented optimisation problem is as follows in (2). However, note that the performance comparison of the ACO, GA and PSO algorithms is not conducted directly on the value of the optimised objective function, B', but on its absolute value, B, as shown in (3).

$$B' = K\left\{\sum_{w=1}^{N}\sum_{s=0}^{l_w} y_w^s \frac{V_w^s}{V_w} T_w\left[\left(\sum_{c=1}^{3} F_w^c\right)\rho_q + \frac{\rho_x}{X_w d_w I_w}\right]\right\} \quad (2)$$

$$B = \left|B'\right| \quad (3)$$

3. Results

3.1. Case Study

The case study includes a network of 19 organic waste generators and three organic waste receptors. These 22 locations (i.e., 19 generators and 3 receptors) are part of the wastewater treatment system managed by Consorci Besòs Tordera (CBT), a public local water administration composed of 64 municipalities in four different regions of Catalonia (Spain) with a population of approximately 470,000 inhabitants. This case study and its anaerobic network system were also considered in [41]. Figure 3 shows the corresponding bipartite graph of the case study.

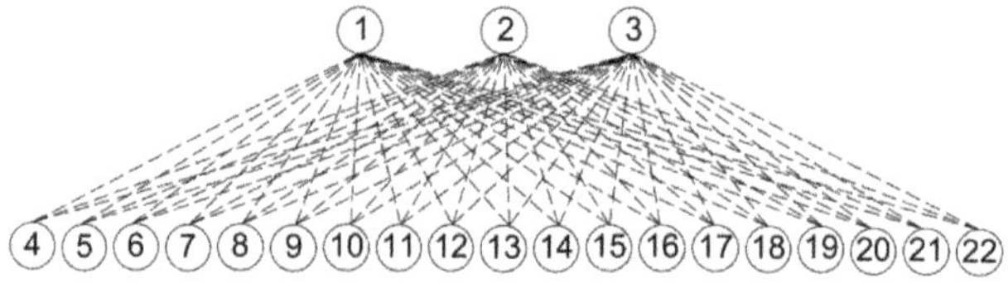

Figure 3. Bipartite graph of the case study.

The three organic waste receptors (R1–R3, or nodes 1–3 of Figure 3) refer to three separate WWTPs that produce their own sewage sludge, but that also have available AnD technology. Due to oversized design, which is a usual practice in WWTP design [71], these AnD systems in R1–R3 have available capacity. This free excess capacity can be used to accept wastes from external sources, such as the undigested sewage sludge of W1–W12 or the industrial substrates from C1–C7.

The 19 waste generators consist of 12 WWTPs that produce undigested sewage sludge (W1–W12, or nodes 4–15 of Figure 3) and seven industrial substrate generators (C1–C7, or nodes 16–22 of Figure 3), which were considered suitable sources of organic waste for the AnD network under study. Each of these locations is a separate and independent system that must manage its own waste produced as best as possible. Additionally, seven industrial substrate generators have been previously verified as feasible substrates for AnD by CBT technical services.

3.2. Simulation Methodology

The algorithms used in this work have been implemented in the MATLAB environment. Simulations were performed with a Lenovo ThinkPad (Lenovo Group, Ltd., Girona, Spain) L14 Gen1-20U10016SP ×64 using the OS Microsoft Windows 10 Pro and an Intel(R) Core(TM) i7-10510U CPU processor (1.80 GHz, 2304 MHz) consisting of four main processors and eight logic processors.

The main optimisation parameters of both GA and PSO algorithms were trimmed in an attempt to select the most suitable array to provide reliable results. Accordingly, the same procedure was already performed for the ACO algorithm to determine the values of its corresponding optimisation parameters in [48], where the same optimisation parameters are used in this work.

For the ACO algorithm, an initial population of 100 individuals (or ants) and 500 iterations per repetition is set, and the values used for the algorithm optimisation parameters are $\alpha = 1$, $\beta = 2$ and $\rho = 0.98$, each corresponding to the importance assigned to the pheromone trail, the importance assigned to the heuristic information and the persistence degree or pheromone evaporation, as explained in [38,58,60]. For GA, the initial population is set to 100, the total number of iterations (or generations) is set to 500, the crossover fraction is set to 0.8 and the fraction of elite children is set to 5% of the corresponding total children. For PSO, the initial population was set to 100, the total number of iterations was set to 500, cognitive attraction was set to 0.8 and the social attraction factor is set to 1.25. Tables 1 and 2 summarise trimming tests for GA and PSO, respectively, where the best results are obtained for higher values of objective index B.

Table 1. Summary of trimming tests for the GA.

Tested Parameters		Best Index (B)	Time(s)
Crossover Fraction	0.2	0.0274	525.73
	0.5	0.0295	574.46
	0.8	**0.0304**	**537.98**
Elite Count	**0.05**	**0.0309**	**541.48**
	0.15	0.0300	536.15
	0.3	0.0291	554.83

Table 2. Summary of trimming tests for PSO.

Tested Parameters		Best Index (B)	Time(s)
Cognitive Attraction	0.2	0.0293	82.86
	0.5	0.0304	62.93
	0.8	**0.0322**	**57.26**
Social Attraction	1.05	0.0301	55.78
	1.25	**0.0329**	**74.98**
	1.95	0.0308	63.45

For the sake of performance comparison, some parameters were fixed for the three algorithms. The fixed parameters are the number of independent simulations (set to 10, the best result is selected), the population (set to 100 individuals), and the maximum number of iterations (set to 500). With these constraints on algorithm trimming, a performance comparison of ACO, GA and PSO was conducted.

The comparison of ACO, GA and PSO performances is based on the value of the fitness function, execution time and an array of technical variables related to the total expected performance of the optimised AnD network: total daily biogas production (in Nm^3), average organic load (in kg of COD per m^3 of volume of the digestion system and day), average carbon to nitrogen ratio (C/N), and average alkalinity (in mg of $CaCO_3$). All algorithms are tested with data from a real case study as the main simulation scenario.

However, other synthetic scenarios are tested to further compare the performance of each algorithm under different scenario conditions.

In the approach presented here, simulated scenarios are based on the waste generator data in Table 3, alongside route distance and receptor system characterisation. For all the 19 waste generators (i.e., the 12 WWTPs without AnD and the seven substrate generators), the addition to the AnD network is optimised. For each of the three AnD systems (i.e., receptors R1, R2 and R3), different volume constraints have been determined, according to operational data and assuming a limit to the hydraulic retention time of 20 days (below that retention time, AnD efficiency is expected to greatly decrease).

Table 3. Waste generator dataset, including distance between waste generators and receptors and characterisation of each receptor of the case study (Baseline Scenario or Scenario 0).

Waste Generator ID	Vw (L by Day)	COD (mg/L)	C/N	Alk (mg/L)	Tw (mg/L)	R1 Distance to R1 (km)	R2 Distance to R2 (km)	R3 Distance to R3 (km)
W1	27,600	19,900	17.8	4300	1.55	5.3	20.5	9.5
W2	47,000	16,900	20.6	3200	1.36	35.9	33.9	45.7
W3	46,300	18,600	19.4	10,100	1.42	21.8	16.6	28.9
W4	20,200	23,400	15.6	3400	1.38	30.4	43.2	19.7
W5	38,400	21,100	17.9	4500	1.35	19.7	24.7	12.4
W6	34,400	18,800	14.0	3800	1.61	14.8	19.9	15.9
W7	13,800	22,600	15.3	2700	1.57	32.1	44.9	18.4
W8	4400	22,100	15.2	1800	2.30	26.5	31.6	27.7
W9	10,800	21,700	15.1	5300	0.93	20.3	33.1	8.8
W10	9500	20,400	15.5	2500	1.28	30	24.8	37.1
W11	17,000	23,300	14.8	7800	0.98	36.9	31.7	44
W12	6500	20,100	16.5	3100	1.40	20.5	33.3	8.7
C1	9000	667,400	42.5	250	0.01	15.9	11.1	23
C2	9000	497,400	461.8	330	0.01	7	12	17.9
C3	9000	155,900	3118.1	60	0.02	27.9	40.7	17.2
C4	9000	459,100	274.1	660	0.10	16.2	11.1	22.9
C5	9000	657,200	2330.6	630	0.01	52.8	65.6	43.8
C6	9000	266,200	2832.4	20	0.01	56.1	33	21
C7	9000	262,100	32,768.4	110	0.01	36.7	24.1	66.4
Maximum Volume (L/day)						122,000	146,000	111,000
COD (mg/L)						18,600	19,100	18,200
C/N						19.1	20.3	18.4
Alk (mg/L)						3100	2900	3400
Tw (mg/L)						1.41	1.68	1.53

Each simulation for ACO, GA and PSO is repeated 10 times since these algorithms have probabilistic, iterative-based search methods. The best solution among these runs is selected for further analysis, although the average fitness function is also registered for discussion.

The data obtained for ACO, GA and PSO (each comprising 10 repetitions of the corresponding algorithm) are compared for every simulated scenario. The baseline scenario (i.e., Scenario 0) corresponds to the real case study, as described in Table 3. Additional synthetic Scenarios 1–4 are simulated, and their corresponding data are created from alterations of the baseline scenario, as described in Table 4.

Table 4. Synthetic scenarios created from original Scenario 0 in Table 3. Description of data alteration procedure.

ID	Description
Baseline (Scenario 0) scenario	Scenario based on data form real case study (see Table 3)
Scenario 1	High COD (×10 COD concentration)
Scenario 2a	Linear modification of distances: ×10 distances
Scenario 2b	Nonlinear modification of distances: square root of original distance
Scenario 3	High volumes (×3 volumes)
Scenario 4	C/N variations (increase of W1–W12 C/N ratio to the 50–60 range)

While for Scenarios 1 and 2 any modification is viable, for Scenario 3, an increase in volume involves a significant increase in execution time. This is because the optimisation problem works around combinations of fixed volumes, and an increase in volume would involve a higher number of possible combinations for the algorithms to consider (i.e., an increase in the search space), hence the expected increase in execution time. Thus, volume modification for Scenario 3 was limited to a triple increase in the baseline scenario volume. Alternatively, for Scenario 4, the C/N ratios were modified while being kept below 60 to facilitate the algorithms in finding a viable solution. This measure was adopted because the C/N ratio was the most limiting optimisation parameter in previous applications of a similar optimisation problem in [41]. Scenario 2 was designed with both linear and nonlinear distance modifications (Scenarios 2a and 2b, respectively) to discuss the effect of distance distribution, as pointed out in [41]. Note that trimming tests were carried out only for GA and PSO using the baseline scenario, assuming that trimmed parameters would suffice for simulation of other synthetic scenarios similar to the baseline scenario.

The optimisation results are presented as a sequence of contributions from all the generators to each anaerobic digester. This optimised contribution sequence can be considered a suggested logistic plan for the co-substrate distribution as follows: once enough substrate has been produced and stocked on a waste generator, a truck of 20 metric tonnes capacity would be fully loaded with substrate from the corresponding waste generator, disregarding the truck waiting time before starting each route; once fully loaded, the truck is assumed to travel to the waste receptor without further stops (assuming it always follows the same route). As long as the cycle of supply routes of all involved waste generators is completed within the AnD retention time of 20 days, the properties of the resulting blending should not vary significantly, especially considering that every waste receptor would have a receiving system for these external organic substrates, where they would be stored and blended before being added to the AnD system. The specific start and finish time for each route along the day have not been considered; this does not affect the optimisation, although it has been noted that it has considerable impact on real-world implementation.

3.3. Algorithm Performance Comparison and Scenario Analysis

The simulation results for each scenario are shown in Table 5. For every scenario and for ACO, GA and PSO, this table shows the best cost index (B) achieved, elapsed optimisation time, and additional parameters related to the performance of the AnD systems: total biogas production, average organic load, average carbon/nitrogen ratio and average alkalinity.

In the baseline scenario, ACO and GA show higher biogas production than PSO (23% and 30% higher, respectively). However, they show a slightly lower *B* index achieved (4% and 11% lower). This result indicates that although one of the main goals of AnD optimisation involves maximising biogas production, it is not all that matters because there are other parameters also subjected to optimisation. PSO appears to find a solution with lower biogas production but better optimises other quality parameters, such as the C/N ratio and alkalinity. However, it is remarkable that out of the three algorithms, ACO and GA find a "similar solution" (prioritising high biogas production), and PSO fins a significantly different solution (prioritising other quality-related parameters).

Table 5. Summary of algorithm performance. The best value *B* is highlighted. Scenario 2a feasible results (*) are associated with a poor solution, so no direct comparison is conducted.

Scenario	**Baseline Scenario**			**Scenario 1**		
Optimisation Method	**ACO**	**GA**	**PSO**	**ACO**	**GA**	**PSO**
Best Index (B)	0.0336	0.0313	**0.0349**	**0.0330**	0.0328	0.0211
Time (seconds)	595.46	325.37	**90.20**	1825.34	1035.53	**221.74**
Total Biogas Production (Nm^3/d)	25,657	**27,133**	20,852	114,870	**198,284**	102,927
Avg Organic Load (kg $COD/m^3 \cdot d$)	2.32	**2.59**	2.09	9.67	**17.46**	9.73
Avg C/N ratio (limited below 60)	50.5	56.1	**46.4**	24.1	54	**32.4**
Avg Alkalinity (g $CaCO_3/m^3$)	3079	3141	**3245**	3183	3193	**3282**
Scenario	**Scenario 2a**			**Scenario 2b**		
Optimisation Method	**ACO**	**GA**	**PSO**	**ACO**	**GA**	**PSO**
Best Index (B)	-	0.0001 *	0.0001 *	0.0287	**0.0333**	0.0319
Time (seconds)	-	62	60	669.56	193	**63.61**
Total Biogas Production (Nm^3/d)	-	14,278	12,325	17,468	**25,404**	19,237
Avg Organic Load (kg $COD/m^3 \cdot d$)	-	1.6	1.4	1.69	**2.5**	2.24
Avg C/N ratio (limited below 60)	-	45.9	32.8	23.4	55.2	**32.4**
Avg Alkalinity (g $CaCO_3/m^3$)	-	3298	3319	**3338**	3174	3221
Scenario	**Scenario 3**			**Scenario 4**		
Optimisation Method	**ACO**	**GA**	**PSO**	**ACO**	**GA**	**PSO**
Best Index (B)	0.0077	**0.0324**	0.0300	0.0339	0.0319	**0.0354**
Time (seconds)	671.03	548.46	**68.10**	1824.94	350.89	**86.47**
Total Biogas Production (Nm^3/d)	17,224	33,657	**35,395**	20,770	**23,326**	19,524
Avg Organic Load (kg $COD/m^3 \cdot d$)	1.70	2.79	**2.92**	2.02	**2.23**	1.90
Avg C/N ratio (limited below 60)	19.1	26.3	**32.9**	35.9	56.4	**37.8**
Avg Alkalinity (g $CaCO_3/m^3$) (limited above 2500)	2985	3020	**3233**	3217	3199	**3272**

In Scenario 1, the COD concentration was increased tenfold. This was done to compare the efficiency of algorithms to optimise substrates with high organic loads, which is especially meaningful for maximisation of biogas production. For this scenario, ACO and GA have better performance, according to the best index *B* achieved. As a natural consequence of substantial COD increases, biogas production also dramatically increases. However, PSO is unable to achieve a competitive solution in relation to both ACO and GA within this scenario and the baseline scenario, respectively.

In Scenario 2a, a tenfold lineal increase in the geographical distances between facilities was conducted. This scenario allows comparing how well each algorithm can handle situations where most substrates have long distances. For this scenario, ACO is unable to find a feasible solution. On the other hand, GA and PSO find a solution, but the corresponding biogas production is far lower than that obtained in the baseline scenario (48% and 41% lower biogas production for GA and PSO, respectively).

Alternatively, Scenario 2b shows the optimisation results when nonlinearly modifying geographical distances between facilities by the square root of the original distances. This modification allows understanding which algorithm would be more favoured by a more equally distributed geographic location of plants. All the algorithms tested are able to find a feasible solution, showing that GA has the best performance (both in terms of best *B* and biogas production). On the other hand, ACO shows the worst best index *B* achieved.

Scenario 3 was modified by a threefold increase in available volume from all sources. The presented modification allows studying the performance of each algorithm when the total number of possible solutions is much greater. ACO shows noticeably poor performance, below the best index *B* achieved by ACO in former scenarios. Although GA shows better performance than PSO in terms of the best index *B* achieved, biogas production appears similar to that in other scenarios.

In Scenario 4, an increase in the C/N ratio for waste generators W1–W12 was conducted. This modification would test the ability of each algorithm when one of the restrictions (i.e., C/N ratio) requires more adjustments. In this case, PSO shows the best index *B* achieved, although it presents the lowest biogas production. Similar to the baseline scenario, ACO shows slightly better performance than GA in terms of the best *B* achieved, but GA still has slightly better biogas production.

Geographical distance modification was performed with two alternative scenarios. Scenario 2a includes a lineal modification of the distance matrix (tenfold), and Scenario 2b considers a distance modified by the square root of the original distance. The relative locations of all involved waste generators and receptors in the case study are shown in Figure 4. For the baseline scenario, waste generators are homogeneously geographically distributed, but receptors are located in a relatively small area—i.e., the geographical distance difference of each receptor from the emitters might be negligible by the optimisation—which may be interpreted as a single receptor with higher volume capacity, caused by the geographical overlapping of waste receptors, or the "big dot" effect. The linear modification of geographical distances in Scenario 2a does not alter this relative distribution, but the nonlinear modification in Scenario 2b does so, avoiding this "big dot" effect by dispersing Receptors A, B, and C in the geographical space. It is important to note that for Scenario 2a, ACO was unable to find a viable solution, and both GA and PSO achieved a relatively poor solution compared to the corresponding solutions for the baseline scenario.

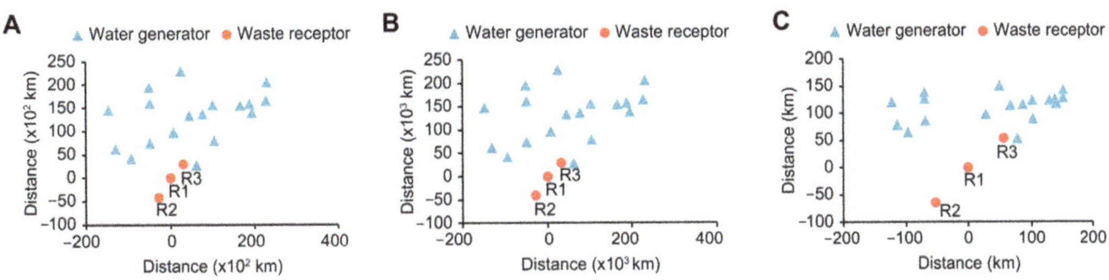

Figure 4. Map of waste generators and waste receptors R1–R3 for Baseline Scenario (**A**), Scenario 2a (**B**) and Scenario 2b (**C**). Distance is expressed as longitudinal distance (X-axis) and latitudinal distance (*Y*-axis) with respect to the R1 plant.

Figure 5 shows the resulting blending profile for the baseline scenario and Scenario 2b. For the baseline scenario, PSO tends to balance the blending of substrates with a low organic load content—i.e., from W1–W12—and selects noticeably lower amounts of high organic load substrates—i.e., from C1–C7— than ACO or GA. This observed behaviour is similar between the three waste receptors A, B and C. On the other hand, ACO and GA tend towards selective blending, showing similar preferences for both receptors B and C. For receptor A, the GA algorithm tends towards slightly more homogeneous blending. In any case, both ACO and GA include more substrates of high organic load—i.e., from C1–C7—except for receptor C.

For Scenario 2b, the ACO blending profiles are similar to those obtained in the baseline scenario—showing a certain tendency to include particular waste generators, —although varying the substrates that the algorithm selects. The GA blending profiles obtained in Scenario 2b are the most affected by geographical distance distortion. GA appears to balance the blending of substrates from all waste generators, much like PSO for both scenarios 2a and 2b. Additionally, the GA blending profile for Scenario 2b accounts for more industrial, high organic load wastes—i.e., from C1–C7—than the PSO blending profile, which remains relatively similar between the baseline scenario and Scenario 2b.

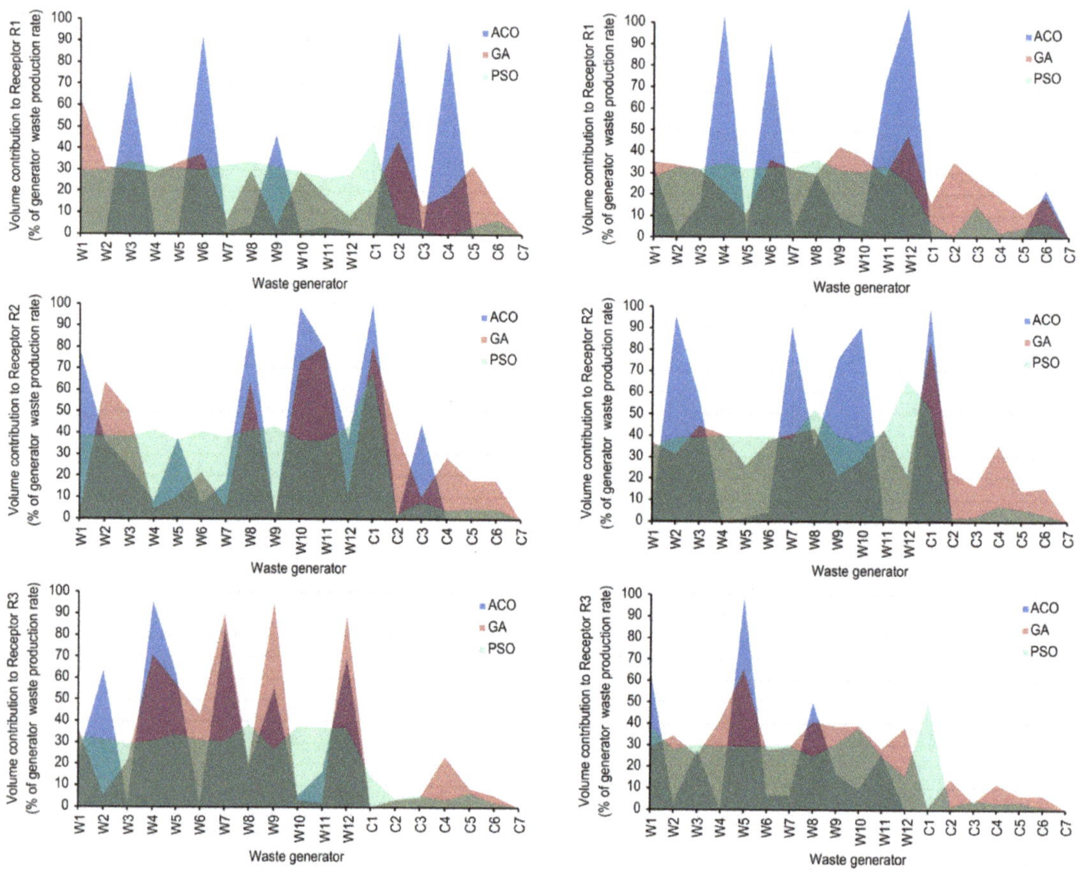

Figure 5. Blending profiles for every waste receptor and ACO, GA and PSO algorithms for the baseline scenario (**left**) and Scenario 2b (**right**).

4. Discussion

Simulations with the optimisation algorithms ACO, GA and PSO were performed, showing successful optimisation results in almost all scenarios. Data from a real case study were used to carry simulations of centralised anaerobic co-digestion blending. As detailed in Section 3, these datasets are composed of 19 organic waste generators and three organic waste receptors within the context of a sanitation network in an area of high industrial activity in Catalonia. This case study composes the baseline scenario. In that previous work, the potential impacts of optimising AnD with wastes from external sources were already demonstrated, bearing up to 77% cost savings regarding waste management. Different modifications were made to this dataset to compare the performance of the ACO, GA, and PSO algorithms under different conditions to assess the performance of each optimisation algorithm in relevant situations. Regarding the optimisation problem, the C/N ratio is the dominant restriction, as was previously seen in [41]. This is the reason why this parameter is included in the discussion of the results, together with biogas production.

For the baseline scenario as seen in Table 5, PSO shows the best index *B* achieved but also the lowest biogas production. However, PSO also shows the lowest C/N ratio, which might play a role in achieving the best solution, compensating for the lack of biogas produced. If biogas production is increased by the design of a particular setup, this could be trimmed by the corresponding weight in the objective function B as a trade-off among the different parameters involved. The results obtained have been considered convenient for the installation under study and improved dramatically performance obtained in the baseline scenario [41]. However, both ACO and GA generally show higher amounts of biogas production, but their best index B values achieved are below that of PSO, and their C/N ratios are above 50.

As shown in Figure 5, ACO and GA show similar behaviours for the baseline scenario, prioritising specific substrates. A first hypothesis suggests that prioritised substrates would be those with higher COD since they would allow higher biogas production. On the other hand, PSO shows a different strategy blending more available substrates and tends to exclude industrial substrates. This trend may point to PSO performing a conservative strategy where it is avoided in all cost situations where the operation of the AnD would be put at risk. Therefore, the general trend is that ACO and GA solve the presented optimisation problem by maximising biogas production and pushing restrictions to the limit, while PSO tends to balance biogas maximisation and the C/N ratio trade-off. In addition, note that PSO has the shortest execution times and ACO the largest, which is observed for all scenarios, indicating PSO to be more computationally efficient, where even here, the execution time is not a drawback for real implementation with the values obtained.

The similarities between ACO and GA and the differences between those and PSO could be partially explained by the nature of these algorithms. Both ACO and GA tend to explore the search space of solutions around the borders, thus increasing the number of non-feasible solutions but also increasing the chances of finding a "rare" solution with a higher best index [72,73]. Thus, these algorithms appear to be based on relatively independent behaviour between particles so that each one can explore separate areas of the border search space and be able to find different non-redundant solutions. On the other hand, PSO algorithm exploration of the search space is based on dependent behaviour between neighbouring particles, which does not encourage particles to explore the limits of the search space. Instead, it promotes the exploration of other mid-term areas between the centre and the borders of the search space. This could help explain why the PSO algorithm attains solutions within shorter execution times but also with generally lower biogas production. Hence, PSO would tend to be a conservative strategy where instead of selecting the most promising solution, single ant or particle, it would prioritise a consensus between the best neighbourhoods.

As detailed in Section 3, Scenario 1 is modified by increasing the organic load of all substrates tenfold. Thus, the dominant condition, in this case, is that organic waste valorisation is fostered, leading to higher biogas production. As observed in Table 5, ACO and GA show better performance than PSO in this scenario, but GA is more efficient since its attained biogas production is noticeably higher than that achieved with ACO.

Additionally, as detailed in Section 3, Scenarios 2a and 2b include a geographical location modification of the involved facilities. As observed in Figure 4, the relative distances between waste generators and receptors (R1, R2, R3) are not modified by lineal modification of the distance when the map plot of the baseline scenario is compared to that of Scenario 2a. However, nonlinear modification of geographical distances in Scenario 2b leads to a different map plot, where waste receptors are more dispersed between them in relation to waste generators (Figure 4). The effect of this distortion of distances is that waste receptors are more separated, thus avoiding geographical overlapping of waste receptors, or the "big dot" effect, i.e., assimilating closer plants as a single centralised plant from a geographical perspective.

Hence, Scenario 2a is modified by tenfold increasing the geographical distances between waste generators and receptors, making geographical distance a dominant condition for optimisation. In this case, ACO is unable to find a solution, and both GA and PSO show extremely poor performance when compared with the baseline scenario, as shown in

Table 5. The linear modification of distances of Scenario 2a shows the performance of each algorithm under the pressure of cases with high geographical distances. This pressure case of Scenario 2a was especially relevant to test because it can significantly impact the logistics processes. On the other hand, Scenario 2b presents a different trend due to the nonlinear modification of distances. As detailed in Table 5, GA shows the best performance, and PSO shows better performance than ACO even in terms of biogas production.

As observed in Figure 5, ACO and PSO maintain similar blending profiles, while GA and PSO also exhibit similar blending profiles, but including GA results in a greater volume of industrial substrates. This shared behaviour between GA and PSO is exclusive to Scenario 2b, but it might indicate that GA behaves similarly to PSO in this case. However, from the operational point of view, GA solutions involve major risks since they tend to include more industrial substrate than PSO solutions.

In Scenario 3, a threefold increase in the volume of all waste generated was performed. First, this result implies that the search space—i.e., the total number of combinations and possible solutions—drastically increases. In this case, Table 5 confirms a similar trend observed for previous scenarios, where GA obtained the best performance and ACO the worst. Again, this finding is consistent with the observation that the ACO algorithm attains weaker performance than GA and PSO for this particular case and that GA and PSO attain similar performance in this study, although GA appears to generally provide better performance than PSO.

Finally, Scenario 4 was composed of increasing the C/N ratio of W1–W12 substrates. These substrates originally conformed to sewage sludge with a low nitrogen load, but in Scenario 4, the drastic increase in the C/N content of sewage sludge was the dominant condition to be tested. Table 5 also presents a summary of the results for Scenario 4, where PSO shows the best performance and GA the worst. The main observation is that the PSO algorithm is more able than the GA to manage situations with high nitrogen loads or major restrictions, while the GA has more potential to maximise biogas production. However, it is more sensitive to high nitrogen loads because it reduces available space to acquire industrial wastes with both high organic loads and high nitrogen loads.

The developed algorithms have successfully optimised the AnD network of the case study, and their performances have been tested under different conditions (i.e., Scenarios 1–4). Simultaneous logistics and quality optimisation of a network of existing waste management facilities is a gap in the current state of the art due to its ad-hoc nature and its interdisciplinarity: there are specialized works for logistics optimisation such as in [45,46], but they do not include process optimisation. The present study implements this logistic optimisation by minimising a cost function designed to this end. The reason for choosing this approach is also based on the need for professionals who manage the AnD network considered here to have a decision support tool capable of integrating logistics and process performance optimisation. And, in addition, to have the ability to handle changing operation conditions and scenarios, as it actually happens in real facilities.

It is also worth noting that it exists a variety of sensors for the determination of physical-chemical parameters that could complement the sensor network considered in this case study, such as a variation of the ones presented in [47]. These sensors could provide additional insight, especially if combined with GIS and process optimisation, and also facilitate the real-time implementation of the presented approach. Additionally, they could also be used as control mechanisms for those cases where ACO and GA optimisation is applied, since attained optimised outcomes pushed quality restrictions of the AnD process close to their thresholds. However, there is a trade-off between information (i.e., data gathered from new sensors) and resources (e.g., implementation, maintenance) which has to be taken into account when considering new sensors.

Overall, this study presents a step forward towards the integrated optimisation of AnD networks, making an innovative attempt to couple logistics and quality optimisation of the centralised digestion process of a real AnD network.

5. Conclusions

In this study, three approaches were developed for the simultaneous optimisation of multiple AnD systems based on ACO, GA and PSO. These methods were applied to a case study based on real data from an AnD network in the area of the Besòs River basin in Catalonia. The performance of each optimisation approach was evaluated. All the approaches successfully optimised biogas production for simulated scenarios while preserving some practical restrictions in optimisation.

For the baseline scenario, ACO and GA allowed maximum biogas production by placing restrictions on the limits of safe operations. On the other hand, PSO solved the optimisation problem with a more conservative strategy where biogas production is lower than that in ACO or GA solutions, in favour of the best AnD operation conditions (i.e., by adjusting the C/N ratio and alkalinity).

In those cases with high opportunities for biogas production (i.e., Scenario 1), GA and ACO would perform the best due to their capabilities of maximising biogas production over that of PSO. GA would perform as the best optimisation algorithm both for cases where distances are significantly different amongst them (i.e., Scenario 2b) and for cases where higher volumes should be handled (i.e., Scenario 3), presumably due to GA's computational potential. Finally, for those cases where other quality-related parameters are restrictions (i.e., Scenario 4), PSO would be the best performing algorithm.

The present study shows an innovative contribution to optimize the performance of centralized AnD systems, combining logistical and quality parameters. To the authors' knowledge, this optimization has not yet been addressed in the literature for an AnD network. In addition, the framework has proven its effectiveness in minimizing the total distance travelled to transport the waste and maximizing biogas production. At the same time, the physical-chemical parameters of the process have been kept within their operational limits.

Further work may include methodologies to improve social impact factor quantification in the optimisation, which might allow better characterisation of the logistic impact of each substrate generator. Additionally, the development of logistic route simulations would be required to enhance real-world distribution planning considering daytime, travel frequency, dynamic waste production-consumption coupled with stocking problems and other time-related issues key to logistic planning.

Author Contributions: Conceptualization, D.P.-H., M.V. and M.À.C.-E.; methodology, D.P.-H., M.V. and M.À.C.-E.; software, D.P.-H.; validation, M.V., M.À.C.-E. and V.P.; formal analysis, M.V., M.P., M.À.C.-E. and V.P.; investigation, D.P.-H., M. Verdaguer; resources, M.P. and M.À.C.-E.; data curation, D.P.-H., M.V., V.P. and M. À. Cugueró-Escofe; writing—original draft preparation, D.P.-H., M.V. and M.À.C.-E.; writing—review and editing, D.P.-H., M.V., V.P. and M.À.C.-E.; supervision, V.P., M.P. and M.À.C.-E.; project administration, M.P. and M.À.C.-E.; funding acquisition, M.P. and M.À.C.-E.. All authors have read and agreed to the published version of the manuscript.

Funding: This research received no external funding.

Institutional Review Board Statement: Not applicable.

Informed Consent Statement: Not applicable.

Data Availability Statement: No new data were created or analyzed in this study. Data sharing is not applicable to this article.

Acknowledgments: The authors would like to acknowledge the support received from Consorci Besòs Tordera. LEQUIA-UdG and SAC-UPC have been recognized as consolidated research groups by the Catalan Government (2017-SGR-1552 and 2017-SGR-482, respectively).

Conflicts of Interest: The authors declare no conflict of interest.

Abbreviations

WWTPs	wastewater treatment plants
AnD	anaerobic digestion
MKP	multidimensional knapsack problem
AnCD	anaerobic co-digestion
ACO	Ant colony optimisation
GA	Genetic algorithm
PSO	Particle swarm optimisation
GUI	graphical user interface
CMAES	covariance matrix adaptation evolution strategy
DE	differential evolution
ANN	perceptron artificial neural network
SGOAs	how stochastic global optimisation algorithms
CO	combinatorial optimisation
EA	evolutionary algorithms
COD	Chemical Oxygen Demand
TN	Total Nitrogen
C/N	Carbon to Nitrogen ratio
CBT	Consorci Besòs Tordera
GIS	geographical information system

References

1. Batstone, D.J.; Virdis, B. The role of anaerobic digestion in the emerging energy economy. *Curr. Opin. Biotechnol.* **2014**, *27*, 142–149. [CrossRef]
2. Bolzonella, D.; Pavan, P.; Battistoni, P.; Cecchi, F. Anaerobic co-digestion of sludge with other organic wastes and phosphorus reclamation in wastewater treatment plants for biological nutrients removal. *Water Sci. Technol.* **2006**, *53*, 177–186. [CrossRef] [PubMed]
3. Melián, J.A.H. Sustainable Wastewater Treatment Systems (2018–2019). *Sustainability* **2020**, *12*, 1940. [CrossRef]
4. Lorick, D.; Macura, B.; Ahlström, M.; Grimvall, A.; Harder, R. Effectiveness of struvite precipitation and ammonia stripping for recovery of phosphorus and nitrogen from anaerobic digestate: A systematic review. *Environ. Evid.* **2020**, *9*, 27. [CrossRef]
5. Namuli, R.; Laflamme, C.B.; Pillay, P. A Computer Program for Modeling the Conversion of Organic Waste to Energy. *Energies* **2011**, *4*, 1973–2001. [CrossRef]
6. Scarlat, N.; Dallemand, J.-F.; Fahl, F. Biogas: Developments and perspectives in Europe. *Renew. Energy* **2018**, *129*, 457–472. [CrossRef]
7. Fouladi, J.; Al-Ansari, T. Conceptualising multi-scale thermodynamics within the energy-water-food nexus: Progress towards resource and waste management. *Comput. Chem. Eng.* **2021**, *152*, 107375. [CrossRef]
8. Baldi, F.; Pecorini, I.; Iannelli, R. Comparison of single-stage and two-stage anaerobic co-digestion of food waste and activated sludge for hydrogen and methane production. *Renew. Energy* **2019**, *143*, 1755–1765. [CrossRef]
9. Mattioli, A.; Gatti, G.; Mattuzzi, G.; Cecchi, F.; Bolzonella, D. Co-digestion of the organic fraction of municipal solid waste and sludge improves the energy balance of wastewater treatment plants: Rovereto case study. *Renew. Energy* **2017**, *113*, 980–988. [CrossRef]
10. Siddique, N.I.; Wahid, Z.A. Achievements and perspectives of anaerobic co-digestion: A review. *J. Clean. Prod.* **2018**, *194*, 359–371. [CrossRef]
11. Villamil, J.; Mohedano, A.F.; Martínez, J.A.V.; Rodriguez, J.; DE LA Rubia, M.A. Anaerobic co-digestion of the process water from waste activated sludge hydrothermally treated with primary sewage sludge. A new approach for sewage sludge management. *Renew. Energy* **2019**, *146*, 435–443. [CrossRef]
12. Díaz-Trujillo, L.A.; Fuentes-Cortés, L.F.; Nápoles-Rivera, F. Economic and environmental optimization for a biogas supply Chain: A CVaR approach applied to uncertainty of biomass and biogas demand. *Comput. Chem. Eng.* **2020**, *141*, 107018. [CrossRef]
13. Mpofu, A.; Kaira, W.; Oyekola, O.; Welz, P. Anaerobic co-digestion of tannery effluents: Process optimisation for resource recovery, recycling and reuse in a circular bioeconomy. *Process Saf. Environ. Prot.* **2021**, *158*, 547–559. [CrossRef]
14. Yağlı, H.; Koç, Y.; Köse, Ö.; Koç, A.; Yumrutaş, R. Optimisation of simple and regenerative organic Rankine cycles using jacket water of an internal combustion engine fuelled with biogas produced from agricultural waste. *Process Saf. Environ. Prot.* **2021**, *155*, 17–31. [CrossRef]
15. Sweelssen, J.; Blokland, H.; Rajamäki, T.; Boersma, A. Capacitive and Infrared Gas Sensors for the Assessment of the Methane Number of LNG Fuels. *Sensors* **2020**, *20*, 3345. [CrossRef] [PubMed]
16. Furst, L.; Feliciano, M.; Frare, L.; Igrejas, G. A Portable Device for Methane Measurement Using a Low-Cost Semiconductor Sensor: Development, Calibration and Environmental Applications. *Sensors* **2021**, *21*, 7456. [CrossRef] [PubMed]

17. Al, R.; Behera, C.R.; Gernaey, K.V.; Sin, G. Stochastic simulation-based superstructure optimization framework for process synthesis and design under uncertainty. *Comput. Chem. Eng.* **2020**, *143*, 107118. [CrossRef]
18. Fouladi, J.; AlNouss, A.; Al-Ansari, T. Sustainable energy-water-food nexus integration and optimisation in eco-industrial parks. *Comput. Chem. Eng.* **2021**, *146*, 107229. [CrossRef]
19. Padrón-Páez, J.I.; Almaraz, S.D.-L.; Román-Martínez, A. Sustainable wastewater treatment plants design through multiobjective optimization. *Comput. Chem. Eng.* **2020**, *140*, 106850. [CrossRef]
20. Dewasme, L.; Sbarciog, M.; Rocha-Cózatl, E.; Haugen, F.; Wouwer, A.V. State and unknown input estimation of an anaerobic digestion reactor with experimental validation. *Control Eng. Pract.* **2019**, *85*, 280–289. [CrossRef]
21. Han, H.-G.; Zhang, H.-J.; Liu, Z.; Qiao, J.-F. Data-driven decision-making for wastewater treatment process. *Control Eng. Pract.* **2020**, *96*, 104305. [CrossRef]
22. Li, Z.; Yan, X. Ensemble model of wastewater treatment plant based on rich diversity of principal component determining by genetic algorithm for status monitoring. *Control Eng. Pract.* **2019**, *88*, 38–51. [CrossRef]
23. Szeląg, B.; Drewnowski, J.; Łagód, G.; Majerek, D.; Dacewicz, E.; Fatone, F. Soft Sensor Application in Identification of the Activated Sludge Bulking Considering the Technological and Economical Aspects of Smart Systems Functioning. *Sensors* **2020**, *20*, 1941. [CrossRef]
24. Ma, Y.; Rose, F.; Wong, L.; Vien, B.; Kuen, T.; Rajic, N.; Kodikara, J.; Chiu, W. Detection of Defects in Geomembranes Using Quasi-Active Infrared Thermography. *Sensors* **2021**, *21*, 5365. [CrossRef] [PubMed]
25. Sánchez, J.B.; Vuono, M.; Dionisi, D. Model-based comparison of sequencing batch reactors and continuous-flow activated sludge processes for biological wastewater treatment. *Comput. Chem. Eng.* **2020**, *144*, 107127. [CrossRef]
26. Naqvi, S.R.; Tariq, R.; Shahbaz, M.; Naqvi, M.; Aslam, M.; Khan, Z.; Mackey, H.; Mckay, G.; Al-Ansari, T. Recent developments on sewage sludge pyrolysis and its kinetics: Resources recovery, thermogravimetric platforms, and innovative prospects. *Comput. Chem. Eng.* **2021**, *150*, 107325. [CrossRef]
27. Babaee, A.; Shayegan, J. Effect of Organic Loading Rates (OLR) on Production of Methane from Anaerobic Digestion of Vegetables Waste. In Proceedings of the World Renewable Energy Congress-Sweden, Linköping, Sweden, 8–13 May 2011; pp. 411–417. [CrossRef]
28. Avilés, A.B.L.; Velázquez, F.D.C.; Del Riquelme, M.L.P. Methodology for Energy Optimization in Wastewater Treatment Plants. Phase III: Implementation of an Integral Control System for the Aeration Stage in the Biological Process of Activated Sludge and the Membrane Biological Reactor. *Sensors* **2020**, *20*, 4342. [CrossRef]
29. Okolie, J.A.; Epelle, E.I.; Tabat, M.E.; Orivri, U.; Amenaghawon, A.N.; Okoye, P.U.; Gunes, B. Waste biomass valorization for the production of biofuels and value-added products: A comprehensive review of thermochemical, biological and integrated processes. *Process Saf. Environ. Prot.* **2021**, *159*, 323–344. [CrossRef]
30. El Ibrahimi, M.; Khay, I.; El Maakoul, A.; Bakhouya, M. Food waste treatment through anaerobic co-digestion: Effects of mixing intensity on the thermohydraulic performance and methane production of a liquid recirculation digester. *Process Saf. Environ. Prot.* **2021**, *147*, 1171–1184. [CrossRef]
31. Fan, Y.; Lv, M.; Niu, X.; Ma, J.; Zhang, D. The key step of gaseous phosphorus release in anaerobic digestion. *Process Saf. Environ. Prot.* **2020**, *137*, 238–245. [CrossRef]
32. Horf, M.; Gebbers, R.; Vogel, S.; Ostermann, M.; Piepel, M.-F.; Olfs, H.-W. Determination of Nutrients in Liquid Manures and Biogas Digestates by Portable Energy-Dispersive X-ray Fluorescence Spectrometry. *Sensors* **2021**, *21*, 3892. [CrossRef]
33. Fréville, A. The multidimensional 0–1 knapsack problem: An overview. *Eur. J. Oper. Res.* **2004**, *155*, 1–21. [CrossRef]
34. Grandón, J.; Derpich, I. A Heuristic for the Multi-knapsack Problem. *WSEAS Trans. Math.* **2011**, *10*, 95–104.
35. Kellerer, H.; Pferschy, U.; Pisinger, D. Multidimensional Knapsack Problems. In *Knapsack Problems*; Springer: Berlin/Heidelberg, Germany, 2004; pp. 235–283. [CrossRef]
36. Glaser, C.; Pavan, A.; Travers, S. The fault tolerance of NP-hard problems. *Inf. Comput.* **2011**, *209*, 443–455. [CrossRef]
37. Somani, D.; Srivastava, H.; Sabumon, P.C.; Anjali, G. A short review of anaerobic co-digestion and feasibility of anaerobic co-digestion of sewage and food waste for sustainable waste management. *Int. J. Earth Sci. Eng.* **2016**, *9*, 55–70.
38. Verdaguer, M.; Clara, N.; Monclús, H.; Poch, M. A step forward in the management of multiple wastewater streams by using an ant colony optimization-based method with bounded pheromone. *Process Saf. Environ. Prot.* **2016**, *102*, 799–809. [CrossRef]
39. Cugueró-Escofet, M.; Puig, V.; Quevedo, J. Optimal pressure sensor placement and assessment for leak location using a relaxed isolation index: Application to the Barcelona water network. *Control Eng. Pract.* **2017**, *63*, 1–12. [CrossRef]
40. García-Gen, S.; Rodríguez, J.; Lema, J. Optimisation of substrate blends in anaerobic co-digestion using adaptive linear programming. *Bioresour. Technol.* **2014**, *173*, 159–167. [CrossRef]
41. Palma-Heredia, D.; Verdaguer, M.; Molinos-Senante, M.; Poch, M.; Cugueró-Escofet, M. Optimised blending for anaerobic co-digestion using ant colony approach: Besòs river basin case study. *Renew. Energy* **2020**, *168*, 141–150. [CrossRef]
42. Rodriguez-Verde, I.; Regueiro, L.; Lema, J.; Carballa, M. Blending based optimisation and pretreatment strategies to enhance anaerobic digestion of poultry manure. *Waste Manag.* **2018**, *71*, 521–531. [CrossRef]
43. Santos, A.T.; Carballa, M.; Morales, N.; Vázquez-Padín, J.R.; Gutierrez, R.; Lema, J. An optimised control system to steer the transition from anaerobic mono- to co-digestion in full-scale plants. *Environ. Sci. Water Res. Technol.* **2019**, *5*, 1004–1011. [CrossRef]
44. Thiriet, P.; Bioteau, T.; Tremier, A. Optimization method to construct micro-anaerobic digesters networks for decentralized biowaste treatment in urban and peri-urban areas. *J. Clean. Prod.* **2019**, *243*, 118478. [CrossRef]

45. Rentizelas, A.; Shpakova, A.; Mašek, O. Designing an optimised supply network for sustainable conversion of waste agricultural plastics into higher value products. *J. Clean. Prod.* **2018**, *189*, 683–700. [CrossRef]
46. Bueno-Delgado, M.-V.; Romero-Gázquez, J.-L.; Jiménez, P.; Pavón-Mariño, P. Optimal Path Planning for Selective Waste Collection in Smart Cities. *Sensors* **2019**, *19*, 1973. [CrossRef] [PubMed]
47. Longo, E.; Sahin, F.A.; Redondi, A.E.C.; Bolzan, P.; Bianchini, M.; Maffei, S. A 5G-Enabled Smart Waste Management System for University Campus. *Sensors* **2021**, *21*, 8278. [CrossRef] [PubMed]
48. Verdaguer, M.; Molinos-Senante, M.; Poch, M. Optimal management of substrates in anaerobic co-digestion: An ant colony algorithm approach. *Waste Manag.* **2016**, *50*, 49–54. [CrossRef]
49. Verdaguer, M.; Clara, N.; Gutierrez, O.; Poch, M. Application of Ant-Colony-Optimization algorithm for improved management of first flush effects in urban wastewater systems. *Sci. Total Environ.* **2014**, *485-486*, 143–152. [CrossRef] [PubMed]
50. Ramachandran, A.; Rustum, R.; Adeloye, A.J. Review of Anaerobic Digestion Modeling and Optimization Using Nature-Inspired Techniques. *Processes* **2019**, *7*, 953. [CrossRef]
51. Zhang, Z.; Liu, Y.; Bo, L.; Yue, Y.; Wang, Y. Economic Optimal Allocation of Mine Water Based on Two-Stage Adaptive Genetic Algorithm and Particle Swarm Optimization. *Sensors* **2022**, *22*, 883. [CrossRef] [PubMed]
52. Marković, D.; Stanković, A.; Petrović, G.; Trajanović, M.; Ćojbašić, Ž. Genetic and Ant Colony Optimization Based Communal Waste Collection Vehicle Routing. *Conf. Inf. Soc. Technol.* **2019**, 209–212.
53. Gaida, D.; Sousa Brito, A.L.; Wolf, C.; Bäck, T.; Bongards, M.; McLoone, S. Optimal Control of Biogas Plants Using Nonlinear MPC. IET Irish Signals Syst. Conf. 2011. Available online: https://www.researchgate.net/publication/224863615_Optimal_Control_of_Biogas_Plants_using_Nonlinear_Model_Predictive_Control (accessed on 26 January 2022).
54. Akbaş, H.; Bilgen, B.; Turhan, A.M. An integrated prediction and optimization model of biogas production system at a wastewater treatment facility. *Bioresour. Technol.* **2015**, *196*, 566–576. [CrossRef]
55. Kana, E.G.; Oloke, J.; Lateef, A.; Adesiyan, M. Modeling and optimization of biogas production on saw dust and other co-substrates using Artificial Neural network and Genetic Algorithm. *Renew. Energy* **2012**, *46*, 276–281. [CrossRef]
56. Alhanjouri, M.; Alfarra, B. Ant Colony versus Genetic Algorithm based on Travelling Salesman Problem. *Int. J. Comput. Technol. Appl.* **2011**, *2*, 570–578.
57. Dorigo, M.; Birattari, M.; Stutzle, T. Ant colony optimization. *IEEE Comput. Intell. Mag.* **2006**, *1*, 28–39. [CrossRef]
58. Dorigo, M.; Stützle, T. The Ant Colony Optimization Metaheuristic: Algorithms, Applications, and Advances. In *Handbook of Metaheuristics. International Series in Operations Research & Management Science*; Glover, F., Kochenberger, G.A., Eds.; Springer: Boston, MA, USA, 2003; Volume 57. [CrossRef]
59. Dorigo, M.; Stützle, T. *Ant Colony Optimization*; MIT Press: Cambridge, MA, USA, 2004.
60. Stützle, T.; Hoos, H. MAX–MIN Ant System. *Futur. Gener. Comput. Syst.* **2000**, *16*, 889–914. [CrossRef]
61. Walters, G.; Savic, D.; Morley, M.; de Schaetzen, W.; Atkinson, R. Calibration of water distribution network models using genetic algorithms. *Trans. Ecol. Environ.* **1998**, *19*, 131–140.
62. Aldosary, A.; Ali, Z.M.; Alhaider, M.M.; Ghahremani, M.; Dadfar, S.; Suzuki, K. A modified shuffled frog algorithm to improve MPPT controller in PV System with storage batteries under variable atmospheric conditions. *Control Eng. Pract.* **2021**, *112*, 104831. [CrossRef]
63. Galuppini, G.; Creaco, E.; Magni, L. Sum-of-delay models for pressure control in Water Distribution Networks. *Control Eng. Pract.* **2021**, *113*, 104844. [CrossRef]
64. Haupt, R.L.; Haupt, S.E. *Practical Genetic Algorithms*; Wiley: Hoboken, NJ, USA, 2004.
65. Gandomi, A.H.; Yang, X.S.; Talatahari, S.; Alavi, A.H. *Metaheuristic Applications in Structures and Infrastructures*; Elsevier Inc.: Amsterdam, The Netherlands, 2013. [CrossRef]
66. Xia, L.; de Jager, B.; Donkers, T.; Willems, F. Robust constrained optimization for RCCI engines using nested penalized particle swarm. *Control Eng. Pract.* **2020**, *99*, 104411. [CrossRef]
67. Aminbakhsh, S.; Sonmez, R. Discrete particle swarm optimization method for the large-scale discrete time–cost trade-off problem. *Expert Syst. Appl.* **2016**, *51*, 177–185. [CrossRef]
68. Ye, C.; Chen, L.; Ni, S.; Zhou, J. Evaluation model of forest eco economic benefits based on discrete particle swarm optimization. *Environ. Technol. Innov.* **2021**, *22*, 101426. [CrossRef]
69. Li, P.; Jiao, X.; Li, Y. Adaptive real-time energy management control strategy based on fuzzy inference system for plug-in hybrid electric vehicles. *Control Eng. Pract.* **2020**, *107*, 104703. [CrossRef]
70. Tuttle, J.F.; Tuttle, J.F.; Vesel, R.; Vesel, R.; Alagarsamy, S.; Alagarsamy, S.; Blackburn, L.D.; Blackburn, L.D.; Powell, K.; Powell, K. Sustainable NOx emission reduction at a coal-fired power station through the use of online neural network modeling and particle swarm optimization. *Control Eng. Pract.* **2019**, *93*, 104167. [CrossRef]
71. Nghiem, L.; Koch, K.; Bolzonella, D.; Drewes, J.E. Full scale co-digestion of wastewater sludge and food waste: Bottlenecks and possibilities. *Renew. Sustain. Energy Rev.* **2017**, *72*, 354–362. [CrossRef]
72. Hibler, D. The GA-ACO Method Applied to Engineering Design. *Procedia Comput. Sci.* **2012**, *12*, 152–157. [CrossRef]
73. Valdez, F.; Moreno, F.; Melin, P. *A Comparison of ACO, GA and SA for Solving the TSP Problem BT—Hybrid Intelligent Systems in Control, Pattern Recognition and Medicine*; Castillo, O., Melin, P., Eds.; Springer International Publishing: Berlin/Heidelberg, Germany, 2020; pp. 181–189. [CrossRef]

sensors

MDPI

Article

Transfer Learning in Wastewater Treatment Plant Control Design: From Conventional to Long Short-Term Memory-Based Controllers

Ivan Pisa [1,2,*], Antoni Morell [1], Ramón Vilanova [2] and Jose Lopez Vicario [1]

1 Wireless Information Networking (WIN) Group, Escola d'Enginyeria, Universitat Autònoma de Barcelona, 08193 Bellaterra, Spain; antoni.morell@uab.cat (A.M.); jose.vicario@uab.cat (J.L.V.)
2 Advanced Systems for Automation and Control (ASAC) Group, Escola d'Enginyeria, Universitat Autònoma de Barcelona, 08193 Bellaterra, Spain; ramon.vilanova@uab.cat
* Correspondence: ivan.pisa@uab.cat

Abstract: In the last decade, industrial environments have been experiencing a change in their control processes. It is more frequent that control strategies adopt Artificial Neural Networks (ANNs) to support control operations, or even as the main control structure. Thus, control structures can be directly obtained from input and output measurements without requiring a huge knowledge of the processes under control. However, ANNs have to be designed, implemented, and trained, which can become complex and time-demanding processes. This can be alleviated by means of Transfer Learning (TL) methodologies, where the knowledge obtained from a unique ANN is transferred to the remaining nets reducing the ANN design time. From the control viewpoint, the first ANN can be easily obtained and then transferred to the remaining control loops. In this manuscript, the application of TL methodologies to design and implement the control loops of a Wastewater Treatment Plant (WWTP) is analysed. Results show that the adoption of this TL-based methodology allows the development of new control loops without requiring a huge knowledge of the processes under control. Besides, a wide improvement in terms of the control performance with respect to conventional control structures is also obtained. For instance, results have shown that less oscillations in the tracking of desired set-points are produced by achieving improvements in the Integrated Absolute Error and Integrated Square Error which go from 40.17% to 94.29% and from 34.27% to 99.71%, respectively.

Keywords: control design; industrial control; transfer learning; WWTP

Citation: Pisa, I.; Morell, A.; Vilanova, R.; Vicario, J.L. Transfer Learning in Wastewater Treatment Plant Control Design: From Conventional to Long Short-Term Memory-Based Controllers. *Sensors* **2021**, *21*, 6315. https://doi.org/10.3390/s21186315

Academic Editors: Miquel À. Cugueró-Escofet and Vicenç Puig

Received: 13 July 2021
Accepted: 18 September 2021
Published: 21 September 2021

1. Introduction

Industrial environments are characterised by running complex and repetitive processes which are sometimes maintained over time. In that sense, control systems are adopted in order to ensure that these processes perform correctly [1]. Most of the times, the development of control strategies can become a complex and time-demanding task since a deep knowledge of the process under control is required. However, the incursion of the Industry 4.0 paradigm and Artificial Neural Network (ANNs) applications are changing the way we control and manage industrial environments. Their main aim is to provide the industries with solutions mainly based on measurements obtained from their systems [2]. Some of these solutions go from basic forecasting systems to more complex solutions, like predictive maintenance ([3], Chapter 9). However, one of the sectors where Industry 4.0 and ANNs are making the point corresponds to the industrial control ([3], Chapter 5). There, ANNs have been adopted for a wide range of tasks, such as the design of soft-sensors or the detection of malfunctions [4–6]. Not only this, but the industrial control domain is experiencing a change in its tendency: ANNs are used more and more as the main control structures than conventional controllers.

One of the industrial sectors where this tendency is observed corresponds to the Wastewater Treatment Plants (WWTPs), which are characterised by running very complex processes where individual operations and actions can change the whole operation of the plant [7]. For that reason, a huge number of control loops are required in order to ensure that each individual operation is correctly performed. Proportional Integral (PI) controllers have mostly been considered as the default and basic controller strategy able to ensure correct WWTP behaviour [8]. However, a complete reduction of the pollutants present in the residual waters cannot be ensured. For that reason, more complex structures have been proposed in order to improve the control performance. Fuzzy and Model Predictive Controllers (MPCs) have been adopted in [9] as the main control strategy to avoid the effluent violations of a WWTP plant, whereas in [10], a hierarchical structure with fuzzy and MPC controllers has also been proposed to determine the control actuation, taking into account the weather and a variable set-point. In this case, the set-point adopted by the MPC controllers is determined by the fuzzy controller whose objective is to maintain the ammonium in the fifth reactor tank of a WWTP at a desired value (please observe Figure 1 to see the distribution of the tanks of a general-purpose WWTP). The problem observed with this kind of structure lies in the fact that they require a model which replicates the relationships between input and output measurements. Besides, most of the time, these relationships consist of non-linear relations which are difficult and tedious to model. This is where ANNs come in, since they are algorithms offering good performance when dealing with these kinds of relationships ([11], Chapter 6). The first approach consists of the adoption of ANNs as elements whose predictions are adopted by conventional control structures. For instance, the solution proposed in [10] has been improved in [12], where Long Short-Term Memory (LSTM) cells have been adopted to predict the WWTP effluent concentrations and determine when and which controller has to actuate. In other cases, neural networks have been considered to directly determine the optimal set-point values adopted by conventional controllers [13] or to implement a Reinforcement Learning (RL) module performing the same task [14]. Moreover, in the last few years, ANNs have been directly considered as the control strategy. In [15], neural networks have been considered to implement an Internal Model Controller (IMC) devoted to managing certain concentrations required in the pollutant reduction tasks performed in the WWTP. This also entails that the control actuation can be decoupled from the physical specifications of the environment [16,17].

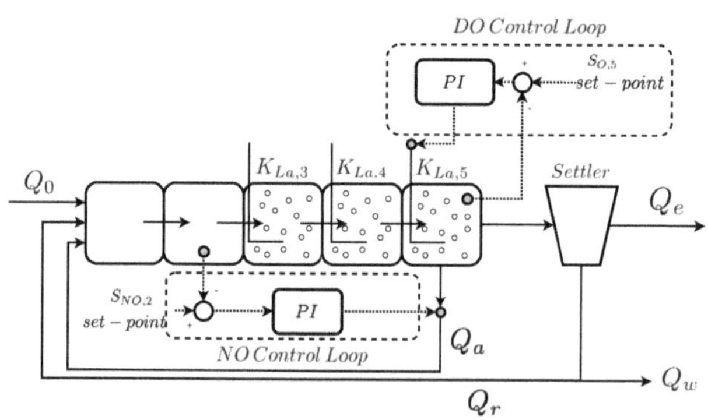

Figure 1. Benchmark Simulation Model No. 1 layout. Q_o, Q_a, Q_r, Q_e, and Q_w are the influent, internal recycle, the external recycle, the effluent, and the wastage flow rates, respectively. Dotted lines correspond to control signals (measured concentrations, desired set-points and actuation signals), while solid lines correspond to process media.

The incursion of ANNs in the industrial control domain presents its own drawbacks that have to be taken into account [18]. The most important one consists in the fact that ANNs have to be designed and trained with amounts of data. This training process is devoted to determining the different hyperparameters of the ANNs, such as the numbers of hidden layers, neurons, learning rates, or even the topology of networks. This has to be performed for each ANN considered, either to complement a conventional controller (PI, MPC, Fuzzy), or to act as the controller as such. Besides, this training process can last hours or even days with regard to the network structure, the hyperparameters, and the amount of data, accordingly [19]. For that reason, transfer learning (TL) methods have been considered to alleviate these tasks.

TL was adopted from image classification tasks, where they were considered to obtain a good image classifier from predesigned and pretrained structures in a source domain [20]. Then, these pretrained structures were retrained with images of the target domain in what is called a fine-tuning process ([21], Chapter 6). In terms of industrial environments, TL was adopted in the design process of soft-sensors, where they are firstly designed and trained in a source domain where a huge number of measurements are available. TL techniques have been adopted mainly to design and implement soft-sensors in those harsh environments, showing a lack of measurements. In [22], TL techniques were considered to design a soft-sensor which would be deployed over a sulphur recovery unit. The problem there is that this environment shows a severe problem of data scarcity; therefore, a traditional ANN training process cannot be performed. To alleviate this, the authors proposed the adoption of TL to design and implement the soft-sensor in an environment without data scarcity problems (the source domain). Then, the obtained soft-sensor was transferred into the environment with the scarcity problem (the target domain) and fine-tuned to adapt its behaviour to this environment [22]. In our case, we propose the adoption of the Transfer Learning-based Control design approach to implement and design the complete control strategy of a general-purpose WWTP. The main idea is to substitute all the PI controllers by LSTM-based PI controllers, where only one is implemented while the others are obtained from transferred versions. The main point here is that instead of training and designing as many LSTM-based PIs as PI controllers, we will implement only a unique LSTM-based structure which will then be transferred into the remaining control loops. In that way, the design of the control loops can be eased at the same time its complexity is reduced. Now, efforts will be focused on designing a unique controller, which will be based on data instead of designing and tuning as many controllers as control loops. In this work, only two control loops have been designed and implemented following this approach. Thus, the benefit of this control approach is not as widely explored as it could be in a scenario where there exists multiple control loops, like in the petrochemical industry [4,23]. However, it has to be taken into account that this approach is mainly based on the adoption of ANNs, which are trained with amounts of data coming from the control loops. Therefore, data have to be accessible in order to adopt this approach; otherwise, the ANNs will not be properly trained, and consequently, the control loops will not act as they should.

This approach has firstly been conducted in [24], where a LSTM-based PI structure has been trained with data from a unique control loop and then transferred into the remaining control loop of a WWTP environment. Notwithstanding this, the structure proposed in [24] considers a unique LSTM cell which requires a total amount of 4 h of WWTP measurements in order to achieve a good control approach. Besides, neither the design of the LSTM-based PI, nor its fine-tuning process is carried out. Therefore, the control performance of the LSTM-based PI can be improved if it is fine-tuned with measurements coming from the target domain, that is, the control loop where the LSTM-based PI is transferred. For that reason, in this manuscript we will continue the work started in [24]. Here, we propose the fine-tuning process and we also analyse the benefits and losses of implementing the LSTM-based PI with data coming from different control loops. Moreover, a new LSTM-based PI structure able to manage the WWTP control loops without requiring 4 h of measurements will be proposed at the same time the control performance will be improved by means of the fine-tuning of this LSTM-based PI controller. Results will show that among all the LSTM-based PI, there exists one able to perform well in the different control loops. Thereby, the fine-tuning process of this LSTM-based controller and its control performance will also be analysed in this work. Besides, the speed-up of the design and implementation process will be explored and analysed in this manuscript as a function of the amount of time required to train the LSTM-based structures. The application where it is tested is specific, but the proposed design approach can be adopted in any kind of industrial environment where measurements are available. In summary, a TL-based design approach is proposed to implement the complete control strategy of a WWTP. The main contributions of this work can be summed up as:

- Conventional PI controllers will be substituted by LSTM-based structures able to improve the conventional controller performance.
- The required knowledge of the process under control will be reduced, since the LSTM-based structures only require input and output measurements of the conventional controllers. Besides, these measurements are easily obtained from a well-known WWTP digital framework: the Benchmark Simulation Model No. 1 [25].
- The design and implementation process of the LSTM-based structures will be sped up, since only a LSTM-based structure will be implemented from scratch. The remaining ones will be obtained through TL approaches.
- A fine-tuning process will be carried out to ensure that the control performance of the control loop is improved with respect to the conventional WWTP controller.

The structure of the manuscript is as follows. The work presented here is introduced in Section 1. The materials and methods adopted in this work are presented in Section 2, especially the Benchmark Simulation Model No. 1 (BSM1), a digital framework which models a general-purpose WWTP. In addition, the LSTM cells, as well as the TL principles are explained in this section. Then, the main contribution of this work, that is, the adoption of TL methods to design and implement the controllers of WWTP control loops are defined and explained in Section 3. The results of the exploratory analyses carried out are reflected in Section 4, while Section 5 concludes the paper.

2. Materials and Methods

2.1. Benchmark Simulation Model No. 1

The Transfer Learning-based Control Design approach proposed here is tested over the Benchmark Simulation Model No. 1 (BSM1). The BSM1 plant is a fictitious WWTP designed by using the engineering principles of an activated sludge process. It characterizes a medium-scale and general-purpose WWTP plant whose main objective is to reduce the nitrogen-derived pollutant products present in residual urban waters [25]. Besides, one of the major aims of BSM1 is to implement a digital framework where different control strategies can be designed and tested before being applied in the real environment. Thus, BSM1 is able to offer generality, easy comparison, and replicability of results in terms of the different control strategies devoted to maintaining certain pollutant components under certain levels or limits [8].

In such a context, BSM1 implements the Activated Sludge Model No. 1 (ASM1) which corresponds to a set of mathematical expressions describing the non-linear and highly complex biological and biochemical processes carried out inside the WWTP plant [26]. These processes mainly consist of the denitrification and nitrification processes where the nitrate and ammonia components are transformed into nitrogen and its derivate products [27]. Notwithstanding this, there are other Activated Sludge Models whose main aim is not only to model the processes carried out to reduce the nitrogen-derived pollutants, but also the phosphorus-derived ones. This is the case for the Activated Sludge Models No. 2, 2d, and 3 [28], which require some updates in the BSM1 framework in order to either consider the phosphorus removal processes like in the phosphorus removal BSM1 framework (BSM1-P), or the sludge treatment, like in the Benchmark Simulation Model No. 2 (BSM2) [29,30]. Nevertheless, the study of these behaviours, as well as the layout of these benchmarks is out of the scope of this work.

2.1.1. BSM1 Layout

The BSM1 layout consists of a set of five reactor tanks and a settler placed just before spilling the clean water into the receiving waters (see Figure 1). The five reactor tanks, where the biological and biochemical processes described in the ASM1 model are carried out, are characterised by their aerated conditions: the first two are anoxic tanks (working with a lack of oxygen), whereas the last three work under aerated conditions [8]. They have a total volume of 6000 m^3 , 1000 m^3 for each anoxic tank and 1333 m^3 for each aerated tank. The settler has a total volume of 6000 m^3. Thus, the total volume of BSM1 equals to 12,000 m^3. Besides, the BSM1 framework has been designed to process an average influent flow rate equal to 18,446 m^3/d and an average biodegradable chemical oxygen demand (COD) of 300 g/m^3. This entails that the BSM1 retention time is equivalent to 14.4 h on average [8,31].

The influent data for municipal WWTP consists of time-series data of the flow and concentrations of the water quality parameters. These influent flow rates depend on many factors: the size of the catchment, the type of the sewer system, and the number of person equivalents, among others. For instance, influent profiles for a WWTP of 100,000 PE are available in [32]. They include dry, rainy, and stormy weather conditions. Besides, these are the usual ones considered when working with BSM1, and therefore the ones considered in the presented work. More information about the BSM1 influent flow and concentrations can be obtained in the BSM1 specifications [8]. Among these 15 variables, the ones of interest in this work are the ones related to the BSM1 default control strategies:

- Nitrate and nitrite nitrogen (NO) control loop: control loop in charge of controlling the nitrate and nitrite nitrogen concentration present in the second reactor tank ($S_{NO,2}$).
- Dissolved oxygen (DO) control loop: control loop in charge of managing the dissolved oxygen present in the fifth reactor tank ($S_{O,5}$).

In the case of the NO control loop, a proportional integral (PI) controller is proposed to manage the internal recycle flow rate (Q_a) in order to ensure that the $S_{NO,2}$ concentration is maintained at the default set-point (1 mg/L). The DO control loop considers another PI structure whose main aim is to maintain the $S_{O,5}$ concentration at the default set-point of 2 mg/L. This is performed by means of varying the oxygen transfer coefficient of the fifth reactor tank ($K_{La,5}$) accordingly to the measured $S_{O,5}$. In that sense, it is worth noting that the two default PI controllers provided in the BSM1 framework have already been tuned, that is, the proportional gain and the integral time parameters are predefined by the BSM1 designers. The control performance of these PI configurations is provided as a start and a baseline with which a new control structure can be compared. Moreover, we have considered the default control strategies, that is, their parameters have been left as the initial configuration proposed by the BSM1 designers.

2.1.2. BSM1 Simulation and Evaluation Protocols

As previously stated, BSM1 has been widely considered as a general-purpose WWTP digital framework offering generality, easy replication, and comparison between different control strategies. In order to ensure a fair comparison in the control performance, BSM1 considers two kind of simulations: (i) a simulation where no variations are produced in the influent, and (ii) a simulation where daily influent and weather variations are produced. In that sense, four influent profiles considering 14 days of influent measurements are provided [25]:

- Constant influent: Influent profile showing constant influent concentrations and flow rates during 14 days.
- Dry influent: Influent profile showing daily variations of the influent concentrations and without any perturbation induced by weather changes.
- Rainy influent: Influent profile showing daily variations of the influent concentrations. Two large rainy perturbations are considered during days 9 and 10.

- Stormy influent: Influent profile showing daily variations of the influent concentrations. Two short but intense stormy perturbations are produced at days 8 and 11.

Thus, the kind of simulation is set accordingly to the influent profile considered in the simulation. However, the BSM1 model has to be previously initialised before performing the simulations. The initialisation process mainly consists in the stabilization of the BSM1 reactor tanks by means of simulating a total amount of 100 days of constant influent ([8], Section 3). Once the model is stabilised, one can perform the desired simulation. From the 14 days simulated, only the last seven days of the simulation, that is, day 7 to day 14 are considered in the performance computation ([8], Section 6). It is also worth noting that only dry, rainy, and stormy influent profiles are considered in this work.

BSM1 also considers its own performance metrics which ease the comparison process among control strategies. They can be divided into two main categories: the environmental metrics and the control ones. The environmental metrics are those showing the improvements achieved in terms of the pollutant reduction when a control strategy is considered instead of another one. Among the different metrics, the two most widely adopted ones are the Overall Cost Index (OCI) and the Effluent Quality Index (EQI). OCI is related to the costs generated in the pollutant reduction process, while the EQI can be understood as a metric telling how clean the water is [8,30]. Nevertheless, we will focus on the control metrics which do not have either an environmental nor a pollutant flavour. In our case, we are going to consider the Integrated Absolute (*IAE*) and Integrated Squared Errors (*ISE*) between the measured variables and their corresponding set-points:

$$IAE = \int_{t=7^{th}\,day}^{t=14^{th}\,day} |r(t) - y(t)|\,dt \tag{1}$$

$$ISE = \int_{t=7^{th}\,day}^{t=14^{th}\,day} (r(t) - y(t))^2\,dt, \tag{2}$$

where $r(t)$ corresponds to the desired set-point, and $y(t)$ to the measured concentration. In this case, $y(t) = \{S_{NO,2}(t), S_{O,5}(t)\}$. Notice that only the control metrics are considered due to the fact that this work is mainly focused on the adoption of transfer learning approaches and ANNs to ease and speed up the design and implementation of the control strategies.

2.2. Long Short-Term Memory Cells

The ANN-based PI controller adopted in this work is mainly based on Long Short-Term Memory (LSTM) cells. They correspond to a type of gated networks which are characterised by their good performance when dealing with time-series signals ([11], Chapter 10). This is possible thanks to the gates that each LSTM cell implements: (i) three sigmoid activation layers, the input gate ($\mathbf{i}(t)$), the forget gate ($\mathbf{f}(t)$), and the output gate ($\mathbf{o}(t)$), and (ii) one hyperbolic tangent layer, the state gate ($\tilde{\mathbf{c}}(t)$) (see Figure 2).

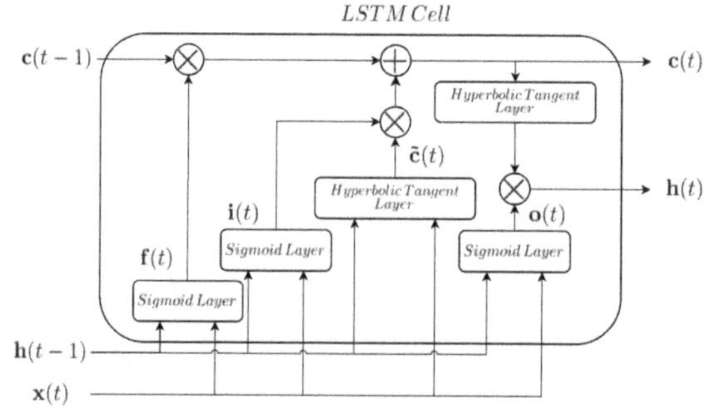

Figure 2. LSTM cell internal structure.

In terms of data, the LSTM cell considers the input data ($\mathbf{x}(t)$) and the output data ($\mathbf{h}(t)$) vectors. Accordingly to them, the forget gate determines the amount of the cell state information that has to be deleted:

$$\mathbf{f}(t) = \sigma(\mathbf{W}_f \cdot \mathbf{x}(t) + \mathbf{U}_f \cdot \mathbf{h}(t-1) + \mathbf{b}_f(t)). \tag{3}$$

Then, the input and state gates determine the new information to be stored in the cell state:

$$\mathbf{i}(t) = \sigma(\mathbf{W}_i \cdot \mathbf{x}(t) + \mathbf{U}_i \cdot \mathbf{h}(t-1) + \mathbf{b}_i(t)) \tag{4}$$

$$\tilde{\mathbf{c}}(t) = tanh(\mathbf{W}_c \cdot \mathbf{x}(t) + \mathbf{U}_c \cdot \mathbf{h}(t-1) + \mathbf{b}_c(t)) \tag{5}$$

$$\mathbf{c}(t) = \mathbf{f}(t) \circ \mathbf{c}(t-1) + \mathbf{i}(t) \circ \tilde{\mathbf{c}}(t). \tag{6}$$

Finally, the output data of the LSTM cell is computed as a function of the input and previous output, as well as the outcome of the output gate:

$$\mathbf{o}(t) = \sigma(\mathbf{W}_o \cdot \mathbf{x}(t) + \mathbf{U}_o \cdot \mathbf{h}(t-1) + \mathbf{b}_o(t)) \tag{7}$$

$$\mathbf{h}(t) = \mathbf{o}(t) \circ tanh(\mathbf{c}(t)). \tag{8}$$

Notice that $\mathbf{W}_x$ and $\mathbf{U}_x$ are the weights of the different gates modifying the input and output data vectors, respectively. $\mathbf{b}_x$ are the biases of the different gates. Finally, $\circ$ is the Hadamard product between two matrices. σ and $tanh$ are the sigmoid and hyperbolic tangent activation functions, respectively. If more information about LSTM cells and their behaviour is required, readers are referred to ([11], Section 10.10).

2.3. Transfer Learning

The main contributions of this work are mainly focused on the adoption of Transfer Learning (TL) techniques to ease and speed up the control design in industrial environments, especially in the WWTPs. In that sense, TL consists of transferring the knowledge obtained in the training process of an ANN structure into another one. For instance, TL techniques have been widely adopted in the design and implementation of image classifiers among others [20]. One clear example is shown in ([21], Chapter 6), where the Inception model, a general-purpose image classifier, is adopted to develop a dog breed classifier. This new classifier is implemented with the Inception classifier without the last layer plus three new convolutional layers connected to the output of the penultimate Inception layer. Therefore, the dog breed classification performance will be derived from the Inception classification one and a new retraining process where a new set of dog breed pictures is considered ([21], Chapter 6). This shows that TL techniques can be considered as techniques which not only obtain ANN models performing well from a source model, but also speed up their designing process since the knowledge of the source model is shared with the new ones ([21], Chapter 4).

In that sense, TL techniques can be categorised into three classes as a function of the data availability in the source and target domains or scenarios ([21] Chapter 4, [22]):

- Inductive Transfer Learning: In inductive transfer learning, the source and target domains do not show data scarcity problems. Therefore, the transfer model can be designed and firstly trained in the source domain and then fine-tuned in the target domain in order to adapt its behaviour to its final application.
- Transductive Transfer Learning: Transductive transfer learning is characterised by the necessity of retraining the transferred model every time a new set of labelled data is available in the target domain. This is motivated by the fact that at the first moment, the target domain has no labelled data.
- Unsupervised Transfer Learning: Unsupervised transfer learning is characterised by the fact that there is no available data neither in the source domain, nor in the target one. Thus, this technique is mainly focused on solving unsupervised tasks like dimensionality reduction.

In our case, we are faced with an Inductive Transfer Learning task, since the source and target domains do not show a data scarcity problem. Here, the source domain consists in the $S_{O,5}$ control loop (DO control loop) or the $S_{NO,2}$ control loop (NO control loop) depending on the base ANN-based controller being implemented. In that sense, if the DO control loop is considered as the source domain, the NO control loop will be considered as the target domain and vice versa.

2.4. Modelling

Three different tools have been considered in this work to implement and test the proposed Transfer Learning-based Control Design approach. They correspond to Simulink and Python. Simulink was adopted due to the fact that the BSM1 model is completely deployed over this simulator. Simulink version 10.1 running over Matlab R2020b was considered. Moreover, all the ANNs involved in the proposed approach are also deployed over the BSM1 model in order to test their behaviour. Thus, they are also implemented in Simulink. In that sense, the ANNs, and especially the LSTM cells have been designed and trained by adopting Python 3.6 with three open-source libraries and a NVIDIA GeForce RTX 2080 Titan GPU memory, which is considered to speed up the LSTM training process:

- NumPy (1.18.1) [33]: library providing a huge amount of tools and operations involving vectors and matrices.
- Scikit-Learn (0.22.1) [34]: library providing most of the functions considered in data preprocessing, cross-validation, and evaluation processes.
- Tensorflow (1.14.0) [35]: library providing lots of ANN structures and techniques. It also implements the Keras API, which offers predefined ANN structures, optimizers, cost functions, or training algorithms. Therefore, nearly any ANN structure can be designed by means of concatenating different predefined Keras structures.

3. TL-Based Control Design

As it has been stated, one of the problems in industrial control is related to the conception and design of the control loop. Most of the times, the design of the controllers can become a tedious and time-consuming process since one has to determine the topology of the controller to be used, as well as the plant or process it is going to manage. In that sense, ANNs have arisen as a possible solution able to alleviate this. They only require pairs of input and output data of the process to be controlled [15]. However, this has its own drawbacks: ANNs have to be correctly trained and designed if a good control performance is required. This can become a time-demanding and computationally expensive process if there are a lot of control loops to design.

For that reason and to alleviate this issue, we propose in this work the TL-based Control Design approach, which is focused on designing and implementing the control strategies of a general purpose WWTP. In this case, the TL-based Control Design approach consists in two stages: (i) the LSTM-based controller, where the design and training of an ANN-based controller is carried out, and (ii) the Control Knowledge Transfer approach, where the transfer of the controller knowledge into the different industrial control loops is performed. The first stage is mainly based on designing an ANN able to manage the signals considered in the control of the industrial process. To achieve this, the proposed ANN-based controller predicts the corresponding actuation signal accordingly to its input measurements, that is, the measured value and its set-point. In our case, the signals involved in the control loops correspond to either the $S_{O,5}$ or the $S_{NO,2}$ concentrations, and their respective actuation signals, the $K_{La,5}$ or Q_a. Besides, the ANN-based controller will be implemented with LSTM cells due to their good performance when dealing with time-series signals, such as the ones obtained from the BSM1 framework. ([11] Section 10.10, [8]). The second stage is mainly focused on transferring the knowledge of the proposed LSTM-based controller into the other control loops. In this case, the LSTM-based controller is considered as the baseline strategy to be transferred (see Figure 3). Thus, the objective is to design only a LSTM-based controller instead of as many LSTM-based structures as control loops present in the WWTP. Then, its knowledge will be transferred into the remaining loops.

It is also important to notice that this transfer approach can be adopted in any industrial scenario. However, there is a requirement that has to be fulfilled; the TL-based Control Design can only be applied among control loops sharing the same control objective. This is motivated by the fact that the ANN-based structure trained in the source control loop will learn how to generate an actuation signal from the controlled ones with the objective of performing certain tasks, for instance, the tracking of a given set-point. Then, the knowledge of this ANN-based structure will be transferred into the target one, which should have the same objective. Otherwise, the target structure would generate actuation signals which do not fulfil the control objective. In the case of this work, the control objective is clear, where both the DO and the NO control loops are designed to track the given set-points regardless of the fact that the involved signals show different values and dynamics [8,25].

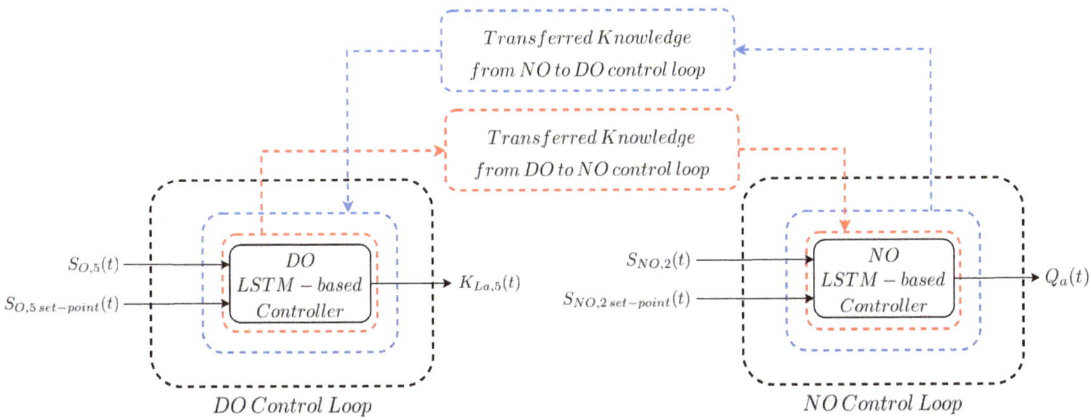

Figure 3. Graphical description of the TL-based Control Design approach. Notice that DO refers to the Dissolved Oxygen ($S_{O,5}$) control loop, whilst NO refers to the nitrate and nitrite ($S_{NO,2}$) control loop.

Once the knowledge of the LSTM-based control structure is transferred, the control performance of the LSTM-based controller can be adjusted through a fine-tuning process which consists in a retraining of the LSTM-based structure. However, this fine-tuning process is different from the usual fine-tuning processes performed in the usual applications of transfer learning, that is, the development of image classifiers. There, the data considered to carry out the fine-tuning process consist in a set of new images where the labels are intrinsically obtained from the same images. Taking up the dog breed classifier, the TL fine-tuning process is performed to the Inception structure with images of different dog breeds where the labels are obviously clear. When talking about industrial processes, the situation completely changes. The new measurements have to be obtained from the control loop where the LSTM-based controller is going to be transferred. In addition, the knowledge about how to control this loop has to also be obtained. For that reason, the data required to perform the fine-tuning processes have to be obtained by means of simulating the behaviour of the industrial process when an existing and conventional controller is applied. If not, the LSTM-based controller will not be able to offer a good control performance.

In this manuscript, two LSTM-based controllers will replicate the behaviour of the default WWTP PIs since they are the ones present in the BSM1 digital framework [8]. This is motivated by the fact that the LSTM-based PI controller can be obtained with data from the DO control loop (DO LSTM-based PI) and then transferred into the NO control loop, or it can be designed considering measurements from the NO control loop (NO LSTM-based

PI) and then transferred into the DO loop. From these two controllers, the one offering the best control performance in both control loops will be fine-tuned.

3.1. LSTM-Based PI

As it has been previously stated, the controller proposed in the TL-based Control Design approach consists in a LSTM-based controller which will act as a PI controller managing either the $S_{O,5}$, or the $S_{NO,2}$ concentrations. Hence, two LSTM-based PI candidates are proposed since there exists two control loops, the DO and the NO control loop. For that reason, we will analyse the control performance of each one in order to determine which LSTM-based PI will be transferred and fine-tuned. The first LSTM-based PI controller corresponds to the DO LSTM-based PI, which is derived from the PI managing the $S_{O,5}$ concentration, while the second one corresponds to the NO LSTM-based PI. It is derived from measurements of the default PI managing the $S_{NO,2}$. Before designing and training the two LSTM-based PIs, one can guess which one will offer the best control performance. If the control performance of the default PI controllers is taken into account (see Figure 4), one can observe that the best PI corresponds to the one managing the $S_{O,5}$, since it is able to maintain the S_O concentration at the desired value (2 mg/L). On the other hand, the PI managing the $S_{NO,2}$ is not able to maintain the desired set-point. Thereby, the control performance of the LSTM-based ones will be similar to the default PI controller from which the data were obtained. In other words, the better the conventional controller performance, the better the LSTM-based one.

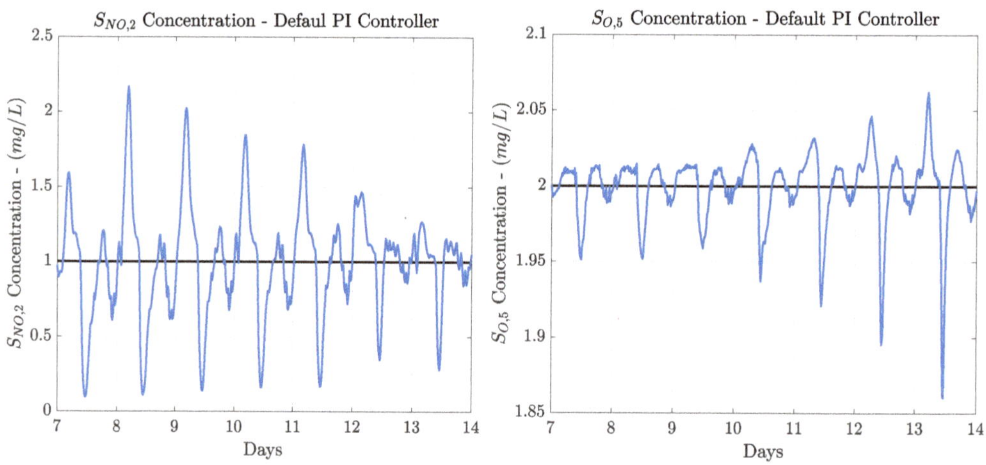

(**a**) $S_{NO,2}$ Default PI control (**b**) $S_{O,5}$ Default PI control

Figure 4. Control performance when the default PI controllers are adopted. Notice that the worst performance is offered by the $S_{NO,2}$ default PI controller.

The DO LSTM-based PI and the NO LSTM-based PI structures are obtained by means of a grid search method where different LSTM-based structures are trained with the same set of measurements. The efforts of the grid search are focused on determining the number of LSTM cells, feedforward layers, and hidden neurons per layer of the LSTM-based structure. Then, the LSTM structure offering the best prediction performance without committing overfitting is the one considered as the main structure in which the LSTM-based PI is based on. In that sense, the grid search is performed instead of finding the parameters characterising the PI controller, that is, the integral time and the proportional gain [36]. This means that a deep knowledge of the process under control is not required. Only pairs of input and output measurements of the existing default PI controllers are needed. To obtain them, a complete year of randomly distributed weather profiles has been simulated in order to achieve a good control performance regardless of the weather conditions. From all the available measurements, the input and output measurements of the LSTM-based PI controller will be determined accordingly to the control loop they will manage:

- DO Control loop: the measurements involved in the DO control loop are the dissolved oxygen ($S_{O,5}$), its desired set-point ($S_{O,5_{set-point}}$), and the oxygen transfer coefficient of the fifth reactor tank of the WWTP ($K_{La,5}$).
- NO Control loop: the measurements involved in the NO control loop are the nitrate and nitrite nitrogen ($S_{NO,2}$), its desired set-point ($S_{NO,2_{set-point}}$) and the internal recirculation flow of the WWTP (Q_a).

These measurements are the ones considered to carry out the grid search method devoted to determining the LSTM-based PI structures. Each one of these measurements is split into three different sets: 70% of the measurements to train the different LSTM-based net configurations, 15% to validate them, and the remaining 15% to test the structures. The grid search process has been carried out adopting the Adam optimizer ([11], Sections 6.5 and 8.5.3) and a total amount of 500 epochs. The initial learning rate value has been set to 1×10^{-3}, however, it is reduced along the process. In addition, LSTM nets are also known to suffer overfitting problems, where they memorise the input and output measurements instead of deriving a model from them ([11], Chapter 7). To avoid this problem, the L2 parameter regularisation technique and early stopping method are considered. The L2 parameter regularisation consists in the addition of extra penalty to the weights of the corresponding layer ([11], Section 7.1.1). This extra penalty is known as the weight decay parameter, which in this case has been set to 5×10^{-4}. On the other hand, early stopping acts as a technique which stops the training process when the validation performance changes its tendency with respect to the training one ([11], Section 7.8). Here, the important point corresponds to the early patience, which determines the amount of epochs that this change of tendency is allowed. In this work, we consider an early patience of five epochs, understanding an epoch as a complete pass over the training dataset ([37], Chapter 2). Both LSTM-based controllers consider the same LSTM-based structure (see Figure 5) which mainly consists in two LSTM cells devoted to extracting and obtaining information from the time correlation between measurements and two feedforward layers which will transform this information into the desired output. Moreover, each structure considers Normalisation and Denormalisation stages in charge of normalising the input measurements towards zero mean and unity variance, and to take them into its natural range, respectively. These two stages are needed since the range of the measurements involved in the control loops are quite different: the mean of the measurements involved in the DO control loop equal to 1.9752 and 144.68 for the $S_{O,5}$ and the $K_{La,5}$, respectively. In the case of the NO loop, the mean values of the variables involved in the control are equal to 0.9937 and 2.1802×10^4 for the $S_{NO,2}$ and Q_a measurements. As a summary, the DO LSTM-based and NO LSTM-based structures are as follows:

- DO LSTM-based PI
  - Input measurements: the dissolved oxygen in the fifth reactor tank ($S_{O,5}(t)$) and its desired set-point ($S_{O,5_{set-point}}(t)$). Besides, the DO LSTM-based net considers the Nonlinear Autoregressive Exogenous principle (NARX) where the output predicted by the net will be considered as an extra input. This extra input provides the LSTM-based structure with information about its performance in the prediction process [38], thus it will be able to correct its predictions as a function of this extra input. In this case, the extra input corresponds to the previously computed actuator signal ($K_{La,5}(t-1)$).
  - Normalisation Stage: stage devoted to normalising the input measurements towards zero mean and unity variance.

- LSTM-based Net: main part of the LSTM-based Controller. It consists of two LSTM cells with 100 and 50 hidden neurons and two feed forward layers with 50 and 25 hidden neurons, respectively.
- Denormalisation Stage: stage devoted to denormalising the actuation signal (DO LSTM-based Net output) towards its real range of values.
- Output: the actuation signal which corresponds to the oxygen transfer coefficient of the fifth reactor tank ($K_{La,5}(t)$).

• NO LSTM-based PI
  - Input measurements: the nitrate and nitrite nitrogen in the second reactor tank ($S_{NO,2}(t)$) and its desired set-point ($S_{NO,2_{set-point}}(t)$). As it happens with the DO LSTM-based PI, the NO LSTM-based controller also considers the NARX principle. In this case, the extra input corresponds to the previously computed actuator signal ($Q_a(t-1)$).
  - Normalisation Stage: stage devoted to normalising the input measurements towards zero mean and unity variance.
  - LSTM-based Net: main part of the LSTM-based Controller. It consists of two LSTM cells with 100 and 50 hidden neurons and two feed forward layers with 50 and 25 hidden neurons, respectively.
  - Denormalisation Stage: stage devoted to denormalising the actuation signal (DO LSTM-based Net output) towards its real range of values.
  - Output: the actuation signal which corresponds to the WWTP internal recirculation flow rate ($Q_a(t)$).

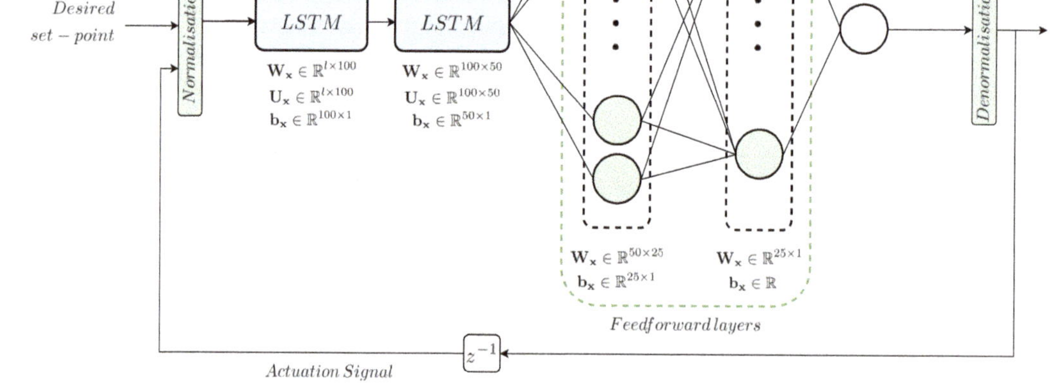

Figure 5. LSTM-based net considered in the LSTM-based Controller. l corresponds to the number of inputs, which in this case is set to three measurements: the measured concentration of interest, $S_{O,5}(t)$ or $S_{NO,2}$, its set-point, $S_{O,5_{set-point}}(t)$ or $S_{NO,2_{set-point}}(t)$, and the actuation variable, $K_{La,5}(t-1)$ or $Q_a(t-1)$.

The prediction performance of both structures has been computed in terms of the difference between the predicted actuation variables and the expected ones (remember that the DO LSTM-based PI predicts the $K_{La,5}$), whereas the NO one predicts the Q_a. Five metrics are adopted, the Root Mean Squared Error (RMSE), the Mean Absolute Error (MAE), the Mean Average Percentage Error (MAPE) the determination coefficient (R^2), and the training time [39]. The RMSE and the MAE tell us how the prediction errors are, that is, if the predictions are close to the expected measurements or not. However, they are absolute metrics in terms of how they do not tell us how big or small these errors are. For that reason, we consider the MAPE, which compares the errors with respect to the expected value. R^2 is considered to determine the correlation between the predicted and expected measurements. Finally, the training time is considered to determine the amount of time to train unique network. Notice that all the prediction metrics are computed considering normalised values, with the exception of the MAPE in order to avoid divisions by zero and the training time. In that sense, the results show that the proposed LSTM-based structures are able to offer a good prediction performance (see Table 1) since both structures yield low RMSE, MAE and MAPE values at the same time they offer a R^2 nearly equal to 1. Therefore, it is corroborated that these structures can be used to implement PI controllers which are mainly based on data.

Table 1. Prediction performance of the DO LSTM-based PI and the NO LSTM-based PI.

	LSTM-Based Prediction Performance				
	RMSE	**MAE**	**MAPE**	R^2	**Training Time**
DO LSTM-based PI	0.026 mg/L	0.018 mg/L	1.347%	0.999	69.91 s
NO LSTM-based PI	0.048 mg/L	0.037 mg/L	6.26%	0.997	98.60 s

3.2. *Control Knowledge Transfer Approach*

The Control Knowledge Transfer approach corresponds to the stage of the TL-based Control Design devoted to transferring the knowledge of the LSTM-based PI structures of one WWTP control loop into the other. The adoption of this stage is motivated by the fact that we looked for the ease and speed-up of the controller design and implementation process, respectively.

In this manuscript, three different TL approaches are considered to achieve the transfer of the control knowledge between control loops. Two of them are considered to determine which controller, the DO or the NO LSTM-based PI, has to be transferred and then fine-tuned. The third approach mainly consists in the adoption of the controller showing the best performance in the source and target domains and its fine-tuning to adapt its behaviour to the dynamics of the target domain, the control loop where it has been transferred. As a summary, the three considered control approaches are:

- Transfer Learning from DO to NO
  The DO LSTM-based PI structure is transferred directly from the DO to the NO control loop. Here, it is important to notice that the structure is not fine-tuned, that is, the LSTM-based PI controller has been trained to manage the $S_{O,5}$ concentration. Besides, only the normalisation and denormalisation stages are adapted to the NO control loop measurements.
- Transfer Learning from NO to DO
  The NO LSTM-based PI structure is directly transferred from the NO to the DO control loop without performing any change, neither in its structure, nor in its weights and biases. Thus, the knowledge on how to manage $S_{NO,2}$ concentration is transferred into the DO control loop. The unique change performed in this transfer approach corresponds to the normalisation and denormalisation stages. They have been adapted to normalise and denormalise the measurements coming from the NO control loop instead of the DO control loop. Following this, the NO LSTM-based PI will be at least equal to the default PI managing the $S_{NO,2}$ concentration, that is, the NO control loop PI. If Figure 4 is taken into account, one can assure that the NO LSTM-based PI controller will not offer such a good control performance as the DO LSTM-based PI derived from the DO control loop.
- LSTM-based controller Fine-tuning & Transfer

This transfer approach is the most important one since it corresponds to the transfer method performing a fine-tuning process and therefore, adapting the behaviour of the transferred controller to the target domain dynamics. In other words, in this transfer learning approach the LSTM-based controller yielding the best control performance between the Transfer Learning from DO to NO and the Transfer Learning from NO to DO will be considered as the candidate to be fine-tuned. Results in Section 4 show that the best control performance is offered by the DO LSTM-based PI. For that reason, this is the LSTM-based controller considered in the fine-tuning process. Nevertheless, this choice can be done at the very beginning if the performance offered by the conventional PI structures is considered (see Figure 4).

In terms of the three TL classes, the LSTM-based controller Fine-tuning and Transfer approach consists in an inductive transfer learning task: data from the source domain, the DO control loop, is considered to firstly obtain the DO LSTM-based PI structure. Then it is fine-tuned (retrained) with data coming from the PI controlling the target domain, the NO control loop. In other words, the default $S_{NO,2}$ controller whose performance is observed in Figure 4a has been considered to perform the fine-tuning process of the DO LSTM-based PI controller. Thus, the obtained controller, the fine-tuned DO LSTM-based PI (FTDO LSTM-based PI) will know how to correctly manage the desired variable, but adapted to the NO control loop. This clearly shows that an existing controller managing the target control loop is compulsory to obtain the measurements considered in the fine-tuning process. This differs from traditional and conventional TL applications, where labelled data are available.

The main point here is that in the fine-tuning process not all the layers of the DO LSTM-based PI controller will be retrained with measurements of the target domain: the weights of the two LSTM cells are blocked whilst the weights and biases of the two feedforward layers (see Figure 5) are modified in the fine-tuning process. The LSTM cells are the ones that are blocked since they are the layers gathering the information about the time-dependence between measurements. The feedforward layers mainly take this information to adapt the output of the controller to the desired control loop. For that reason, these are the layers which will be retrained just to adapt the outcomes of the LSTM layers to the new domain.

The measurements of the target domain are again obtained by performing a whole-year simulation of the BSM1 behaviour when the three weather profiles, dry, rainy, and stormy, are randomly distributed. The weights and biases of the two retrained feedforward layers are obtained considering the same training parameters as in the case of the DO LSTM-based PI training process: initial learning rate equals to 1×10^{-3}, the weight decay equals to 5×10^{-4} and the early patience is set to 5 epochs.

4. Results

4.1. TL-Based Control Design Results

The performance of the TL-based Control Design approach is determined by means of analysing the control performance of each one of the proposed TL approaches: (i) the Transfer Learning from DO to NO, (ii) the Transfer Learning from NO to DO, and (iii) the LSTM-based controller Fine-tuning and Transfer. In that sense, the two first results will determine which controller, the DO LSTM-based PI or the NO LSTM-based PI, is performing better in both control loops when no fine-tuning process is carried out. Finally, the one performing better is fine-tuned and its control performance is computed in the last TL approach. Results will show which is the best option not only to obtain a complete and good control approach mainly based on data, but also to speed-up the design process of the complete WWTP control strategy.

The control performance has been computed in terms of fix and variable set-points in order to determine if the TL-based Control Design approach is suitable for both types of set-points. Fix set-points are considered since the default control strategy considers them in order to assure that the nitrification and denitrification processes, the ones performing the pollutant reduction task, are correctly performed [8,27]. They have been set to 2 mg/L and 1 mg/L for the $S_{O,5}$ and $S_{NO,2}$ control loops, respectively. Notwithstanding, variable set-points are the ones of most interest since most of the times the set-points are computed by means of other control strategies or are varied in order to optimise the pollutant reduction process [10,14,40,41]. In this case, the variable DO set-point has been computed accordingly to the Fuzzy Logic adopted in [10], where the Fuzzy controller is considered to determine the $S_{O,5}$ set-point generating the lower $S_{NH,5}$. Moreover, the three different BSM1 weather profiles have also been simulated to determine if the control design approach can be considered regardless of the weather conditions.

4.2. Transfer Learning from DO to NO

The first computed control performance corresponds to the situation where the DO LSTM-based PI is obtained with data from the DO control loop. Then, it is transferred into the NO control loop without performing any fine-tuning process. Results are shown in Table 2, where the first important effect that one can notice is that the control performance in the DO control loop, that is, in the management of $S_{O,5}$, is even better than the control offered by the default PI. This effect is motivated by two situations: (i) the fact that the DO LSTM-based PI has been trained through the simulation of the control strategy when random variations in the set-point are provided, and (ii) the NARX principle which provides the LSTM-based structure with information about the previous predicted outcomes. Thus, the LSTM-based structure has learnt how to correct variations present either in the set-point, or in the measured concentration.

Table 2. Control performance when the DO LSTM-based PI derived from the DO control loop is transferred into the NO control loop.

Transfer Learning from DO to NO Control Loop						
Fix Set-point						
	Dry Weather		Rainy Weather		Stormy Weather	
Structure	*IAE*	*ISE*	*IAE*	*ISE*	*IAE*	*ISE*
PI—$S_{O,5}$	0.148	0.007	0.143	0.007	0.158	0.007
DO LSTM-based PI—$S_{O,5}$	0.006	9.98×10^{-6}	0.006	1.12×10^{-5}	0.006	1.29×10^{-5}
PI—$S_{NO,2}$	1.594	0.691	1.922	0.951	1.874	0.977
DO LSTM-based PI—$S_{NO,2}$	1.008	0.290	1.401	0.578	1.033	0.357
Variable Set-point						
PI—$S_{O,5}$	0.185	0.016	0.155	0.014	0.206	0.020
DO LSTM-based PI—$S_{O,5}$	0.013	2.34×10^{-4}	0.016	4.48×10^{-4}	0.016	4.05×10^{-4}
PI—$S_{NO,2}$	1.792	0.858	2.132	1.089	1.884	0.989
DO LSTM-based PI—$S_{NO,2}$	1.271	0.503	1.672	0.758	1.358	0.593

In terms of the *IAE* and *ISE* metrics, one can observe that they are improved with respect to the default PI control performance when a fixed set-point is considered. In addition, these improvements are achieved regardless of the weather profile. In other words, the *IAE* and *ISE* values were improved by around a 95.98% and a 99.84% in average with respect to the default PI controller, respectively. For instance, the highest *IAE* improvement is achieved when the stormy influent profile is simulated. The *IAE* offered by the default $S_{O,5}$ PI controller is equivalent to 0.158, while it is reduced until 0.006 when the DO LSTM-based PI is considered. This entails that the difference between the measured and the $S_{O,5}$ controller is minimal. In terms of the *ISE*, the highest improvement is obtained when the dry weather is simulated. The achieved improvement equals to 99.86% with respect to the default PI controller. Notwithstanding, this improvement equals to 99.84% and 99.82% when the rainy and stormy weathers are considered, respectively. These results show that the DO LSTM-based PI is able to be highly improved when a fixed set-point is considered. However, the important results are the ones obtained when a variable set-point is considered, since it corresponds to the most frequently adopted set-point topology.

In such a context, the same effect is observed when a $S_{O,5}$ variable set-point is considered. In this case, the average improvement in terms of the *IAE* and *ISE* equals to 91.67% for the *IAE* and a 97.77% for the *ISE*. Now, the highest improvement is achieved when the dry weather is considered: the *IAE* and the *ISE* are improved by 92.97% and 98.54% with respect to the default PI controller performance. This is motivated by the fact that rainy and stormy influents are derived from the dry weather where the rainy and stormy episodes are included. For that reason, the LSTM-based structure has more often observed the effects of the PI controlling the $S_{O,5}$ when dry episodes are observed rather than stormy or rainy ones. In addition, the control performance clearly shows that the DO LSTM-based PI controller can be adopted as the main controller in the DO control loop (see Figure 6). As it is observed, the output of the controller is much closer to the given set-point of 2 mg/L than the default PI output.

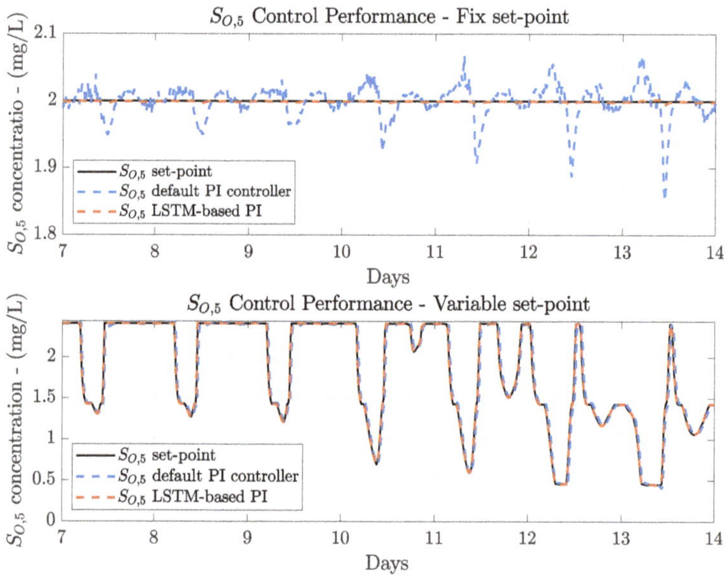

Figure 6. Control performance for the DO control loop when the LSTM-based PI is considered.

However, the most important point is to determine the control performance of the NO control loop, since in this case the DO LSTM-based PI is directly transferred into the NO control loop. The changes performed in the control structure correspond to the normalisation and denormalisation stages, which have been adapted to the range of values involved in the control of $S_{NO,2}$. Results show that the control performance of the DO LSTM-based PI controller can be improved by, on average, 33.07% and 42.94% in the case of the *IAE* and the *ISE*, respectively, when it is managing the $S_{NO,2}$ and considering a fixed set-point. For instance, the highest improvement with respect to the default PI structure is achieved when the stormy weather is considered. The *IAE* and *ISE* obtained in such a situation equal to 1.033 and 0.357, respectively, which in percentage values equal to an improvement of a 44.88% and a 63.46% for the *IAE* and *ISE* respectively. At the same time, this represents a reduction of the *IAE* and *ISE* improvement of 51.32 and 36.36 percentage points with respect to the improvement achieved when the DO LSTM-based PI is managing the $S_{O,5}$. This is clearly motivated by the fact that the DO LSTM-based PI is designed to offer its best performance when managing the DO control loop. When a $S_{O,5}$ variable set-point is chosen, one can observe that the average improvements in the NO control loop and in terms of the *IAE* and *ISE* are equal to 26.19% and 37.27%, respectively, being the dry weather the one showing the highest improvement (see Figure 7). The *IAE* values go from 1.792 to 1.271 while the *ISE* values go from 0.858 to 0.503, respectively.

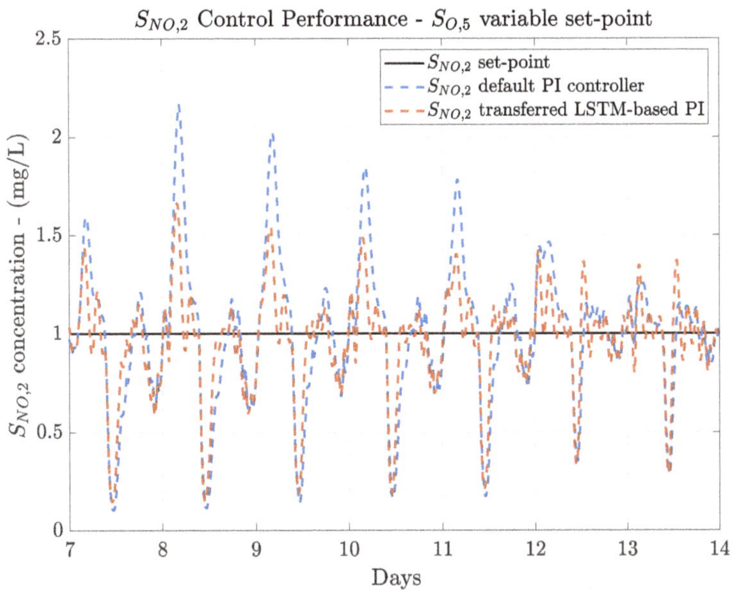

Figure 7. Control performance for the NO control loop when the LSTM-based PI derived from the DO control loop is transferred into it.

These results show that the DO LSTM-based PI controller is able to improve the default PI controllers performance. For that reason, it is considered as a candidate to be fine-tuned in order to adapt its behaviour to the $S_{NO,2}$ control management and therefore, achieve a better improvement in the management of this loop.

4.3. Transfer Learning from NO to DO

Before performing the fine-tuning process, the control performance of the NO LSTM-based PI is also computed to determine its behaviour when managing the NO control loop (its source domain) and its performance when managing the DO loop (its target domain). Results are shown in Table 3 where at first sight it is clearly observed that the *IAE* and *ISE* metrics are improved with respect to the default $S_{NO,2}$ PI controller. When a $S_{O,5}$ fix set-point is considered, the NO control loop *IAE* is improved in average a 24.32% while the corresponding *ISE* is improved around a 39.03% in average. Both with respect to the default NO control loop PI controller. The *ISE* improvement shows that the proposed NO LSTM-based PI controller, which has been derived from the NO control loop, is able to reduce the highest errors between the measured $S_{NO,2}$ and its set-point, with respect to the default PI controller. However, the control performance can be still improved since the improvement achieved in terms of the *IAE* error is still low. For instance, the best improvement is observed when the stormy weather is considered. There, the obtained *IAE* goes from 1.874 to 1.360, whereas the *ISE* goes from 0.977 to 0.543. These values represent an improvement around a 27.43% and a 44.42% when the obtained *IAE* and *ISE* values are compared to the default PI control metrics. In terms of the $S_{O,5}$ control performance, the transferred NO LSTM-based PI shows that the *IAE* performance is degraded instead of improved. For instance, when the NO LSTM-based PI is adopted, the *IAE* is increased from 0.148 to 0.158 when the dry weather is considered. This effect is motivated by the fact that the default PI of the NO control loop is not offering such a good control performance as the default PI of the DO control loop. Thus, the control performance will not be improved if data from the NO control loop is obtained to derived the NO LSTM-based PI and then transfer it into the DO control loop.

Table 3. Control performance when the NO LSTM-based PI derived from the NO control loop is transferred into the DO control loop.

Transfer Learning from NO to DO Control Loop						
	Fix Set-point					
	Dry Weather		Rainy Weather		Stormy Weather	
Structure	*IAE*	*ISE*	*IAE*	*ISE*	*IAE*	*ISE*
PI—$S_{NO,2}$	1.594	0.691	1.922	0.951	1.874	0.977
NO LSTM-based PI—$S_{NO,2}$	1.302	0.486	1.399	0.542	1.360	0.543
PI—$S_{O,5}$	0.148	0.007	0.143	0.007	0.158	0.007
NO LSTM-based PI—$S_{O,5}$	0.158	0.004	0.146	0.004	0.160	0.004
	Variable Set-point					
PI—$S_{NO,2}$	1.792	0.858	2.132	1.089	1.884	0.989
NO LSTM-based PI—$S_{NO,2}$	1.266	0.464	1.574	0.662	1.372	0.557
PI—$S_{O,5}$	0.185	0.016	0.155	0.014	0.206	0.020
NO LSTM-based PI—$S_{O,5}$	0.288	0.030	0.239	0.022	0.385	0.049

Visually, one can observe that the $S_{NO,2}$ control performance is slightly improved with respect to the default PI (see Figure 8). The peaks of $S_{NO,2}$ concentration are reduced, however, the desired set-point is not achieved. In terms of the $S_{O,5}$, the control performance is even slightly degraded with respect to the default PI controller. As it can be observed, the measured $S_{O,5}$ does not show variations as the default PI controller, however, there exists an offset which produces the *IAE* increment. For that reason, the *ISE* metric in terms of the $S_{O,5}$ is still reduced, it now equals to 0.004 in average instead of 0.007. Notice that the *ISE* tells if there exists a huge difference between the measured and the desired concentration, whereas the *IAE* tells if the difference is maintained over time.

When a variable set-point is considered, one can observe that the control performance is only improved in terms of the NO control loop. The *IAE* and *ISE* metrics are improved in averages with respect to the default PI controller of 27.56% and 42.94%, respectively. In terms of the DO control loop performance, results show that transferring the NO LSTM-based PI controller derived from the NO loop into the DO loop is not an option, since all the control metrics are degraded. For instance, the *IAE* and *ISE* metrics are nearly doubled with respect to the default PI controller when the stormy weather profile is simulated. These results entail that the NO LSTM-based PI cannot be considered as a candidate to be fine-tuned since it does not improve the control performance of target domain at the same time that the improvement achieved in the source domain is much lower than the one achieved by the DO LSTM-based PI. In addition, this also corroborates one of the main ideas stated before: the better the conventional control performance, the better the LSTM-based one. For that reason, the DO LSTM-based PI is the one considered to perform the fine-tuning process. It is important to notice that the initial training of both structures, the DO LSTM-based PI and the NO LSTM-based PI is not compulsory. In Figure 4b it is clearly observed that the control loop offering the best performance corresponds to the PI managing the DO control loop. Hence, the DO LSTM-based PI can be initially adopted to be trained. Then, it will be transferred into the NO control loop and fined-tuned. As a consequence, there is no need to train or even implement the NO LSTM-based PI.

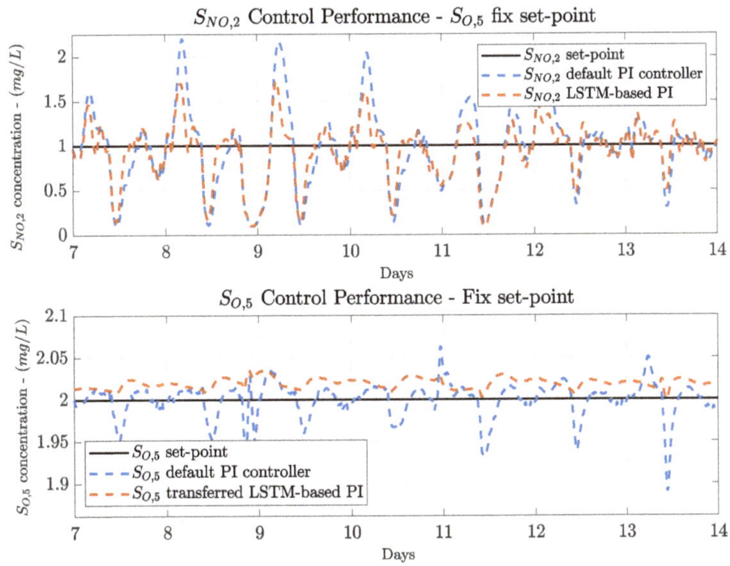

Figure 8. Control performance for the NO and DO control loops when the stormy weather is considered. The LSTM-based PI managing the DO control loop is derived from the NO control loop and transferred into the DO one.

4.4. LSTM-Based Controller Fine-Tuning & Transfer

Once the control performance of the DO and NO LSTM-based PI controllers is computed one can clearly observe that the DO LSTM-based PI is the controller offering the best control performance in both control loops. For that reason, the fine-tuning of the DO LSTM-based PI controller is proposed. To perform this tasks, data coming from the default $S_{NO,2}$ PI controller is considered. In that sense, information about how to control and manage the $S_{NO,2}$ concentration is provided to the DO LSTM-based PI. Thus, the fine-tuned version of the controller, the FTDO LSTM-based PI, should be able to improve a better control performance in terms of the $S_{NO,2}$ managing process.

Now, the prediction performance of the FTDO LSTM-based PI equals to a RMSE of 0.095 mg/L, a MAE of 0.067 mg/L, a MAPE of 6.24% and a R^2 of 0.991. Its training time equals to 20.27 s. At first sight one can observe that prediction performance is degraded with respect to the DO and NO LSTM-based PI controllers. However, this degradation is motivated by the fact that the proposed FTDO LSTM-based PI controller has learnt how to correctly manage the $S_{O,5}$ and $S_{NO,2}$ concentration instead of a unique one. In addition, the training time in this occasion equals to 20.27 s, which means that the time spent in the fine-tuning process is largely reduced with respect to training the LSTM-based structure from scratch. This effect is motivated by the information already present in the LSTM structure, that is, the weights and biases of the blocked LSTM cells. This corroborates that TL techniques can be adopted to simplify and speed up the control design process. Let's suppose that instead of transferring the knowledge of the DO LSTM-based PI into the NO control loop and performing a fine-tuning process, we decide to control each loop with its corresponding LSTM-based PI structure. The amount of time devoted to training the networks correspond to 69.91 and 98.60 s for the DO and NO control loops, respectively. This equals to a total time of 168.51 s only in terms of the training time. Although this time is affordable, if the DO LSTM-based PI is transferred into the NO control loop, only 69.91 s plus the time spent in the fine-tuning process, no more than 21 s is required. Thus, the total amount of time invested in the design process equals to 90.18 s, which represents a reduction of 78.33 s with respect to training two individual nets. Therefore, the reduction of the training time is clearly observed. In addition, it is important to notice that the WWTP we are dealing with only considers two control loops. However, this reduction of time will be higher in these situations where the number of control loops to design is larger. In that sense, an estimation of the training time reduction can be performed. If we suppose that the training time of the baseline LSTM-based PIs (from scratch) correspond to $t_{baseline}$ and that the time spent in the fine-tuning process on average equals to t_{ft}, the reduction of time (Δt) provided by our approach can be computed as:

$$\Delta t = N \cdot t_{baseline} - [t_{baseline} + (N-1) \cdot t_{ft}] = (N-1)[t_{baseline} - t_{ft}], \tag{9}$$

where $t_{ft} \ll t_{baseline}$. N equals to the number of control loops where the baseline LSTM-based PI is the transfer. As it is observed, the higher the number of control loops to design, the higher the reduction of time and the higher the benefit of the proposed methodology. Not only this, but this methodology can also be applied in those situations where the control of a new WWTP scenario has to be designed. In such a context, the new control structure can be derived by transferring the knowledge of the control structure of an already controlled WWTP. This would involve an even higher reduction of the complexity and time required in the development of the control strategy. All these facts motivate us to consider the TL methods in the design of the WWTP control loops.

In terms of the control performance, results of the FTDO LSTM-based PI control are shown in Table 4, where the *IAE* and *ISE* values are computed for different weather profiles and set-points. It is worth noticing that the $S_{NO,2}$ is now managed by the fine-tuned and transferred DO LSTM-based PI, that is, the FTDO LSTM-based PI, whereas the $S_{O,5}$ concentration is managed by the DO LSTM-based PI.

Table 4. Control performance when the DO LSTM-based PI derived from the DO control loop is transferred into the NO control loop. Then, the NO controller is fine-tuned with data from the default PI controller managing the $S_{NO,2}$.

LSTM-Based Controller Fine-Tuning & Transfer						
Fix Set-point						
	Dry Weather		Rainy Weather		Stormy Weather	
Structure	*IAE*	*ISE*	*IAE*	*ISE*	*IAE*	*ISE*
PI—$S_{O,5}$	0.143	0.007	0.143	0.007	0.158	0.007
DO LSTM-based PI—$S_{O,5}$	0.004	5.12×10^{-6}	0.008	2.43×10^{-5}	0.006	1.76×10^{-5}
PI—$S_{NO,2}$	1.594	0.691	1.922	0.951	1.874	0.977
FTDO LSTM-based PI—$S_{NO,2}$	0.091	0.002	1.150	0.625	0.357	0.151
Variable Set-point						
PI—$S_{O,5}$	0.185	0.016	0.155	0.014	0.206	0.020
DO LSTM-based PI—$S_{O,5}$	0.013	1.99×10^{-4}	0.017	3.91×10^{-4}	0.017	3.72×10^{-4}
PI—$S_{NO,2}$	1.792	0.858	2.132	1.089	1.884	0.989
FTDO LSTM-based PI—$S_{NO,2}$	0.129	0.004	0.643	0.261	0.324	0.122

When a fixed set-point is considered, one can observe that the control performance is hugely improved not only in terms of the $S_{O,5}$, but also in terms of the $S_{NO,2}$. The improvement of the DO control loop with respect to the default PI controller is translated into an average reduction of the *IAE* around a 95.94% and an average reduction in the *ISE* around a 99.78%. Thereby, this is translated in a better tracking process of the $S_{O,5}$ and consequently, a better management of this concentration. In terms of the NO control loop, one can observe that the *IAE* and *ISE* are hugely improved as well. However, there is an exception with the rainy weather. In this case, the $S_{NO,2}$ *IAE* and *ISE* are only improved a 40.17% and a 34.27%, respectively. This is motivated by the fact that the rainy weather profile shows two large perturbation during days 9 and 11. Besides, the fine-tuning process is performed with measurements obtained from the $S_{NO,2}$ default PI controller when a whole year of randomly distributed weathers is simulated. Thus, this entails that most of the knowledge provided to the DO LSTM-based PI consists in the control actuations to manage the $S_{NO,2}$ concentration when the dry weather is considered (remember that rainy and stormy weathers are equal to the dry weather with the exception of the two rainy and the two stormy episodes). On average, the NO control loop *IAE* and *ISE* are reduced by 73.47% and 72.84% with respect to the default $S_{NO,2}$ PI control performance. The greatest improvement is observed when the dry weather is considered. The *IAE* is reduced from 1.594 to 0.091, whereas the *ISE* is decreased from 0.691 to 0.002 (see Figure 9). In the case of the rainy weather, the reduction of the *IAE* and *ISE* is lower, the *IAE* changes from 1.922 to 1.150 and the *ISE* from 0.951 to 0.625. Nevertheless, this *IAE* value corresponds to the lowest NO control loop *IAE* value of the three TL approaches considered in this work.

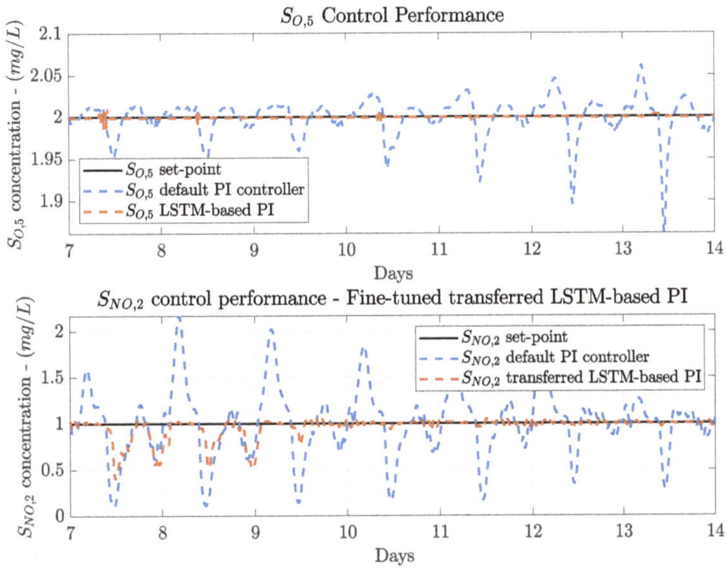

Figure 9. Control performance for the NO and DO control loops when a $S_{O,5}$ fix set-point and dry weather are considered. The LSTM-based PI managing the NO loop is transferred from the DO control loop and fined-tuned with data from the NO control loop.

Results of the control performance when a variable set-point is considered show the same tendency as the fix set-point ones. The *IAE* and *ISE* metrics have been improved for all the weather profiles. Again, the most important results are the ones corresponding to the NO control loop, which is the controller whose control performance improvement is sought with the fine-tuning process. In that sense, the best improvement is now observed when the dry weather is simulated. The *IAE* has been decreased from 1.792 to 0.129, which equals to an improvement of 92.80%. In terms of the *ISE*, it is decreased from 0.858 to 0.004, which represent an improvement of a 99.53%. It is important to notice that the lowest control performance is obtained when the rainy weather is considered, the *IAE* deceases from 2.132 to 0.643 while the *ISE* is reduced from 1.089 to 0.261. Although these improvements are not so high as the ones achieved with the dry weather, they are still much better than the performance obtained when the fine-tuning process is not carried out. For instance, the *IAE* has been improved a 69.84% whilst the *ISE* has been improved a 76.03%. The *IAE* improvement represents an increase of 48.26 and 42.63 percentage points with respect to the improvements achieved in the Transfer Learning from DO to NO and from NO to DO. In terms of the *ISE*, these increments equal to 45.64 and 36.82 percentage points, respectively. Visually, we can observe in Figure 10 that the $S_{NO,2}$ desired value of 1 mg/L is obtained at the same time the $S_{O,5}$ variable set-point is correctly tracked. In addition, the rainy episodes are plotted to show that the FTDO LSTM-based PI controller requires some more knowledge to finally learn how to manage these events.

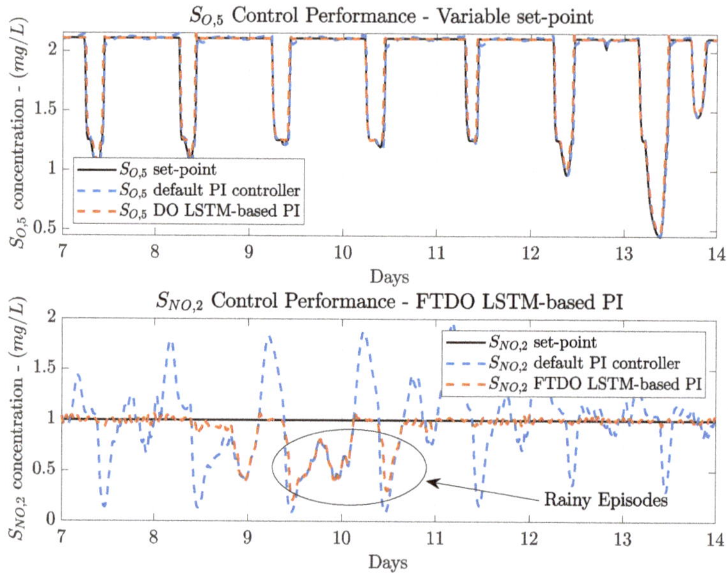

Figure 10. Control performance for the NO and DO control loops when a $S_{O,5}$ variable set-point and rainy weathers are considered. The LSTM-based PI managing the NO loop is transferred from the DO control loop and fined-tuned with data from the NO control loop.

As a summary, the control performance is improved in all terms regardless of the set-point topology and the weather profiles. This entail that the best option to design or improve a control strategy of an industrial plant, and especially a WWTP, is to obtain a first baseline controller, the DO LSTM-based PI, and then transfer its knowledge to the rest of control loops. The main point is to design the baseline controller with data coming from the controller performing better. In our case, this controller corresponds to the $S_{O,5}$ default PI controller. Then, the obtained DO LSTM-based PI is transferred into the remaining control loops and fine-tuned with data coming from controllers actuating in the target domain. Moreover, this approach entail that control loops can be designed without requiring a high knowledge of the different processes carried out in the plant. Only input and output measurements of a control strategy performing well are required. The rest of the control loops will be derived from the implemented one. Thus, the higher the number of control loops, the higher the benefit offered with this design approach. In our case, this benefit is not widely exploded since we have only transferred the LSTM-based PI between two control loops. However, this approach can be adopted in other scenarios where the number of control loops is largely higher than the ones managed here. In that sense, the benefit of this approach should be much higher than the one observed here.

Finally, the results observed in this manuscript motivates us to open a new research line where the transfer learning approach presented here is considered as the initial process of a reinforcement learning based control design. Then, instead of performing the fine-tuning process with measurement coming from a conventional control approach, it could be performed following a reinforcement leaning process. In that sense, the controller would be fine-tuned over time, adapting its output to the incoming measurements.

5. Conclusions

In this work, we presented a new industrial control design process which involves the application of LSTM-based neural networks and transfer learning approaches. The main purpose was to design and implement the control loops managing a general-purpose WWTP. The application is specific; however, the design approach can be adopted in any kind of industrial environment. The main idea is that TL techniques allow us to derive new control strategies from a baseline one without performing a deep tuning process of each control structure. This reduces the control design complexity, as well as the time invested in the training process of each data-based control structure. Thus, the higher the number of control loops, the higher the improvement achieved.

In our case, the proposed control design approach consists in two main processes: (i) the design of a neural network-based controller with data obtained from an existing control loop and (ii) the transfer of the controller knowledge into the remaining loops. To achieve that, three different design approaches were proposed, two of them mainly consisting in the design of the LSTM-based controller with data from a control loop and then transferring it to the others without retraining the net structure. In that sense, we considered the development of the LSTM-based PI controller either with measurements from the $S_{O,5}$ control loop or from the $S_{NO,2}$ one, both from a general-purpose WWTP. The third option considers the development of the LSTM-based PI with data from the $S_{O,5}$ control loop and the fine-tuning of its transferred version. Results show that there exists a trade-off between deriving the LSTM-based PI with measurement from the $S_{NO,2}$ or the $S_{O,5}$ control loops. If the LSTM-based PI controller is derived with measurements from the $S_{O,5}$ control loop, one can observe that the $S_{O,5}$ control performance is highly improved with respect to the default PI controller regardless of the weather influent and the considered set-point. In addition, the $S_{NO,2}$ control performance experiences a slight improvement as well. On the other hand, the NO control performance experienced an improvement at expense of degrading the $S_{O,5}$ control performance when the LSTM-based PI transferred into the DO control loop is implemented with measurements from the NO control loop. To solve this trade-off, we considered the third option, where the LSTM-based PI derived from the $S_{O,5}$ control loop was adopted and transferred into the DO control loop. Its transferred version was fine-tuned with measurements coming from the default PI controller managing the $S_{NO,2}$ control loop.

Results show that a high improvement is achieved in the $S_{O,5}$ control loop as well as in the $S_{NO,2}$ one. Besides, the lowest *IAE* and *ISE* improvements in terms of the $S_{NO,2}$ when compared to the default $S_{NO,2}$ PI controller equalled to 69.84% and 76.03% for the *IAE* and *ISE*, respectively, which are even higher improvements than in the cases where the fine-tuning process is not considered. This clearly shows that designing a LSTM-based PI in a control loop, transferring it to another different one, and then performing a fine-tuning process is the best option if a high level of improvement of the control performance is sought. Besides, this also entails a speed-up and a complexity reduction of the control design process since only the design and training of one control loop has to be performed. Again, the higher the number of control loops to design, the higher the benefit obtained following this design approach.

Author Contributions: I.P. has designed and trained the artificial neural networks considered in the TL-based methodology adopted in this work. He has also implemented and tested the proposed structures over the BSM1 framework. The manuscript has been also written by I.P., A.M., R.V. and J.L.V. supervised the work. All authors have read and agreed to the published version of the manuscript.

Funding: The APC was funded by the Spanish Government under Project TEC2017-84321-C4-4-R co-funded with European Regional Development Funds of the European Union.

Institutional Review Board Statement: Not applicable.

Informed Consent Statement: Not applicable.

Data Availability Statement: The data considered in this study correspond to the influent and effluent data generated and available in the same BSM1 framework defined in [8,25].

Acknowledgments: This work has received support from the Catalan Government under Projects 2017 SGR 1202 and 2017 SGR 1670, and from the Spanish Government through the MICINN projects PID2019-105434RB-C33 and TEC2017-84321-C4-4-R.

Conflicts of Interest: The authors declare no conflict of interest.

Abbreviations

The following abbreviations are used in this manuscript:

ANN	Artificial Neural Network
ASM1	Activated Sludge Model N.1
BSM1	Benchmark Simulation Model No. 1
BSM1-P	Benchmark Simulation Model No. 1 with Phosphorus processing
BSM2	Benchmark Simulation Model No. 2
$\mathbf{b}_x$	Biases of the xth hidden layer
DO	Dissolved Oxygen in the fifth reactor tank ($S_{O,5}$) control loop
DO LSTM-based PI	LSTM-based PI controller trained with data from the $S_{O,5}$ control loop
FTDO LSTM-based PI	DO LSTM-based PI fine-tuned with data from the $S_{NO,2}$ control loop
IAE	Integrated Absolute Error
ISE	Integrated Squared Error
$K_{La,x}$	Oxygen Transfer Coefficient of the xth reactor tank measured in day^{-1}
LSTM	Long Short-Term Memory
MAE	Mean Absolute Error
MAPE	Mean Average Percentage Error
MLP	Multilayer Perceptron
MPC	Model Predictive Controller
NO	Nitrate and nitrite nitrogen in the second reactor tank ($S_{NO,2}$) control loop
NO LSTM-based PI	LSTM-based PI controller trained with data from the $S_{NO,2}$ control loop
PI	Proportional Integral controller
PID	Proportional Integral Derivative controller
Q_0	Influent flow rate
Q_a	Internal recirculation flow rate
Q_r	External recirculation flow rate
R^2	Determination coefficient
RMSE	Root Mean Squared Error
$S_{NO,x}$	Nitrate and nitrite nitrogen in the xth reactor tank measured in mg/L
$S_{NH,x}$	Ammonium concentration in the xth reactor tank measured in mg/L
$S_{O,x}$	Dissolved oxygen concentration in the xth reactor tank measured in mg/L
TL	Transfer Learning
$\mathbf{U}_x$	Weights affecting the previous output data of the xth hidden layer
WWTP	Wastewater Treatment Plant
$\mathbf{W}_x$	Weights affecting the input data of the xth hidden layer

References

1. Ogata, K. *Modern Control Engineering*; Prentice Hall: Upper Saddle River, NJ, USA, 2010.
2. Wollschlaeger, M.; Sauter, T.; Jasperneite, J. The future of industrial communication: Automation networks in the era of the internet of things and industry 4.0. *IEEE Ind. Electron. Mag.* **2017**, *11*, 17–27. [CrossRef]
3. Ustundag, A.; Cevikcan, E. *Industry 4.0: Managing the Digital Transformation*; Springer: Cham, Switzerland, 2017.
4. Rani, A.; Singh, V.; Gupta, J. Development of soft sensor for neural network based control of distillation column. *ISA Trans.* **2013**, *52*, 438–449. [CrossRef]
5. Pisa, I.; Santín, I.; Vicario, J.L.; Morell, A.; Vilanova, R. ANN-based soft sensor to predict effluent violations in wastewater treatment plants. *Sensors* **2019**, *19*, 1280. [CrossRef] [PubMed]
6. Thennadil, S.N.; Dewar, M.; Herdsman, C.; Nordon, A.; Becker, E. Automated weighted outlier detection technique for multivariate data. *Control Eng. Pract.* **2018**, *70*, 40–49. [CrossRef]
7. Zejda, V.; Máša, V.; Václavková, Š.; Skryja, P. A Novel Check-List Strategy to Evaluate the Potential of Operational Improvements in Wastewater Treatment Plants. *Energies* **2020**, *13*, 5005. [CrossRef]
8. Alex, J.; Benedetti, L.; Copp, J.; Gernaey, K.V.; Jeppsson, U.; Nopens, I.; Pons, M.N.; Rieger, L.; Rosen, C.; Steyer, J.P.; et al. *Benchmark Simulation Model No. 1 (BSM1)*; Technical Report; Department of Industrial Electrical Engineering and Automation, Lund University: Lund, Sweden, 2008.

9. Santin, I.; Pedret, C.; Vilanova, R. Fuzzy control and model predictive control configurations for effluent violations removal in wastewater treatment plants. *Ind. Eng. Chem. Res.* **2015**, *54*, 2763–2775. [CrossRef]
10. Santin, I.; Pedret, C.; Vilanova, R. Applying variable dissolved oxygen set point in a two level hierarchical control structure to a wastewater treatment process. *J. Process Control* **2015**, *28*, 40–55. [CrossRef]
11. Goodfellow, I.; Bengio, Y.; Courville, A. *Deep Learning*; MIT Press: Cambridge, MA, USA, 2016.
12. Pisa, I.; Santín, I.; Morell, A.; Vicario, J.L.; Vilanova, R. LSTM based Wastewater Treatment Plants operation strategies for effluent quality improvement. *IEEE Access* **2019**, *7*, 159773–159786. [CrossRef]
13. Sadeghassadi, M.; Macnab, C.J.; Gopaluni, B.; Westwick, D. Application of neural networks for optimal-setpoint design and MPC control in biological wastewater treatment. *Comput. Chem. Eng.* **2018**, *115*, 150–160. [CrossRef]
14. Hernández-del Olmo, F.; Gaudioso, E.; Dormido, R.; Duro, N. Tackling the start-up of a reinforcement learning agent for the control of wastewater treatment plants. *Knowl.-Based Syst.* **2018**, *144*, 9–15. [CrossRef]
15. Pisa, I.; Morell, A.; Vicario, J.L.; Vilanova, R. Denoising Autoencoders and LSTM-Based Artificial Neural Networks Data Processing for Its Application to Internal Model Control in Industrial Environments—The Wastewater Treatment Plant Control Case. *Sensors* **2020**, *20*, 3743. [CrossRef]
16. Kasireddy, I.; Nasir, A.W.; Singh, A.K. IMC based Controller Design for Automatic Generation Control of Multi Area Power System via Simplified Decoupling. *Int. J. Control Autom.* **2018**, *16*, 994–1010. [CrossRef]
17. Wang, C.; Zhao, W.; Luan, Z.; Gao, Q.; Deng, K. Decoupling control of vehicle chassis system based on neural network inverse system. *Mech. Syst. Signal Prossess.* **2018**, *106*, 176–197. [CrossRef]
18. Souza, F.A.; Araújo, R.; Mendes, J. Review of soft sensor methods for regression applications. *Chemom. Intell. Lab. Syst.* **2016**, *152*, 69–79. [CrossRef]
19. Amirabadi, M.A.; Kahaei, M.H.; Nezamalhosseini, S.A. Novel suboptimal approaches for hyperparameter tuning of deep neural network [under the shelf of optical communication]. *Phys. Commun.* **2020**, *41*, 101057. [CrossRef]
20. Zhuang, F.; Qi, Z.; Duan, K.; Xi, D.; Zhu, Y.; Zhu, H.; Xiong, H.; He, Q. A Comprehensive Survey on Transfer Learning. *Proc. IEEE* **2021**, *109*, 43–76. [CrossRef]
21. Sarkar, D.; Bali, R.; Ghosh, T. *Hands-On Transfer Learning with Python: Implement Advanced Deep Learning and Neural Network Models Using TensorFlow and Keras*; Packt Publishing Ltd.: Birmingham, UK, 2018.
22. Curreri, F.; Patanè, L.; Xibilia, M.G. RNN-and LSTM-Based Soft Sensors Transferability for an Industrial Process. *Sensors* **2021**, *21*, 823. [CrossRef]
23. Miccio, M.; Cosenza, B. Control of a distillation column by type-2 and type-1 fuzzy logic PID controllers. *J. Process Control* **2014**, *24*, 475–484. [CrossRef]
24. Pisa, I.; Morell, A.; Vicario, J.L.; Vilanova, R. Transfer Learning Approach for the Design of Basic Control Loops in Wastewater Treatment Plants. In Proceedings of the 26th IEEE International Conference on Emerging Technologies and Factory Automation (ETFA), Vasteras, Sweden, 7–10 September 2021.
25. Copp, J.B. *The Cost Simulation Benchmark: Description and Simulator Manual (Cost Action 624 and Action 682)*; Office for Official Publications of the European Union: Luxembourg, 2002.
26. Henze, M.; Grady, L., Jr.; Gujer, W.; Marais, G.; Matsuo, T. Activated Sludge Model No 1. *IAWPRC Scientific and Technical Reports*; IAWPRC: London, UK, 1987; Volume 29.
27. Halling-Sørensen, B.; Jorgensen, S.E. *The Removal of Nitrogen Compounds from Wastewater*; Elsevier: Amsterdam, The Netherlands, 1993.
28. Henze, M.; Gujer, W.; Mino, T.; van Loosdrecht, M.C. *Activated Sludge Models ASM1, ASM2, ASM2d and ASM3*; IWA Publishing: London, UK, 2000.
29. Gernaey, K.V.; Jørgensen, S.B. Benchmarking combined biological phosphorus and nitrogen removal wastewater treatment processes. *Control Eng. Prac.* **2004**, *12*, 357–373. [CrossRef]
30. Gernaey, K.V.; Jeppsson, U.; Vanrolleghem, P.A.; Copp, J.B.; International Water Association. Task Group on Benchmarking of Control Strategies for Wastewater Treatment Plants. In *Benchmarking of Control Strategies for Wastewater Treatment Plants*; IWA Publishing: London, UK, 2014.
31. Martin, C.; Vanrolleghem, P.A. Analysing, completing, and generating influent data for WWTP modelling: A critical review. *Environ. Model. Softw.* **2014**, *60*, 188–201. [CrossRef]
32. Modelling & Integrated Assessment, Benchmarking. Available online: http://iwa-mia.org/benchmarking/ (accessed on 9 November 2020).
33. Oliphant, T.E. *A Guide to NumPy*; Trelgol Publishing: Spanish Fork, UT, USA, 2006; Volume 1.
34. Virtanen, P.; Gommers, R.; Oliphant, T.E.; Haberland, M.; Reddy, T.; Cournapeau, D.; Burovski, E.; Peterson, P.; Weckesser, W. SciPy 1.0: Fundamental Algorithms for Scientific Computing in Python. *Nat. Methods* **2020**, *17*, 261–272. [CrossRef] [PubMed]
35. Abadi, M.; Agarwal, A.; Barham, P.; Brevdo, E.; Chen, Z.; Citro, C.; Corrado, G.S.; Davis, A.; Dean, J.; Devin, M.; et al. TensorFlow: Large-Scale Machine Learning on Heterogeneous Systems. 2015. Available online: https://www.tensorflow.org/ (accessed on 5 October 2020).
36. Vilanova, R.; Visioli, A. *PID Control in the Third Millennium—Lessons Learned and News Approaches*; Springer: London, UK, 2012.
37. Da Silva, I.N.; Spatti, D.H.; Flauzino, R.A.; Liboni, L.H.B.; dos Reis Alves, S.F. *Artificial Neural Networks*; Springer International Publishing: Cham, Switzerland, 2017.

38. Boussaada, Z.; Curea, O.; Remaci, A.; Camblong, H.; Mrabet Bellaaj, N. A nonlinear autoregressive exogenous (NARX) neural network model for the prediction of the daily direct solar radiation. *Energies* **2018**, *11*, 620. [CrossRef]
39. Manu, D.S.; Thalla, A.K. Artificial intelligence models for predicting the performance of biological wastewater treatment plant in the removal of Kjeldahl Nitrogen from wastewater. *Appl. Water Sci.* **2017**, *7*, 3783–3791. [CrossRef]
40. Qiao, J.F.; Hou, Y.; Zhang, L.; Han, H.G. Adaptive fuzzy neural network control of wastewater treatment process with multiobjective operation. *Neurocomputing* **2018**, *275*, 383–393. [CrossRef]
41. Santín, I.; Barbu, M.; Pedret, C.; Vilanova, R. Dissolved Oxygen Control in Biological Wastewater Treatments with Non-Ideal Sensors and Actuators. *Ind. Eng. Chem. Res.* **2019**, *58*, 20639–20654. [CrossRef]

sensors

MDPI

Article

Water Quality Indicator Interval Prediction in Wastewater Treatment Process Based on the Improved BES-LSSVM Algorithm

Meng Zhou [1], Yinyue Zhang [1], Jing Wang [1], Yuntao Shi [1,*] and Vicenç Puig [2]

1 School of Electrical and Control Engineering, North China University of Technology, Beijing 100144, China; zhoumeng@ncut.edu.cn (M.Z.); zyy17718301596@163.com (Y.Z.); jwang@ncut.edu.cn (J.W.)

2 Advanced Control Systems Research Group at Institutde Robòtical, CSIC-UPC, Universitat Politècnica de Catalunya-BarcelonaTech, 08028 Barcelona, Spain; vicenc.puig@upc.edu

* Correspondence: shiyuntao@ncut.edu.cn

Abstract: This paper proposes a novel interval prediction method for effluent water quality indicators (including biochemical oxygen demand (BOD) and ammonia nitrogen (NH3-N)), which are key performance indices in the water quality monitoring and control of a wastewater treatment plant. Firstly, the effluent data regarding BOD/NH3-N and their necessary auxiliary variables are collected. After some basic data pre-processing techniques, the key indicators with high correlation degrees of BOD and NH3-N are analyzed and selected based on a gray correlation analysis algorithm. Next, an improved IBES-LSSVM algorithm is designed to predict the BOD/NH3-N effluent data of a wastewater treatment plant. This algorithm relies on an improved bald eagle search (IBES) optimization algorithm that is used to find the optimal parameters of least squares support vector machine (LSSVM). Then, an interval estimation method is used to analyze the uncertainty of the optimized LSSVM model. Finally, the experimental results demonstrate that the proposed approach can obtain high prediction accuracy, with reduced computational time and an easy calculation process, in predicting effluent water quality parameters compared with other existing algorithms.

Keywords: water quality monitoring; data pre-processing; improved IBES-LSSVM algorithm; interval prediction method

Citation: Zhou, M.; Zhang, Y.; Wang, J.; Shi, Y.; Puig V. Water Quality Indicator Interval Prediction in Wastewater Treatment Process Based on the Improved BES-LSSVM Algorithm. *Sensors* **2022**, *22*, 422. https://doi.org/10.3390/s22020422

Academic Editor: Assefa M. Melesse

Received: 30 October 2021
Accepted: 23 December 2021
Published: 6 January 2022

1. Introduction

Nowadays, freshwater is considered one of the most critical resources for humans, since it can ensure the availability of an acceptable quantity of water for livelihoods, health, ecosystems and production. Hence, freshwater plays a key role in poverty and disease burden reduction, economic growth and environmental sustainability [1,2]. This fact has long been acknowledged all over the world. However, due to industrial pollution, rapid population growth and farmland sewage caused by the extensive use of chemical fertilizers, pesticides and herbicides, the shortage of freshwater sources is a serious and challenging issue [3,4].

Wastewater treatment is one key technology to potentially provide additional water supplies, and it is very important for the functioning of the economy and society. Wastewater treatment has been attracting a lot of attention, since it can not only remove organic wastes to reduce the environmental burden, but also offer the advantage of producing a renewable source of water [5,6]. Wastewater treatment is a very complex process with a variety of physical and biochemical reactions since it presents nonlinear dynamic behavior, time delay and uncertainty [7]. In wastewater treatment plant processes, effluent water quality monitoring is an important task that involves measuring the evolution of the quality parameters in time.

Note that most traditional methods of measuring these quality indicators for wastewater treatment processes are based on manual lab-based monitoring approaches, with manual sample collection, long-time transportation and biological/microbial testing in a laboratory, which is cumbersome and time-consuming. Usually, the testing equipment is very expensive and cannot be used online. In addition, since the process of wastewater treatment is complex, some control strategies are necessary and required to be deployed to guarantee that effluent quality indicators behave normally. In recent decades, water quality monitoring has been evolving to the latest wireless sensor networks [8], such that most of the important indicators of effluent water (pressure, pH, level and so on) can be measured by their corresponding sensors online. However, there are still some parameters that cannot be measured quickly due to high costs and the limitations of sensors, such as BOD and NH3-N. Usually, the concentration of the BOD/NH3-N effluent associated with a wastewater treatment process is an important factor to measure the water quality since the discharge of a large amount of NH3-N and BOD wastewater will lead to water eutrophication, which can affect human health. In China's "Pollutant Discharge Standard for Urban Wastewater Treatment Plants (*GB*18918-2002)", the Class A standard stipulates that the maximum discharge for NH3-N is 5 mg/L, while for BOD, it is 10 mg/L. Thus, measuring these effluent quality indicators with high accuracy is an important issue.

Researchers have focused on soft-sensing methods to predict these effluent quality indicators and the prediction task is addressed combining data analytics and water quality control. Soft-sensing methods aim to find some certain relationships between easy-to-measure variables and difficult-to-measure variables in the sewage treatment process. Then, a suitable model is established based on these relationships, and difficult-to-measure variables can be predicted based on the soft-sensing models.

Machine learning approaches are usually considered a subset of artificial intelligence. They focus on some statistical models and algorithms to extract patterns from data so that useful inferences can be used to predict new data. Recently, with the development of machine learning, artificial neural network (ANN), support vector machine (SVM), decision tree, random forest, ensemble learning and many other methods have been researched in depth and have a wide range of applications, including text processing, computer vision, healthcare, finance and robotics. They can also be used for socio-economic and environmental studies [9–12]. In [12], the impacts of flood protection in Bangladesh were evaluated by machine learning methods. In [13], a gray model and ANN method were investigated to predict suspended matter and chemical oxygen demand in the wastewater treatment process. Cong et al. proposed a mixed soft sensor model based on a wavelet neural network and adaptive weighted fusion for the online prediction of effluent COD [14]. M. Hamada carried out the assessment of a wastewater treatment plant's performance based on ANN and a multiple linear regression method [15]. M. Zeinolabedini et al. proved that applying various parent wavelet functions to the neural network structure can improve the accuracy of predicting the wastewater sludge volume [16]. A. K. Kadam et al. used ANN and multiple linear regression to model and predict water quality parameters in river basins [17]. S. Heddam et al. investigated a generalized regression neural network model to predict the BOD of effluent in wastewater treatment plants [18]. Tan et al. predicted the first weighting from the working face roof in a coal mine based on a GA-BP neural network [19]. V. Nourani et al. proved that the prediction ability of a neural network ensemble is more reliable [20].

Compared with the ANN method, SVM is another important prediction technique, which can effectively solve the problem of high-dimensional data model construction under the condition of limited samples, and has strong generalization ability. Hence, many scholars have carried out a lot of research on SVM-based prediction. Cheng et al. proposed a variety of kernel single-class SVMs to monitor and predict the intake conditions of wastewater treatment plants [21]. Han et al. developed a neural network model for predicting the sludge volume index based on information transfer strength and adaptive second-order algorithms [22]. Wu et al. proposed an adaptive multi-output soft sensor model for monitoring wastewater treatment and made several simulation comparisons to prove the superiority of the algorithm [23]. K. Lotfi et al. used a linear–nonlinear hybrid method to predict the effluent index of a wastewater treatment plant, which improves the prediction ability of the single method [24]. Han et al. proposed a data-based predictive control strategy and proved its superiority through several simulations [25]. In [26], the total solid content of a wastewater treatment plant was predicted by an SVM model, which can enhance performance and durability.

Although SVM is a small-sample learning method and has been widely used to solve the wastewater prediction problem, the calculation process is multifarious, which is difficult to implement for large-scale training samples [27]. To overcome these disadvantages, the least-squares support vector machine (LSSVM) has been proposed. LSSVM improves the performance of the SVM algorithm by solving linear programming rather than quadratic programming. In this way, the calculation process can be reduced and the computation speed greatly improved [28]. Zhang et al. proposed an improved LSSVM model based on SVM to predict river flow [29]. Fei Luo et al. integrated the Gustafson-Kessel algorithm and least-squares support vector machine for line prediction of [30]. D. S. Manu et al. combined SVM and an adaptive neuro-fuzzy reasoning system model to predict the effluent nitrogen content of wastewater treatment plants [31]. Liu et al. investigated the online prediction of effluent COD in an anaerobic wastewater treatment system based on principal component analysis and the LSSVM algorithm [32].

Note that there are some unknown parameters in the kernel functions of LSSVM that need to be selected in advance. Generally, these parameters are determined according to experience, which may be time-consuming, and it is difficult to find the optimal parameters. Nowadays, swarm intelligence optimization algorithms are researched extensively, since the optimal solution can be found by swarm intelligence to perform a collaborative search mechanism. The results of the combination of swarm intelligence optimization algorithms and machine learning methods can be found in a large number of references. In [33], a hybrid model of particle swarm optimization (PSO) and support vector machine is proposed to predict the turbidity and pH value of sand filtered water in irrigation systems. Han et al. use an adaptive PSO algorithm to design self-organizing radial basis function neural networks to improve the accuracy and save time [34]. Chen et al. study the artificial bee colony optimization back-propagation network to predict the water quality of a water diversion project [35]. Fan et al. use the LSSVM model to improve the performance of predicting the safety factor of a circular slope [36]. Mahdi Shariati et al. use the gray wolf algorithm to optimize ELM model parameters to predict the compressive strength of partially replaced cement concrete [37]. However, to the best of the authors' knowledge, these swarm intelligence methods may fall into local optima and do not find the global optimal solutions.

Most of the above-mentioned methods only focus on point prediction, without providing information regarding accuracy. The prediction results have strong uncertainty that affects the decision-making process, increasing the risk of not making good decisions. Prediction interval (PI) is a standard tool for quantifying prediction uncertainty. PI not only provides the range where the target value is most likely to exist, but also indicates its accuracy. Yao et al. combined the mean variance estimation (MVE) method with a recurrent neural network to measure the uncertainty in prediction [38]. Yuan et al. combined beta distribution with the PSO-LSTM model to obtain the wind power prediction interval with high reliability and a narrow interval width, so as to provide decision support for the safe and stable operation of power systems [39]. Liao et al. combined the bootstrap method with the long and short memory network to realize the uncertain prediction of the remaining service life of the machine [40]. Marin et al. obtained the prediction interval of power consumption by combining the delta method with a fuzzy prediction model [41]. Sun et al. constructed a high-quality prediction interval based on the two-step method of dual ELM and applied it to the scheduling of a gas system [42]. In recent years, a direct interval prediction method called upper and lower bound estimation (LUBE) has been proposed. The main idea of this method is to directly construct the upper and lower bounds of PI by optimizing the coefficients of the neural network according to the interval quality evaluation index. This approach can provide good performance and does not consider strict data distribution assumptions, such that it can provide more information about the prediction results, which motivates the work of this paper.

The main objective of this paper is to obtain a soft-sensor-based interval prediction method with high prediction accuracy and less computational time to predict the effluent water quality parameters, which is significant for water quality monitoring and control. Aiming at the online prediction of BOD/NH3-N effluent in a wastewater treatment plant within a smart data-driven framework, the main contributions of this paper are the following:

- Data pre-processing methods, i.e., abnormal data elimination and normalization, are taken into consideration after the data and their related auxiliary variables are collected. Then, some key factors of the wasterwater quality indicators are selected based on the gray correlation analysis algorithm.
- In order to improve the prediction accuracy of BOD/NH3-N effluent, a novel IBES-LSSVM algorithm is proposed, in which an improved bald eagle search (IBES) optimization algorithm is used to find the optimal parameters of the least-squares support vector machine (LSSVM). The superiority of the proposed method is verified by comparing it with the existing soft-sensing models (such as GWO, WOA, PSO, SSA) using some benchmark functions and providing higher prediction accuracy.
- In order to estimate the uncertainty of the model prediction results and make better decisions, after obtaining the point prediction results, the interval prediction bounds of effluent quality are also generated. Compared with some existing soft-sensing models, the proposed interval prediction method can obtain a more accurate prediction range.

The structure of this paper is as follows: In Section 2, the problem description is given, including the real data collection, data pre-processing and gray-correlation-analysis-based data selection. Section 3 describes the model uncertainty analysis by using the proposed IBES-LSSVM algorithm and LUBE algorithm. In Section 4, the simulation examples are depicted, demonstrating the effectiveness of the proposed method based on the BOD and NH3-N data. Section 5 draws the main conclusions of this paper.

2. Problem Description

In this paper, a soft-sensing-based method is investigated to analyze and predict the water quality indicators, including three main aspects: data collection, data pre-processing and data interval prediction. The main steps of the approach presented in this paper are shown in Figure 1.

Under a smart data-driven framework, in order to predict water quality tendencies and analyze the mechanisms behind the considered data sources, enough relevant experimental data in real time must be collected based on the prediction quality indicators. Most collected data may present several issues, such as data sparsity and data synchronization, among others. After the data are collected, they must be pre-processed in advance by applying several procedures, such as data cleaning, abnormal data elimination or normalization. Then, correlation analysis from different dimensions of water quality indicators should be considered to extract the relations between these auxiliary variables and find the key factors.

2.1. Data Collection

Due to the complexity of the wastewater treatment process and the large number of parameters that need to be set, it is necessary to determine the characteristic variables related to the water quality to be determined as auxiliary variables. The data that can evaluate the quality or impact of water quality in wastewater treatment plants are mainly divided into the following four categories [43]:

- **Physical data:** Physical properties are the ones that must be monitored throughout the treatment process, including total suspended solids, temperature, conductivity, transparency, total dissolved solids, etc.

- **Chemical data:** Chemical water quality indices of the national comprehensive discharge standard for water pollutants, including: pH, biochemical oxygen demand, biochemical oxygen consumption, heavy metals, nitrates, etc.
- **Biological data:** Biomarkers include a variety of microorganisms in the water, such as mayflies, E. coli, etc.
- **Environmental data:** Environmental data cover the whole process of water supply, including indexes of weather, hydrology, soil or ecology.

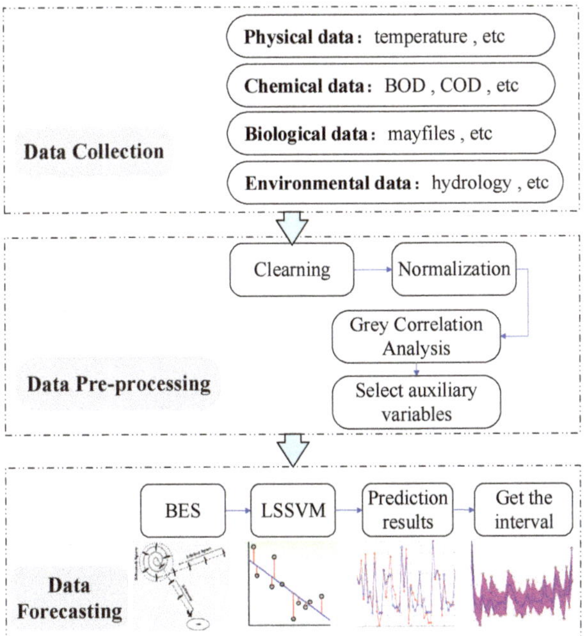

Figure 1. Main steps of the proposed approach.

This paper focuses on a real wastewater treatment plant in Beijing, China, from August 2014 to September 2014 [7,44]. Two data sets are collected first, which are used to predict the BOD/NH3-N effluent, separately. (1) BOD data set: containing 360 batches of data with 23 variables (including the BOD effluent parameters)—the detailed information is shown in Table 1; (2) NH3-N data set: including 10 characteristic variables related to NH3-N effluent parameters, as shown in Table 2.

2.2. Elimination of Abnormal Data

Data collected from wastewater treatment plants can contain erroneous values because of improper instrument operation, human or environmental interference and other factors. As a result, we need to analyze the collected data first, and eliminate some abnormal or meaningless data.

In this paper, we use the 3σ criterion to handle the abnormal data of the two collected data sets. The sample data are denoted as $x_1, x_2, \cdots, x_n$. η_i is used to represent the data residual error. Then, the standard deviation is calculated as follows:

$$\sigma = \sqrt{\frac{\sum\limits_{i=1}^{n} {\eta_i}^2}{n-1}} \tag{1}$$

$$\eta_i = x_i - \bar{x} \tag{2}$$

where n represents the number of elements in the data set, and $\bar{x}$ is the data average.

If the residual error of particular data sample x_i satisfies

$$|\eta_i| > 3\sigma \tag{3}$$

this means that it corresponds to an abnormal sample and needs to be eliminated. Otherwise, x_i is accepted.

Table 1. Effluent BOD data set.

Number	Auxiliary Variable	
01	Influent pH (IPH)	
02	Effluent pH (EPH)	
03	Influent SS	(mg/L)
04	Effluent SS (ESS)	(mg/L)
05	Influent BOD (IBOD)	(mg/L)
06	Influent COD (ICOD)	(mg/L)
07	Effluent COD (ECOD)	(mg/L)
08	Sludge settling ratio of biochemical tank	(mg/L)
09	MLSS in biochemical tank (MLSS)	(mg/L)
10	Biochemical pool Do	(mg/L)
11	Influent oil (IOil)	(mg/L)
12	Effluent oil (EOil)	(mg/L)
13	Influent NH3-N (INH3-N)	(mg/L)
14	Effluent NH3-N	(mg/L)
15	Influent Chroma (IC)	(d)
16	Effluent Chroma (EC)	(d)
17	Influent TN (IT)	(mg/L)
18	Effluent TN	(mg/L)
19	Influent phosphate concentration (IPC)	(mg/L)
20	Effluent phosphate concentration	(mg/L)
21	Inlet water temperature	(°C)
22	Outlet water temperature	(°C)
23	Effluent BOD (EBOD)	(mg/L)

Table 2. Effluent NH3-N data set.

Number	Auxiliary Variable	
01	Effluent TP	(mg/L)
02	Influent TP (ITP)	(mg/L)
03	Temperature (T)	(°C)
04	Anaerobic terminal ORP (ATORP)	(mv)
05	Aerobic front end DO	(mg/L)
06	Aerobic terminal DO	(mg/L)
07	Total suspended solids TTS (TTS)	(mg/L)
08	Effluent PH (EPH)	
09	Effluent ORP (EORP)	(mL)
10	Effluent nitrate (EN)	(mg/L)
11	Effluent NH3-N (ENH3-N)	(mg/L)

2.3. *Data Normalization*

Different variables often have different dimensions and dimensional units. In order to eliminate the dimensional influence between indicators, it is necessary to normalize the data to achieve uniformity among the different data indicators. There are four classes of normalization methods, i.e., rescaling, mean normalization, standardization and scaling to unit length. In this paper, the rescaling method is selected. The normalization formula is as follows:

$$\tilde{x}_i = \frac{x_i - x_{i\,\min}}{x_{i\,\max} - x_{i\,\min}} \tag{4}$$

where x_i is any value of a variable; $x_{i\,\min}$ and $x_{i\,\max}$ are, respectively, the minimum and maximum value of the variable.

After this kind of normalization, all the values of the data are set in the range of $[0, 1]$.

2.4. Correlation Degree Analysis

Since different characteristic variables will have different influences on the predicted variables, to obtain a soft-sensing model with a simpler structure, it is necessary to choose the quality indicators with high correlations. Selecting $\acute{m}$ auxiliary variables from m variable, it has $\acute{m} < m$. In practice, the larger m is, the smaller $\acute{m}$ is compared to m.

In this paper, the gray relational degree analysis method is investigated to select the characteristic variables of BOD and NH3-N effluents. Gray relational degree analysis is a multi-factor statistical method, which describes the strength of the relationship between various factors according to the gray relational degree. This method looks for the inconsistency between quantitative results and quantitative analysis in the traditional mathematical statistics method and reduces the amount of calculation.

The gray correlation coefficient is formulated as follows:

$$\beta = |x_0(k) - x_j(k)| \tag{5}$$

$$\mu_j(k) = \frac{\min\limits_j \min\limits_k \beta + \rho \cdot \max\limits_j \max\limits_k \beta}{\beta + \rho \cdot \max\limits_j \max\limits_k \beta} \tag{6}$$

where j means the j-th variable, k is the k-th iteration, $x_0(k)$ is the output variable, $x_j(k)$ is the input variable, μ_j is the gray correlation coefficient and ρ is the resolution coefficient. If ρ is smaller, the difference between correlation coefficients is larger, and the distinguishing ability is stronger.

Then, the gray correlation degree can be calculated as follows:

$$\gamma_j = \frac{1}{n} \sum_{k=1}^{n} \mu_j(k) \tag{7}$$

where n is the number of variables.

If the gray correlation degree is larger, this means that the corresponding variable has a higher correlation with the effluent quality indicators. Then, according to the gray correlation degree, the characteristic variables are sorted from front to back. Usually, a threshold is determined in advance as $\hbar$, and then the key indicators can be selected as the input of the soft-sensing model if

$$\gamma_j > \hbar \tag{8}$$

is satisfied.

3. Methodology

In this section, a novel IBES-LSSVM method is proposed to find the optimal kernel function parameters of the LSSVM in Figure 2.

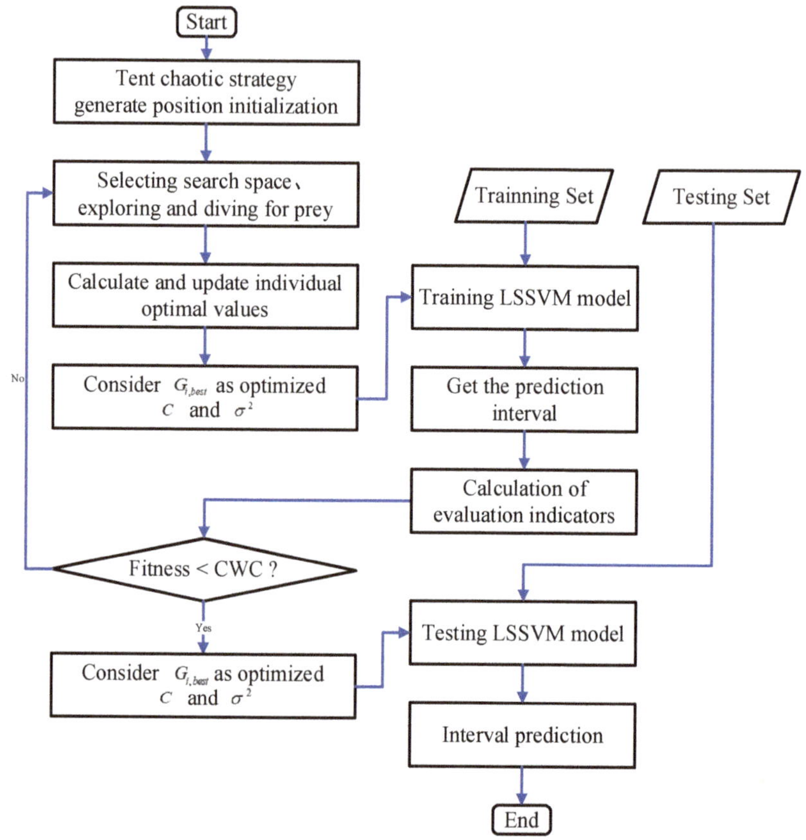

Figure 2. Flow chart of IBES-LSSVM model.

3.1. LSSVM Algorithm

The theory of LSSVM was first proposed by Suykens in 1994. LSSVM is a kernel learning machine following the principle of structural risk minimization and is suitable for analyzing the issue of sample classification and regression estimation [45].

In LSSVM theory, firstly, the sample data are mapped to higher dimensions through nonlinear changes, and linear functions are used for fitting in this high-dimensional feature space:

$$y(x) = w \cdot \phi(x) + b \tag{9}$$

where $y(x)$ is the output variable, x is the input variables, and w and b are weight and bias terms, respectively.

The optimization objectives of the LSSVM regression algorithm can be formulated as

$$\begin{aligned} &\min J(w, \xi_i) = \frac{1}{2} w^T w + \frac{C}{2} \sum_{i=1}^{n} \xi_i^2 \\ &s.t. \\ &y_i = w \cdot \phi(x) + b + \xi_i, \quad i = 1, 2, \cdots, n \end{aligned} \tag{10}$$

where C is the regularization coefficient, ξ_i is the relaxation variable, and $\sum_{i=1}^{n} \xi_i^2$ is the experience risk.

By means of Lagrange multipliers α_i, (10) can be expressed as:

$$
\begin{aligned}
L(w,b,\xi_i,\alpha_i) = \frac{1}{2}w^T w + \frac{C}{2}\sum_{i=1}^{N}\xi_i^2 \\
-\sum_{i=1}^{n}\alpha_i[w\cdot\phi(x)+b+\xi_i - y_i]
\end{aligned}
\tag{11}
$$

According to Karush–Kuhn–Tucker (KKT) optimization conditions:

$$
\begin{cases}
\frac{\partial L}{\partial \mathrm{b}} = 0 \Rightarrow \sum_{i=1}^{n}\alpha_i = 0 \\
\frac{\partial L}{\partial w} = 0 \Rightarrow w = \sum_{i=1}^{n}\alpha_i\phi(x_i) \\
\frac{\partial L}{\partial \xi_i} = 0 \Rightarrow \alpha_i = C\xi_i \\
\frac{\partial L}{\partial a} = 0 \Rightarrow w\cdot\phi(x_i) + b + \xi_i - y_i
\end{cases}
\tag{12}
$$

By defining kernel functions, the optimization problem (11) can be transformed into a linear solution issue:

$$
\begin{pmatrix}
0 & 1 & \cdots & 1 \\
1 & \frac{K(x_1,x_1)+1}{C} & \cdots & K(x_1,x_n) \\
\vdots & \vdots & & \vdots \\
1 & K(x_n,x_1) & \cdots & \frac{K(x_n,x_n)+1}{C}
\end{pmatrix}
\begin{pmatrix} b \\ \alpha_1 \\ \vdots \\ \alpha_n \end{pmatrix}
=
\begin{pmatrix} 0 \\ y_1 \\ \vdots \\ y_n \end{pmatrix}
\tag{13}
$$

where $K(x,x_i)$ is the kernel function.

The Lagrange multiplier and its parameters can be obtained from (13). Therefore, the output of LSSVM can be obtained:

$$
\hat{y}(x) = \sum_{i=1}^{n}\alpha_i K(x,x_i) + b
\tag{14}
$$

For LSSVM, there are many different types of kernel functions, such as linear function, polynomial kernel function, radial basis function (RBF), sigmoid kernel function, etc. Different kernel functions will produce difference types of LSSVM. In this paper, we select RBF as the kernel function of the model:

$$
K(x,x_i) = \exp(-\frac{\|x-x_i\|^2}{2\sigma^2})
\tag{15}
$$

where σ is the variance of RBF.

Through the aforementioned analysis, LSSVM has two tunable parameters (regularization coefficient C and variance of radial basis kernel function σ with RBF), which are important and need to be determined. To obtain the optimal two parameters, the next step is to use an improved PSO algorithm to optimize them.

3.2. *IBES-LSSVM Algorithm*

The BES algorithm is an optimization algorithm that simulates the hunting strategy of vultures when looking for fish. It can obtain a single optimal solution through multiple iterations and finally obtain the overall optimal solution, such that the position of the optimal solution corresponds to the optimal parameter value.

BES hunting is divided into three stages. In the first stage (selection space), the eagle selects the space with the largest prey number. In the second stage (spatial search), the eagle moves in the selected space to find the prey. In the third stage (dive), the eagle swings from the best position determined in the second stage and determines the best hunting.

In the selection stage, firstly, this paper optimizes the initial prey position and adopts the tent chaos strategy, which has the advantages of simple structure and strong ergodicity. Then, the linear decreasing method is used to improve the control parameters of the vulture iterative update position. The optimal model parameters of the model can be found that improve the quality of the fitting. The tent chaotic mapping function is described as:

$$P_{i+1} = \begin{cases} P_i/\lambda, & P_i \in [0,\lambda) \\ (1-P_i)/(1-\lambda), & P_i \in [\lambda,1] \end{cases} \tag{16}$$

where λ is $[0,1]$.

Then, the vultures hunt for food. The formula is:

$$P_{\text{new},i} = P_{\text{best}} + R_1 \cdot C_1 \cdot (P_{mean} - P_i) \tag{17}$$

where R_1 is a parameter controlling the position change, and C_1 is a random number between $(0,1)$. P_{best} is the current optimal location. P_{mean} is the average distribution location of vultures after the previous search. P_i is the location of the i-th vulture.

In the search phase, vultures search for prey in the selected search space and move in different directions in the spiral space to speed up the search. The best position for subduction is:

$$P_{i,\text{new}} = P_i + b(i) \cdot (P_i - P_{i+1}) + a(i) \cdot (P_i - P_{mean}) \tag{18}$$

where:

$$a(i) = \frac{ar(i)}{\max(|ar|)} \tag{19}$$

$$b(i) = \frac{br(i)}{\max(|br|)} \tag{20}$$

$$ar(i) = r(i) \cdot \sin[(\theta(i))] \tag{21}$$

$$\text{br}(i) = r(i) \cdot \cos[(\theta(i))] \tag{22}$$

$$r(i) = \theta(i) + R_2 \cdot C_3 \tag{23}$$

$$\theta(i) = \pi \cdot \omega \cdot C_2 \tag{24}$$

$$\omega = (1 - \frac{i}{i_{max}})^2 \cdot (\omega_{max} - \omega_{min}) + \omega_{min} \tag{25}$$

where $\theta(i)$ and $r(i)$ are the polar angle and polar diameter of the spiral equation, respectively. ω and R_2 are the parameters controlling the spiral trajectory. C_2 and C_3 are a random number within $(0,1)$. The $a(i)$ and $b(i)$ represent the position of the vulture in polar coordinates, and the values are $(-1,1)$.

During the dive phase, vultures swing from the best position in the search space to their target prey. All points also move towards the best point according to

$$\begin{aligned} P_{i,\text{new}} = {} & C_4 \cdot P_{best} + a_1(i) \cdot (P_i - R_3 \cdot P_{man}) \\ & + b_1(i) \cdot (P_i - R_4 \cdot P_{best}) \end{aligned} \tag{26}$$

where:

$$a_1(i) = \frac{ar(i)}{\max(|ar|)} \tag{27}$$

$$b_1(i) = \frac{br(i)}{\max(|br|)} \tag{28}$$

$$ar(i) = r(i) \cdot \sinh[(\theta(i))] \tag{29}$$

$$\text{br}(i) = r(i) \cdot \cosh[(\theta(i))] \tag{30}$$

$$r(i) = \theta(i) \tag{31}$$

$$\theta(i) = \pi \cdot \omega \cdot C_5 \tag{32}$$

where R_3 and R_4 represent the moving speed of the vulture to the optimal point. C_4 and C_5 are random numbers within $(0,1)$.

3.3. *Interval Prediction*

The traditional point prediction cannot deal with the uncertainty in the operation of the system. In order to obtain the numerical estimation and its reliability, the practical application requires the calculation of the prediction interval. Interval prediction indicates the estimation interval of the range of predicted values in a certain confidence interval. Therefore, the prediction interval is composed of the upper and lower line of prediction, which provides its accuracy within a certain confidence level. Assuming that the confidence level is $(1-\mu)\%$, l and u are the lower and upper limits, respectively, when $P(l < y < u) = 1 - \mu\%$, and PI can be expressed as $[l, u]$. For a given confidence interval, the smaller the range of prediction interval, the smaller the uncertainty of prediction and the higher the accuracy.

The evaluation indexes of interval prediction are as follows [46].

PICP: The ratio of the real value to the upper and lower bounds of the prediction interval

$$PICP = \frac{1}{n}\sum_{i=1}^{n} c_i \tag{33}$$

If the predicted value is within the $[l_i, u_i]$ range, c_i is 1. Otherwise, c_i is 0. If all predicted values are included in the prediction interval, $PICP$ = 100%. n is the number of prediction points. In theory, $PICP \geqslant (1-\mu)\%$; otherwise, PI is invalid or unreliable. When comparing the PIs by the model, the other indexes should be as small as possible under the condition that the $PICP$ is as close to the confidence level as possible.

PINAW: The narrow PI has more information and practical value than the wide PI according to

$$PINAW = \frac{1}{nR}\sum_{i=1}^{n}(u_i - l_i) \tag{34}$$

where R is the range of predicted values, respectively.

PINRW: Represents the standard square root width of the predicted interval. The expression is:

$$PINRW = \frac{1}{R}\sqrt{\frac{1}{n}\sum_{i=1}^{n}(u_i - l_i)^2} \tag{35}$$

CWC: In practical application, it is often hoped that a narrow prediction interval width can still be obtained under the condition of high prediction probability, i.e., the prediction interval range probability and interval width will conflict. Therefore, the comprehensive index CWC is proposed:

$$CWC = PINAW\left(1 + \varrho(PICP) \cdot e^{-\tau \cdot (PICP - (1-\mu))}\right) \tag{36}$$

where τ and μ are constants.

When working with training data, the set $\varrho(PICP)$ is 1. In addition, in data verification, $\varrho(PICP)$ is a step function:

$$\varrho = \begin{cases} 0 & PICP \geq 1-\mu \\ 1 & PICP < 1-\mu \end{cases} \tag{37}$$

LUBE is a method based on neural networks to directly calculate the lower and upper bound of the prediction interval. Assuming that the two node values of the output layer of the neural network are the upper and lower limits of the interval, respectively, all the predicted values are included in this range at the confidence level $(1-\mu)\%$. The training purpose of a neural network is to minimize the objective function CWC. In this way, the probability and width of the prediction interval are considered at the same time, and the advantages and disadvantages of the prediction interval PI can be comprehensively evaluated.

The flow-chart of the proposed IBES-LSSVM algorithm is shown in Figure 2, which mainly includes the procedure presented in Algorithm 1.

Algorithm 1 LUBE interval prediction based on IBES-LSSVM model.

Input: Measured data of wasterwater treatment plant.
Output: Prediction interval of BOD/NH3-N effluent.
Step 1: Abnormal data elimination, normalization of the data according to Equations (1)–(4).
Step 2: Analyzing and selecting the key indicators with high correlation degree by Equations (5)–(8).
Step 3: The bald eagle population is initialized by tent chaos strategy based on Equation (16).
Step 4: Local optimal solution.

for all X_i do:
    for all X_i do:
        Obtain predicted value by means of Equations (9)–(15), (17).
    end for
    Using confidence, mean, standard deviation and other parameters, the prediction interval is obtained according to *norminv*() formula.
    Evaluate interval fitness by means of Equations (33)–(37).
end for
Obtain the local optimal solution.

Step 5: Global optimal solution.

While $t \leq iter$ do:
    for all X_i do:
        Update parameter X, C, σ by using Equations (18)–(25).
        Obtain different predictions by using Equations (9)–(15).
    end for
    Using confidence, mean, standard deviation and other parameters, the prediction interval is obtained according to *norminv*() formula.
    Judge and update by Equations (33)–(37).
    for all X_i do:
        Update parameter X, C, σ by using Equations (26)–(32).
        Obtain different predictions by using Equations (9)–(15).
        Using confidence, mean, standard deviation and other parameters, the prediction interval is obtained according to *norminv*() formula.
        Judge and update by means of Equations (33)–(37).
    end for
    $t = t + 1$
end while
Obtain the global optimal solution.

Step 6: Return the global optimal prediction interval.
Step 7: Output C, σ, fitness and other index values by using Equations (33)–(37), (38)–(41).

4. Simulation Results

In this section, the data sets of BOD/NH3-N effluents are collected from a wastewater treatment plant in Beijing and are used to verify the effectiveness of the proposed approach.

The following evaluation indices of several certainty point predictions are evaluated as follows:

$$MSE = \frac{1}{n}\sum_{i=1}^{n}(\hat{y}_i - y_i)^2 \quad (38)$$

$$RMSE = \sqrt{\frac{1}{n}\sum_{i=1}^{n}(\hat{y}_i - y_i)^2} \quad (39)$$

$$MAE = \frac{1}{n}\sum_{i=1}^{n}|\hat{y}_i - y_i| \tag{40}$$

$$R^2 = 1 - \frac{\sum_{i=1}^{n}(\hat{y} - y_i)^2}{\sum_{i=1}^{n}(\hat{y} - \bar{y})^2} = \frac{\sum_{i=1}^{n}(y_i - \bar{y})^2}{\sum_{i=1}^{n}(\hat{y} - \bar{y})^2} \tag{41}$$

4.1. Experiment of Benchmark Functions

The proposed approach is based on the six functions listed in Table 3 with the corresponding ranges and parameters. The range is the boundary of the function search space.

In order to verify the superiority of the proposed approach, it is compared with the WOA, GWO, PSO and SSA algorithms. Statistical results are presented in Table 4. Moreover, the iteration process is depicted in Figures 3–8. From the results, we can see that the convergence rate of IBES is better than that of the other algorithms and the proposed IBES method is able to provide competitive results on the benchmark functions.

Table 3. Benchmark functions.

	Function	Range	Parameters
F1	$F(x) = -\sum_{i=1}^{10}\left[(X-a_i)(X-a_i)^T + c_i\right]^{-1}$	$[1,10]$	dim = 4 popsize = 100 iteration = 300
F2	$F(x) = -\sum_{i=1}^{7}\left[(X-a_i)(X-a_i)^T + c_i\right]^{-1}$	$[1,10]$	dim = 4 popsize = 100 iteration = 300
F3	$F(x) = -\sum_{i=1}^{5}\left[(X-a_i)(X-a_i)^T + c_i\right]^{-1}$	$[1,10]$	dim = 4 popsize = 100 iteration = 300
F4	$F(x) = -\sum_{i=1}^{4} c_i \exp\left(-\sum_{j=1}^{6} a_{ij}(x_j - p_{ij})^2\right)$	$[0,1]$	dim = 6 popsize = 100 iteration = 200
F5	$F(x) = -\sum_{i=1}^{4} c_i \exp\left(-\sum_{j=1}^{3} a_{ij}(x_j - p_{ij})^2\right)$	$[1,3]$	dim = 3 popsize = 100 iteration = 120
F6	$F(x) = \left(\frac{1}{500} + \sum_{j=1}^{25}\frac{1}{j+\sum_{i=1}^{2}(x_i - a_{ij})^6}\right)^{-1}$	$[-65,65]$	dim = 2 popsize = 100 iteration = 180

Table 4. Simulation results of algorithms.

	GWO	PSO	WOA	SSA	IBES	Theoretical Value
F1	−10.5364	−105364	−10.5364	−10.5364	**−10.5364**	−10
F2	−10.4042	−10.4029	−10.4029	−10.4029	**−10.4029**	−10
F3	−10.1561	−10.1532	−10.1576	−10.1532	**−10.1532**	−10
F4	−3.3220	−3.3311	−3.3231	−3.3220	**−3.3220**	−3
F5	−3.8628	−3.8628	−3.8627	−3.8628	**−3.8628**	−3
F6	0.9980	0.9980	0.9980	2.9821	**0.9980**	1

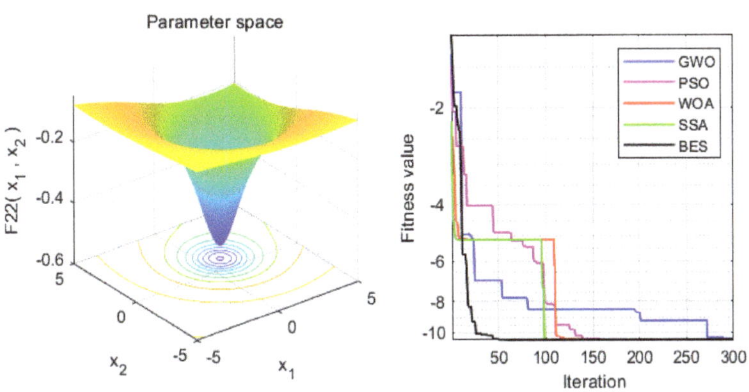

Figure 3. The result of F1.

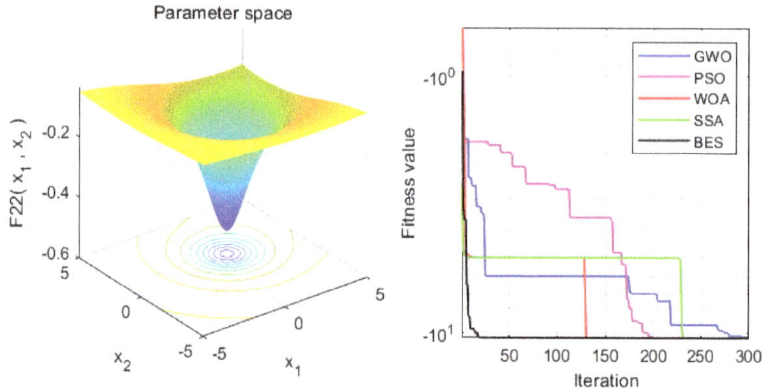

Figure 4. The result of F2.

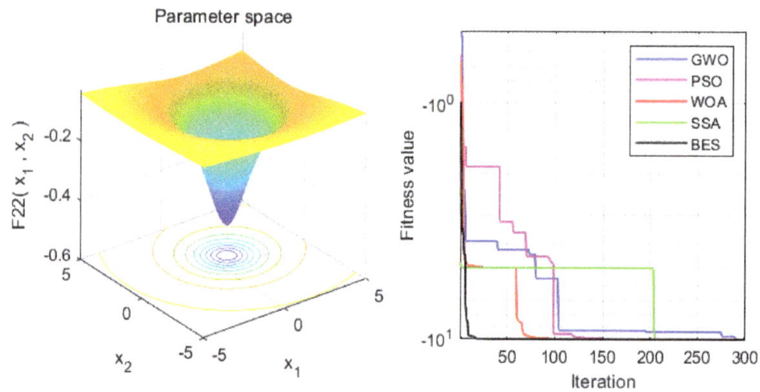

Figure 5. The result of F3.

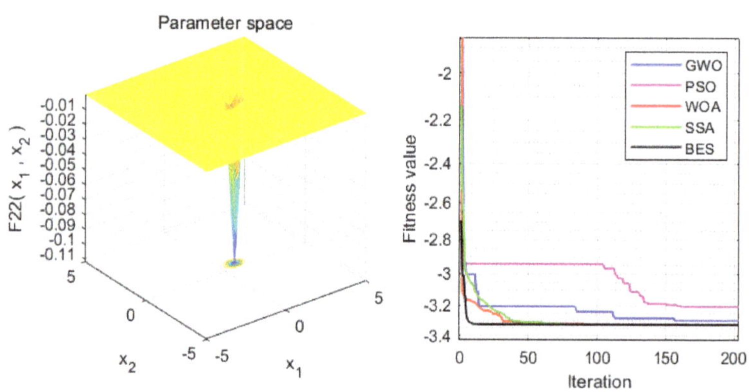

Figure 6. The result of F4.

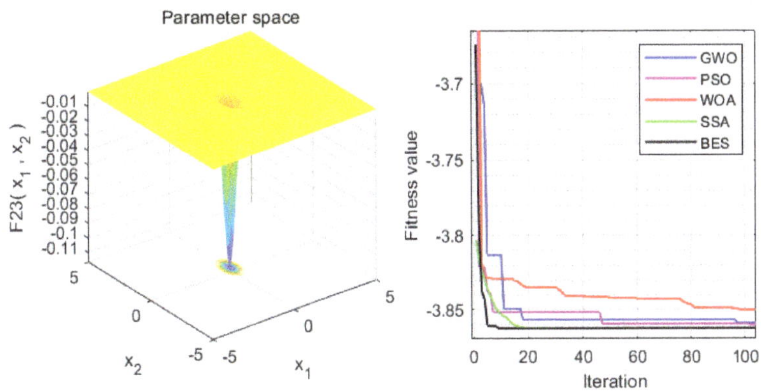

Figure 7. The result of F5.

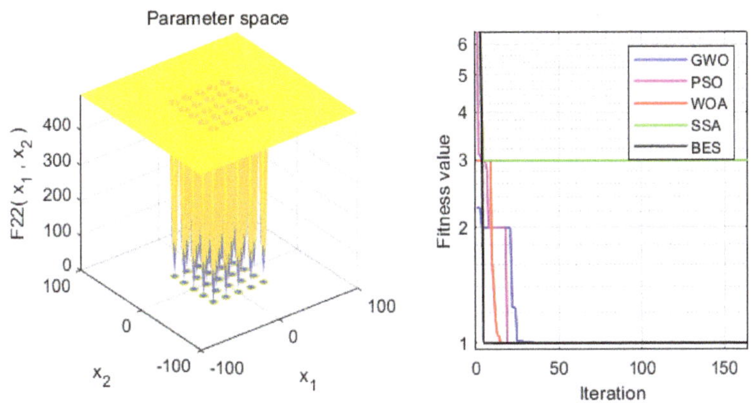

Figure 8. The result of F6.

4.2. Experiment of BOD Data

BOD is one of the most important effluent quality indexes and can reflect the water pollution situation [7]. First, the key auxiliary variables are selected for the BOD effluent data set by calculating the gray correlation degree based on (7). The threshold of the gray correlation degree is chosen as 0.8. Hence, 14 auxiliary variables (as shown in Table 5) are selected as the soft measurement model inputs. Including the output effluent BOD, there are 15 key indicators; the detailed information is shown in Figure 9. Moreover, the description of each datum is given in Figure 10.

In this paper, the BOD effluent data set has 365 sets of data; among them, 335 sets of data are randomly selected as training samples, and the remaining 30 sets of data are treated as the prediction samples. In order to demonstrate the superiority of the proposed IBES-LSSVM method, it is compared with some existing results, i.e., CNN, LSTM, ELMAN, WOA-LSSVM, GWO-LSSVM, PSO-LSSVM and SSA-LSSVM. In the experiments, the initialization conditions are set as: *iter* is 50, $n = 30$, $\omega_{max} = 10$, $\omega_{min} = 0$, $R_1 = 1.8$, $R_2 = 1$, $R_3 = 1.5$, $R_4 = 1.5$.

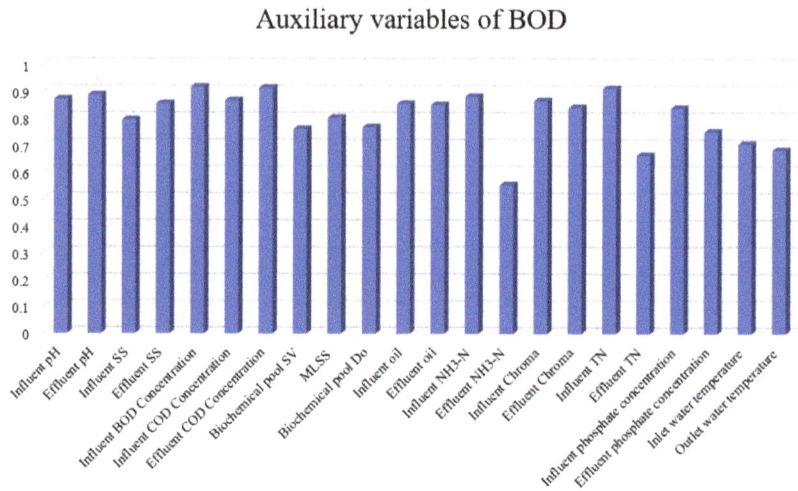

Figure 9. Auxiliary variables of BOD.

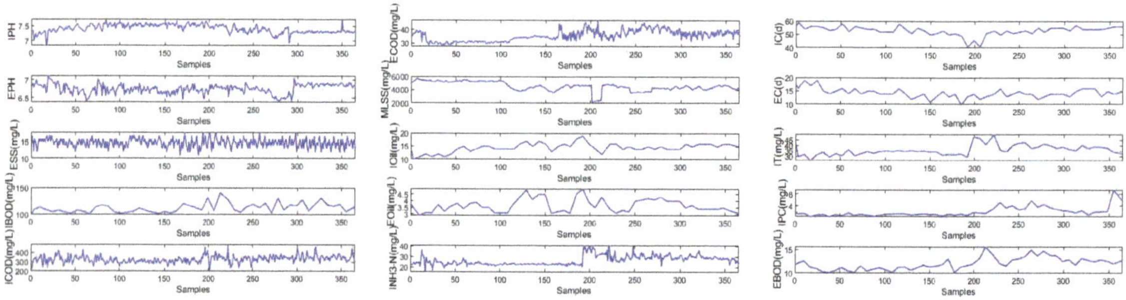

Figure 10. Original data of BOD.

Table 5. Data after processing.

Number of Coefficient	Auxiliary Variable	Correlation
1	Influent BOD	0.9179
2	Effluent COD	0.9151
3	Influent TN	0.9119
4	Effluent pH	0.8878
5	Influent NH3-N	0.8826
6	Influent pH	0.8716
7	Influent COD	0.8676
8	Influent Chroma	0.8669
9	Influent oil	0.8562
10	Effluent SS	0.8556
11	Effluent oil	0.8519
12	Effluent Chroma	0.8415
13	Influent phosphate	0.8397
14	MLSS in biochemical tank	0.8037

From Tables 6 and 7 and Figures 11–13, we can see that, compared with the existing CNN model, LSTM model, ELMAN model, WOA-LSSVM model, GWO-LSSVM model, PSO-LSSVM model and SSA-LSSVM model, the prediction accuracy of the proposed method is better, demonstrating its effectiveness.

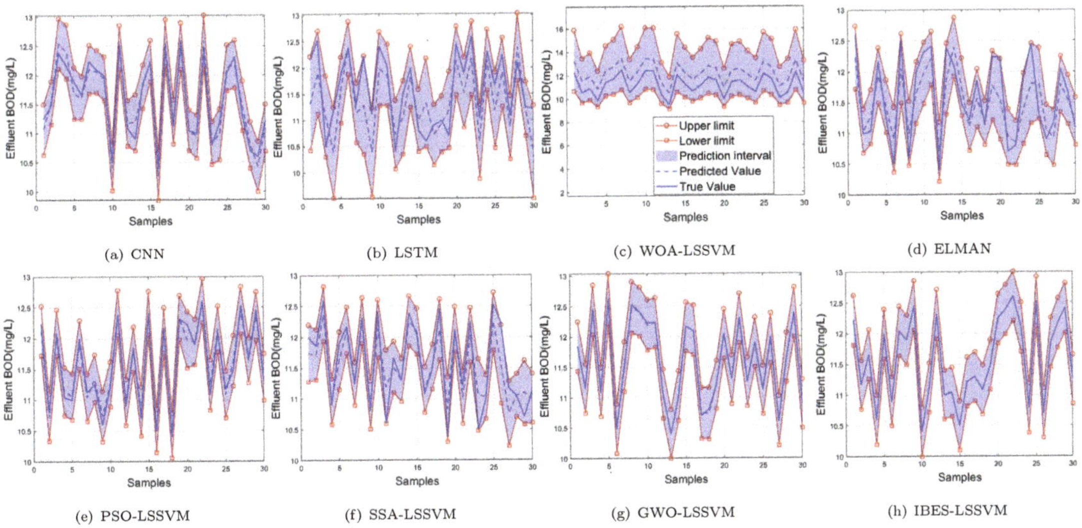

(a) CNN (b) LSTM (c) WOA-LSSVM (d) ELMAN

(e) PSO-LSSVM (f) SSA-LSSVM (g) GWO-LSSVM (h) IBES-LSSVM

Figure 11. 99% of BOD.

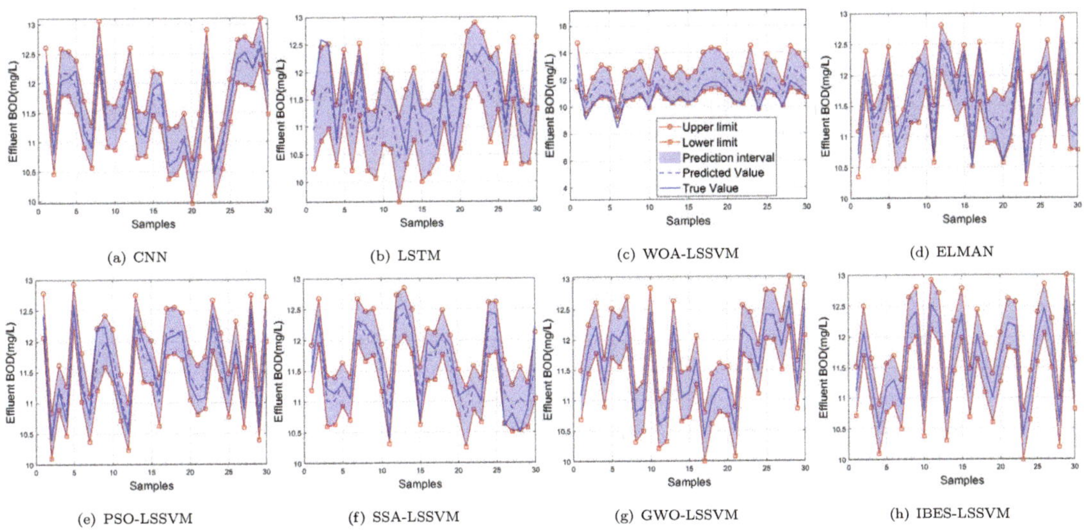

(a) CNN (b) LSTM (c) WOA-LSSVM (d) ELMAN

(e) PSO-LSSVM (f) SSA-LSSVM (g) GWO-LSSVM (h) IBES-LSSVM

Figure 12. 95% of BOD.

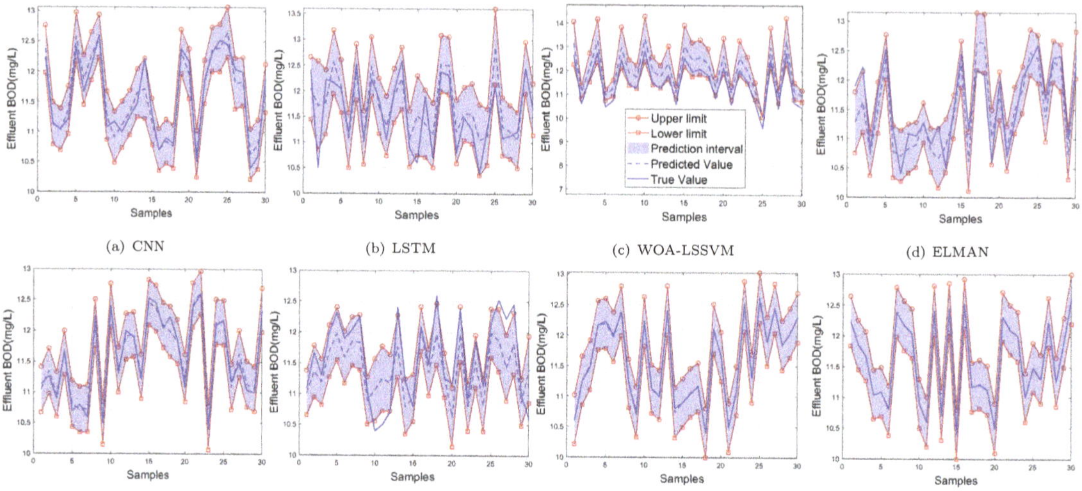

(a) CNN (b) LSTM (c) WOA-LSSVM (d) ELMAN

(e) PSO-LSSVM (f) SSA-LSSVM (g) GWO-LSSVM (h) IBES-LSSVM

Figure 13. 90% of BOD.

Table 6. Predictive index of BOD.

Model	MSE	RMSE	MAE	R^2
CNN	0.0847	0.1500	0.1115	0.9503
LSTM	0.1310	0.2985	0.2330	0.8132
ELMAN	0.2425	0.3120	0.2523	0.7849
GWO-LSSVM	0.0659	0.0217	0.0182	0.9889
WOA-LSSVM	0.0711	0.1831	0.1521	0.9693
PSO-LSSVM	0.0587	0.1049	0.0851	0.9757
SSA-LSSVM	0.0726	0.2371	0.1707	0.9758
IBES-LSSVM	**0.0201**	**0.0104**	**0.0103**	**0.9911**

Table 7. PI of BOD.

	μ = 90%					μ = 95%					μ = 99%				
	PICP	PINRW	CWC	PINAW	Time	PICP	PINRW	CWC	PINAW	Time	PICP	PINRW	CWC	PINAW	Time
CNN	0.9298	0.2731	0.2731	0.2348	41.489	0.9617	0.3848	0.3848	0.3325	42.940	0.9911	0.2841	0.2841	0.2413	46.076
LSTM	0.9124	0.3632	0.3632	0.3112	27.486	0.9609	0.3796	0.3796	0.3254	27.731	0.9913	0.3554	0.3554	0.3020	27.821
ELMAN	0.9073	0.2978	0.2978	0.2474	316.316	0.9549	0.2573	0.2573	0.2202	241.446	0.9909	0.2571	0.2571	0.2132	90.582
WOA-LSSVM	0.9104	0.2663	0.2663	0.2325	1.686	0.9633	0.2697	0.2697	0.2346	1.873	0.9909	0.2673	0.2673	0.2245	1.654
GWO-LSSVM	0.9099	0.2557	0.2557	0.2241	1.396	0.9587	0.2668	0.2668	0.2355	1.389	0.9911	0.2689	0.2689	0.2254	2.012
PSO-LSSVM	0.9111	0.2519	0.2519	0.2198	1.029	0.9544	0.2596	0.2596	0.2155	0.967	0.9908	0.2773	0.2773	0.2277	0.963
SSA-LSSVM	0.9072	0.2901	0.2901	0.2543	1.428	0.9563	0.3178	0.3178	0.2613	1.410	0.9907	0.2961	0.2691	0.2245	1.599
IBES-LSSVM	**0.9053**	**0.2468**	**0.2468**	**0.2007**	**1.406**	**0.9531**	**0.2569**	**0.2569**	**0.2064**	**1.432**	**0.9907**	**0.2569**	**0.2569**	**0.2111**	**1.207**

4.3. *Experiment of NH3-N Data*

In this experiment, the NH3-N effluent data set is considered, which has been described in [44]. First, the gray correlation degree is calculated from (7), and the results are presented in Figure 14. In addition, each selected auxiliary datum of the NH3-N data set is shown in Figure 15.

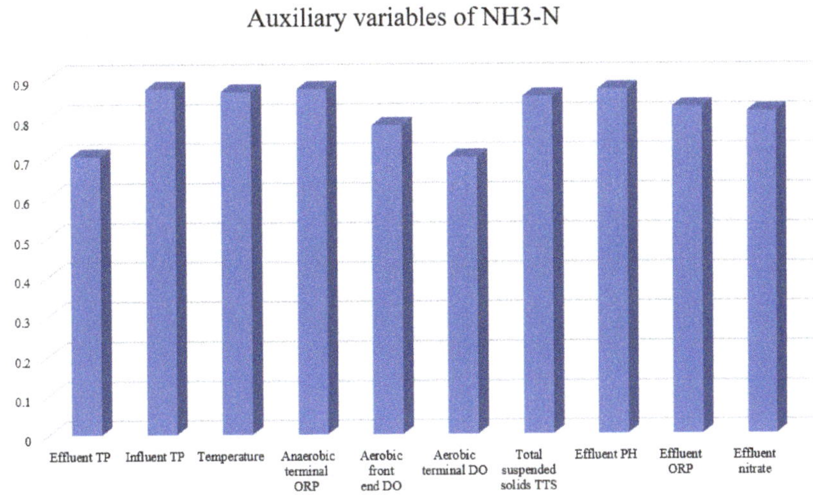

Figure 14. Auxiliary variables of NH3-N.

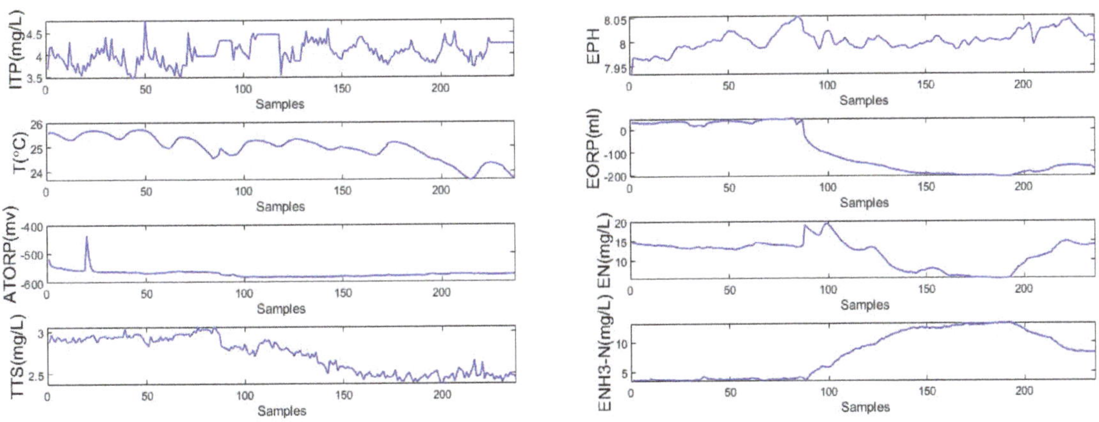

Figure 15. Original data of NH3-N.

In this example, the threshold of the gray correlation degree is also chosen as 0.8; hence, 7 auxiliary variables (as shown in Table 8) are selected as the soft measurement model input. The experimental data of effluent NH3-N used in this paper are from a sewage treatment plant in Beijing. In total, 237 sets of data were obtained, including 200 sets of data that were randomly selected as training samples, and the remaining 37 sets of data were treated as the prediction samples.

Table 8. Data after processing.

Number of Coefficient	Auxiliary Variable	Correlation
1	Influent TP	0.8730
2	Anaerobic terminal ORP	0.8726
3	Effluent PH	0.8693
4	Temperature	0.8659
5	Total suspended solids TTS	0.8525
6	Effluent ORP	0.8257
7	Effluent nitrate	0.8143

In order to demonstrate the superiority of the proposed BES-LSSVM method, it is compared with some existing approaches, i.e., CNN, LSTM, ELMAN, WOA-LSSVM, GWO-LSSVM, PSO-LSSVM and SSA-LSSVM. In the experiments, the parameters are set as follows: *iter* is 50, $n = 30$, $\omega_{max} = 10$, $\omega_{min} = 0$, $R_1 = 1.8$, $R_2 = 1.2$, $R_3 = 1.8$, $R_4 = 1.8$.

From Tables 9 and 10 and Figures 16–18, we can see that, compared with the existing CNN model, LSTM model, ELMAN model, WOA-LSSVM model, GWO-LSSVM model, PSO-LSSVM model and SSA-LSSVM model, the prediction accuracy of the proposed method is the best, demonstrating its effectiveness.

Table 9. PI of NH3-N.

	μ = 90%					μ = 95%					μ = 99%				
	PICP	PINRW	CWC	PINAW	Time	PICP	PINRW	CWC	PINAW	Time	PICP	PINRW	CWC	PINAW	Time
CNN	0.9231	0.53951	0.53951	0.50111	29.991	0.9619	0.49776	0.49776	0.46854	32.446	0.9919	0.52063	0.52063	0.48445	31.703
LSTM	0.9182	0.49437	0.49437	0.44235	22.176	0.9588	0.42320	0.42320	0.37824	22.637	0.9921	0.53185	0.53185	0.50111	21.272
ELMAN	0.9066	0.38637	0.38637	0.34255	6.661	0.9580	0.37625	0.37625	0.32142	3.175	0.9912	0.42032	0.42032	0.38764	3.120
WOA-LSSVM	0.9197	0.49711	0.49711	0.45739	1.547	0.9581	0.46106	0.46106	0.42131	1.711	0.9913	0.47562	0.47562	0.41121	1.584
GWO-LSSVM	0.9227	0.51067	0.51067	0.46174	1.346	0.9601	0.51117	0.51117	0.47894	1.166	0.9913	0.51776	0.51776	0.45669	1.163
PSO-LSSVM	0.9241	0.48209	0.48209	0.45394	0.959	0.9604	0.47815	0.47815	0.42756	0.797	0.9917	0.49209	0.49209	0.46401	0.801
SSA-LSSVM	0.9112	0.40579	0.40579	0.35752	1.363	0.9574	0.38947	0.38947	0.34556	1.184	0.9909	0.38777	0.38777	0.36454	1.142
IBES-LSSVM	**0.9037**	**0.34531**	**0.34531**	**0.30989**	**1.354**	**0.9556**	**0.34906**	**0.34906**	**0.31128**	**1.181**	**0.9907**	**0.34677**	**0.34677**	**0.31001**	**1.366**

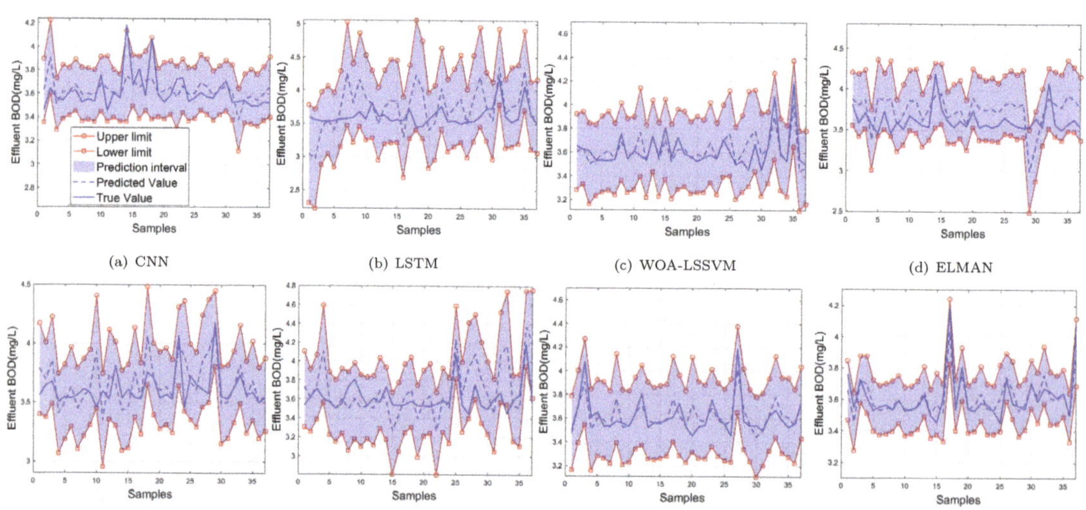

(a) CNN (b) LSTM (c) WOA-LSSVM (d) ELMAN
(e) PSO-LSSVM (f) SSA-LSSVM (g) GWO-LSSVM (h) IBES-LSSVM

Figure 16. 99% of NH3-N.

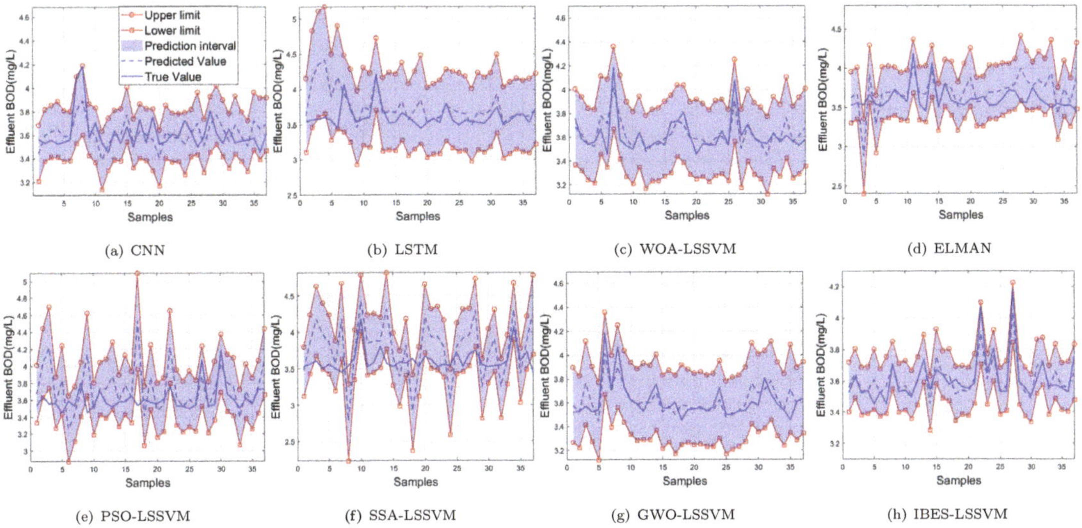

(a) CNN (b) LSTM (c) WOA-LSSVM (d) ELMAN

(e) PSO-LSSVM (f) SSA-LSSVM (g) GWO-LSSVM (h) IBES-LSSVM

Figure 17. 95% of NH3-N.

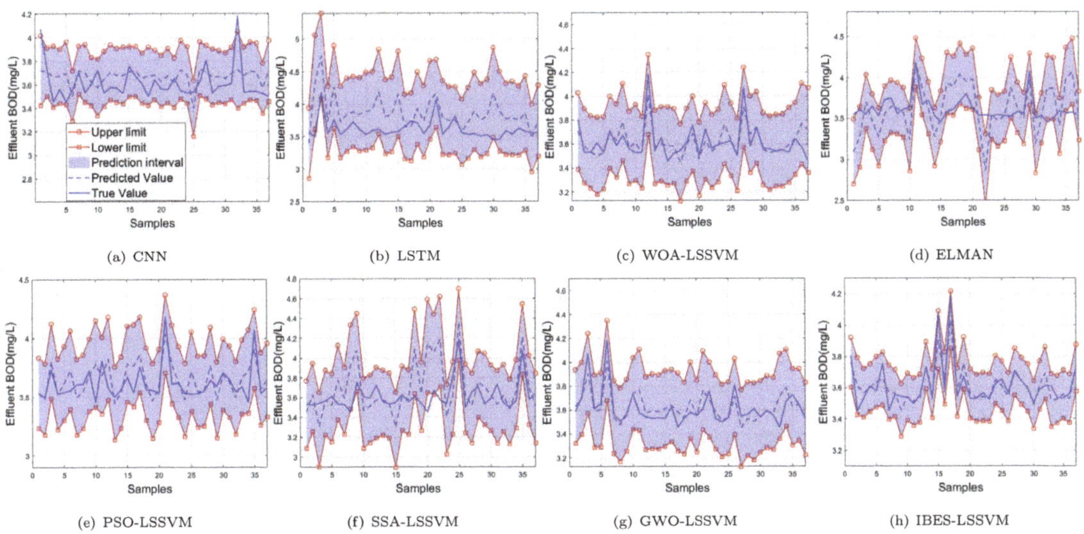

(a) CNN (b) LSTM (c) WOA-LSSVM (d) ELMAN

(e) PSO-LSSVM (f) SSA-LSSVM (g) GWO-LSSVM (h) IBES-LSSVM

Figure 18. 90% of NH3-N.

Table 10. Predictive index of NH3-N.

Model	MSE	RMSE	MAE	R^2
CNN	0.1874	0.1711	0.1450	0.8932
LSTM	0.1138	0.2131	0.1663	0.7666
ELMAN	0.0954	0.1846	0.1564	0.7872
GWO-LSSVM	0.0997	0.0895	0.0628	0.7280
WOA-LSSVM	0.1929	0.2371	0.1709	0.8959
PSO-LSSVM	0.1312	0.1722	0.1247	0.8922
SSA-LSSVM	0.1196	0.1958	0.2037	0.8117
IBES-LSSVM	**0.0917**	**0.0645**	**0.0450**	**0.8967**

5. Conclusions

This paper investigates an improved IBES-LSSVM algorithm to predict the effluent water quality indicators of a wastewater treatment plant, in which an improved BES method is proposed to find the optimal LSSVM parameters. To deal with the uncertainties of the data, the prediction interval is generated within a certain confidence level, which could provide the upper and lower bounds of the prediction results. Compared with other existing methods, the proposed approach demonstrates high prediction accuracy, with reduced computational time and an easy calculation process, in predicting effluent water quality parameters. Note that the proposed results can only predict the water quality indicators, but this is not the end work for a wastewater treatment plant process. The application of this work to reliable decision-making and the generation of a suitable control strategy will be our future work.

Author Contributions: Conceptualization M.Z., V.P.; methodology M.Z., J.W., Y.S., V.P.; resources Y.S.; writing-review and editing M.Z., V.P., Y.Z.; supervision M.Z., J.W.; investigation J.W.; formal analysis M.Z.; software and data curation Y.Z. All authors have read and agreed to the published version of the manuscript.

Funding: This work was partially supported by the National Key Research and Development Program (Grant No. 2018YFC1602704), National Natural Science Foundation of China (Grant No. 51805021,61973023), the Fundamental Research Funds for Beijing (No. 110052972027/015) and the Research foundation for Talents of NCUT (No. 213051360020XN173/017), "Science and Technology Innovation" Special Fund Project in Shijingshan District: 401053762117.

Institutional Review Board Statement: Not applicable.

Informed Consent Statement: Not applicable.

Data Availability Statement: Data are available upon request.

Acknowledgments: This work was carried out with the help of Dingyuan Chen at Beijing University of Technology.

Conflicts of Interest: The authors declare no conflict of interest.

References

1. Julio, N.; Figueroa, R.; Ponce Oliva, R.D. Water resources and governance approaches: Insights for achieving water security. *Water* **2021**, *13*, 3063. [CrossRef]
2. Abrams, A.L.; Carden, K.; Teta, C.; Wågsæther, K. Water, sanitation, and hygiene vulnerability among rural areas and small towns in south Africa: Exploring the role of climate change, marginalization, and inequality. *Water* **2021**, *13*, 2810. [CrossRef]
3. Shannon, M.A.; Bohn, P.W.; Elimelech, M.; Georgiadis, J.G.; Mariñas, B.J.; Mayes, A.M. Science and technology for water purification in the coming decades. *Nature* **2008**, *452*, 301–310. [CrossRef]
4. Olsson, G. ICA and me—A subjective review. *Water Res.* **2012**, *46*, 1585–1624. [CrossRef]
5. Thürlimann, C.M.; Dürrenmatt, D.J.; Villez, K. Soft-sensing with qualitative trend analysis for wastewater treatment plant control. *Control Eng. Pract.* **2018**, *70*, 121–133. [CrossRef]
6. Vorosmarty, C.J.; Hoekstra, A.Y.; Bunn, S.E.; Conway, D.; Gupta, J. Fresh water goes global. *Science* **2015**, *349*, 478–479. [CrossRef] [PubMed]
7. Qiao, J.; Wang, L.; Yang, C.; Gu, K. Adaptive Levenberg-Marquardt algorithm based echo state network for Chaotic time series prediction. *IEEE Access* **2018**, *6*, 10720–10732. [CrossRef]
8. Chen, Y.; Han, D. Water quality monitoring in smart city: A pilot project. *Autom. Constr.* **2018**, *89*, 307–316. [CrossRef]
9. Lu, H.; Ma, X. Hybrid decision tree-based machine learning models for short-term water quality prediction. *Chemosphere* **2020**, *249*, 126169. [CrossRef]
10. Zhou, P.; Li, Z.; Snowling, S.; Baetz, B.W.; Na, D.; Boyd, G. A random forest model for inflow prediction at wastewater treatment plants. *Stoch. Environ. Res. Risk Assess.* **2019**, *33*, 1781–1792. [CrossRef]
11. Liu, H.; Zhang, H.; Zhang, Y.; Zhang, F.; Huang, M. Modeling of Wastewater Treatment Processes Using Dynamic Bayesian Networks Based on Fuzzy PLS. *IEEE Access* **2020**, *8*, 92129–92140. [CrossRef]
12. Manandhar, A.; Fischer, A.; Bradley, D.J.; Salehin, M.; Islam, M.S.; Hope, R.; Clifton, D.A. Machine learning to evaluate impacts of flood protection in Bangladesh, 1983–2014. *Water* **2020**, *12*, 483. [CrossRef]

13. de Canete, J.F.; Orozco, P.D.S.; Baratti, R.; Mulas, M.; Ruano, A.; Garcia-Cerezo, A. Soft-sensing estimation of plant effluent concentrations in a biological wastewater treatment plant using an optimal neural network. *Expert Syst. Appl.* **2016**, *63*, 8–19. [CrossRef]
14. Cong, Q.; Yu, W. Integrated soft sensor with wavelet neural network and adaptive weighted fusion for water quality estimation in wastewater treatment process. *Measurement* **2018**, *124*, 436–446. [CrossRef]
15. Hamada, M.; Adel Zaqoot, H.; Abu Jreiban, A. Application of artificial neural networks for the prediction of Gaza wastewater treatment plant performance-Gaza strip. *J. Appl. Res. Water Wastewater* **2018**, *5*, 399–406. [CrossRef]
16. Zeinolabedini, M.; Najafzadeh, M. Comparative study of different wavelet-based neural network models to predict sewage sludge quantity in wastewater treatment plant. *Environ. Monit. Assess.* **2019**, *191*, 1–25. [CrossRef]
17. Kadam, A.; Wagh, V.; Muley, A.; Umrikar, B.; Sankhua, R. Prediction of water quality index using artificial neural network and multiple linear regression modelling approach in Shivganga River basin, India. *Model. Earth Syst. Environ.* **2019**, *5*, 951–962. [CrossRef]
18. Heddam, S.; Lamda, H.; Filali, S. Predicting effluent biochemical oxygen demand in a wastewater treatment plant using generalized regression neural network based approach: A comparative study. *Environ. Process.* **2016**, *3*, 153–165. [CrossRef]
19. Tan, T.; Yang, Z.; Chang, F.; Zhao, K. Prediction of the First Weighting from the Working Face Roof in a Coal Mine Based on a GA-BP Neural Network. *Appl. Sci.* **2019**, *9*, 4159. [CrossRef]
20. Nourani, V.; Elkiran, G.; Abba, S.I. Wastewater treatment plant performance analysis using artificial intelligence—An ensemble approach. *Water Sci. Technol.* **2018**, *78*, 2064–2076. [CrossRef]
21. Cheng, T.; Dairi, A.; Harrou, F.; Sun, Y.; Leiknes, T. Monitoring influent conditions of wastewater treatment plants by nonlinear data-based techniques. *IEEE Access* **2019**, *7*, 108827–108837. [CrossRef]
22. Han, H.; Liu, H.; Liu, Z.; Qiao, J. Fault detection of sludge bulking using a self-organizing type-2 fuzzy-neural-network. *Control Eng. Pract.* **2019**, *90*, 27–37. [CrossRef]
23. Wu, J.; Cheng, H.; Liu, Y.; Liu, B.; Huang, D. Modeling of adaptive multi-output soft-sensors with applications in wastewater treatments. *IEEE Access* **2019**, *7*, 161887–161898. [CrossRef]
24. Lotfi, K.; Bonakdari, H.; Ebtehaj, I.; Mjalli, F.S.; Zeynoddin, M.; Delatolla, R.; Gharabaghi, B. Predicting wastewater treatment plant quality parameters using a novel hybrid linear-nonlinear methodology. *J. Environ. Manag.* **2019**, *240*, 463–474. [CrossRef]
25. Han, H.; Zhang, L.; Qiao, J. Data-based predictive control for wastewater treatment process. *IEEE Access* **2017**, *6*, 1498–1512. [CrossRef]
26. Mateo Pérez, V.; Mesa Fernández, J.M.; Ortega Fernández, F.; Villanueva Balsera, J. Gross Solids Content Prediction in Urban WWTPs Using SVM. *Water* **2021**, *13*, 442. [CrossRef]
27. Ribeiro, D.; Sanfins, A.; Belo, O. Wastewater Treatment Plant Performance Prediction with Support Vector Machines. In *Advances in Data Mining. Applications and Theoretical Aspects*; Springer: Berlin/Heidelberg, Germany, 2013; pp. 99–111. [CrossRef]
28. Baghban, A.; Sasanipour, J.; Habibzadeh, S.; Zhang, Z. Sulfur dioxide solubility prediction in ionic liquids by a group contribution—LSSVM model. *Chem. Eng. Res. Des.* **2019**, *142*, 44–52. [CrossRef]
29. Zhang, N.; Kamaha, R.; Behera, P. Prediction of Surface Water Supply Sources for the District of Columbia Using Least Squares Support Vector Machines (LS-SVM) Method. *Adv. Comput. Sci. Int. J.* **2015**, *4*, 1–9.
30. Luo, F.; Qiao, X.; Liao, W. Soft-sensing modeling based on GK-LSSVM method for online predictions of BOD in activated sludge process. In Proceedings of the 2017 International Conference on Robotics and Automation Sciences (ICRAS), Hong Kong, China, 26–29 August 2017; pp. 134–138.
31. Manu, D.; Thalla, A.K. Artificial intelligence models for predicting the performance of biological wastewater treatment plant in the removal of Kjeldahl Nitrogen from wastewater. *Appl. Water Sci.* **2017**, *7*, 3783–3791. [CrossRef]
32. Liu, Z.; Wan, J.; Ma, Y.; Wang, Y. Online prediction of effluent COD in the anaerobic wastewater treatment system based on PCA-LSSVM algorithm. *Environ. Sci. Pollut. Res.* **2019**, *26*, 12828–12841. [CrossRef] [PubMed]
33. Nieto, P.G.; García Gonzalo, E.; Arbat, G.; Duran Ros, M.; de Cartagena, F.R.; Puig Bargués, J. A new predictive model for the filtered volume and outlet parameters in micro-irrigation sand filters fed with effluents using the hybrid PSO-SVM-based approach. *Comput. Electron. Agric.* **2016**, *125*, 74–80. [CrossRef]
34. Han, H.; Lu, W.; Hou, Y.; Qiao, J. An adaptive-PSO-based self-organizing RBF neural network. *IEEE Trans. Neural Netw. Learn. Syst.* **2016**, *29*, 104–117. [CrossRef] [PubMed]
35. Chen, S.; Fang, G.; Huang, X.; Zhang, Y. Water quality prediction model of a water diversion project based on the improved artificial bee colony–backpropagation neural network. *Water* **2018**, *10*, 806. [CrossRef]
36. Zeng, F.; Amar, M.N.; Mohammed, A.S.; Motahari, M.R.; Hasanipanah, M. Improving the performance of LSSVM model in predicting the safety factor for circular failure slope through optimization algorithms. *Eng. Comput.* **2021**, 1–12. [CrossRef]
37. Shariati, M.; Mafipour, M.S.; Ghahremani, B.; Azarhomayun, F.; Ahmadi, M.; Trung, N.T.; Shariati, A. A novel hybrid extreme learning machine–grey wolf optimizer (ELM-GWO) model to predict compressive strength of concrete with partial replacements for cement. *Eng. Comput.* **2020**, 1–23. [CrossRef]
38. Yao, W.; Zeng, Z.; Lian, C. Generating probabilistic predictions using mean-variance estimation and echo state network. *Neurocomputing* **2017**, *219*, 536–547. [CrossRef]
39. Yuan, X.; Chen, C.; Jiang, M.; Yuan, Y. Prediction interval of wind power using parameter optimized Beta distribution based LSTM model. *Appl. Soft Comput.* **2019**, *82*, 105550. [CrossRef]

40. Liao, Y.; Zhang, L.; Liu, C. Uncertainty prediction of remaining useful life using long short-term memory network based on bootstrap method. In Proceedings of the 2018 IEEE International Conference on Prognostics and Health Management (ICPHM), Seattle, WA, USA, 11–13 June 2018; pp. 1–8.
41. Marín, L.G.; Cruz, N.; Sáez, D.; Sumner, M.; Núñez, A. Prediction interval methodology based on fuzzy numbers and its extension to fuzzy systems and neural networks. *Expert Syst. Appl.* **2019**, *119*, 128–141. [CrossRef]
42. Sun, X.; Wang, Z.; Hu, J. Prediction interval construction for byproduct gas flow forecasting using optimized twin extreme learning machine. *Math. Probl. Eng.* **2017**, *2017*, 5120704. [CrossRef]
43. Wu, D.; Wang, H.; Seidu, R. Smart data driven quality prediction for urban water source management. *Future Gener. Comput. Syst.* **2020**, *107*, 418–432. [CrossRef]
44. Chen, Z.; Yang, C.; Qiao, J. The optimal design and application of LSTM neural network based on the hybrid coding PSO algorithm. *J. Supercomput.* **2021**, *5*. [CrossRef]
45. Su, Y.; Han, L.; Wang, J.; Wang, H. Quantum-behaved RS-PSO-LSSVM method for quality prediction in parts production processes. In *Concurrency and Computation Practice and Experience*; John Wiley & Sons: Hoboken, NJ, USA, 2019; pp. 1–15.
46. Li, R.; Jin, Y. A wind speed interval prediction system based on multi-objective optimization for machine learning method. *Appl. Energy* **2018**, *228*, 2207–2220. [CrossRef]

sensors

MDPI

Article

Estimation of Infiltration Volumes and Rates in Seasonally Water-Filled Topographic Depressions Based on Remote-Sensing Time Series

Pavel P Fil [1,2,*], Alla Yu Yurova [1], Alexey Dobrokhotov [3] and Daniil Kozlov [1]

1 V. Dokuchaev Soil Science Institute, 109017 Moscow, Russia; alla.yurova@gmail.com (A.Y.Y.); daniilkozlov@gmail.com (D.K.)
2 Department of Geography, Lomonosov Moscow State University, 119991 Moscow, Russia
3 Agrophysical Research Institute, 195220 St. Petersburg, Russia; dobralexey@gmail.com
* Correspondence: philpaulp@gmail.com

Abstract: In semi-arid ecoregions of temperate zones, focused snowmelt water infiltration in topographic depressions is a key, but imperfectly understood, groundwater recharge mechanism. Routine monitoring is precluded by the abundance of depressions. We have used remote-sensing data to construct mass balances and estimate volumes of temporary ponds in the Tambov area of Russia. First, small water bodies were automatically recognized in each of a time series of high-resolution Planet Labs images taken in April and May 2021 by object-oriented supervised classification. A training set of water pixels defined in one of the latest images using a small unmanned aerial vehicle enabled high-confidence predictions of water pixels in the earlier images (Cohen's K = 0.99). A digital elevation model was used to estimate the ponds' water volumes, which decreased with time following a negative exponential equation. The power of the exponent did not systematically depend on the pond size. With adjustment for estimates of daily Penman evaporation, function-based interpolation of the water bodies' areas and volumes allowed calculation of daily infiltration into the depression beds. The infiltration was maximal (5–40 mm/day) at onset of spring and decreased with time during the study period. Use of the spatially variable infiltration rates improved steady-state shallow groundwater simulations.

Keywords: closed depressions; temporary water bodies; remote sensing; infiltration

Citation: Fil, P.P.; Yurova, A.Y.; Dobrokhotov, A.; Kozlov, D. Estimation of Infiltration Volumes and Rates in Seasonally Water-Filled Topographic Depressions Based on Remote-Sensing Time Series. *Sensors* **2021**, *21*, 7403. https://doi.org/10.3390/s21217403

Academic Editors: Miquel Àngel Cugueró-Escofet and Vicenç Puig

Received: 14 September 2021
Accepted: 3 November 2021
Published: 7 November 2021

1. Introduction

Shallow groundwater is present in many semi-arid landscapes across the world either intermittently or permanently, depending on the lithological profile, topography, and water balance. Unlike in wetter environments with diffuse groundwater recharge, recharge in these environments is primarily focused (local) in areas of excess water input [1]. In such environments, where moisture deficits in upland soils are high, groundwater recharge will only occur if there is sufficient infiltration of converging flow to overcome the deficits. One mechanism involved is a localized recharge process that routes surface water runoff within the landscape to topographically low areas (depressions), allowing infiltration of water through ephemeral seasonal ponds [2–4]. Moreover, depression-focused recharge driven by snowmelt is a major annual hydrological event in cold semi-arid regions such as the Pothole Prairie Region of North America. In recent decades, there has been an accelerated increase in process understanding of the contributions of prairie potholes to surface runoff [5,6] and depression-focused groundwater recharge [3] in this part of North America. The knowledge has been acquired through studies involving conceptual and mathematical modeling of hydrological processes of surface flows [5–7], subsurface flows and combinations of the two [8,9], applications of isotopic and environmental tracers [3], digital elevation model (DEM)-based delineations of depressions and their watersheds [10–14], assessments of hydrologic connectivity [6,10,15–17], and remote sensing with high and intermediate resolution [18,19]. Studies in various catchments have shown that both horizontal and vertical connectivity in pothole hydrological systems are very site-specific and no model can be applied to a new system without validation.

In Russia, areas rich in pothole-like systems of depressions ("zapadiny") in interfluves of forest-steppe catchments cover a larger region than in North America, extending across much of European Russia and into Siberia. However, after an initial period of intensive hydrological research in the 1960s to the 1980s there was very little study of depression-focused groundwater recharge despite advances in GIS-facilitated simulation and remote sensing. Moreover, there is increasing societal need for such studies to enhance the understanding of key landscape functions related to water storage or movement, e.g., water capacitance, carbon sequestration, and both nutrient retention and cycling [17,20] and precision agricultural management. With some justification, early studies noted similarities between prairie potholes and forest-steppe zapadiny. However, before applying tools developed in North American research to Russian systems, there is a need for quantitative evaluation of concepts that emerged in earlier local studies.

One of the key hypotheses developed during the 1960s is that the major source of recharge for shallow groundwater in areas such as the Oka-Don Lowland of the Tambov region in European Russia is depression-focused infiltration during snowmelt [21]. In a very recent study an indirect method was used to calibrate the groundwater recharge to hydraulic conductivity ratio for application in an analytical steady-state solution of the 2D shallow groundwater flow equation using soil redoximorphic features of typical classified catenas of the Samovetc catchment in this lowland [22]. In the cited study, the same recharge rate was prescribed for all points along a topographical transect. In contrast, in the study presented here, the spatial variability of depression-focused groundwater recharge along the transect was studied in a field campaign in spring 2021 during, immediately after snowmelt, and several weeks later.

There is no single method for classifying remote-sensing data for the ponds' retrieval. The methods and materials used vary greatly depending on the region of study, season of the year, image resolution or type of the pond. In terms of wavelengths used in the electromagnetic spectrum, they are visible (RGB), near infrared (NIR), shortwave infrared (SWIR) and thermal infrared (TIR) [23,24]. In addition to optical methods, data from RADAR and LIDAR are also used [25]. Methods for extraction of small water bodies are divided into four groups. The first group is the threshold methods, the essence of which is the discretization of individual spectral channels or spectral indicators based on expert or experimental threshold values [26,27]. The second group of methods are statistical methods, such as those using multivariate regression or discriminant analysis. Classification methods (the third group) are a matrix of combinations of different methods—this is a pixel or object-oriented approach, classifications with or without training, various classification machines; for example: a random forest or support vector machine, neural algorithms [28–32]. There are also various special techniques (group four) such as entropy-based computer vision techniques [33].

In this work, remote-sensing data were used to construct a mass balance and estimate volumes of ephemeral ponds by object-oriented supervised classification of high-resolution Planet Labs images of the Tambov area acquired from April to May 2021. The data acquired on dynamic changes in delineated ponds, in combination with a DEM, observations using an unmanned aerial vehicle (UAV), a widely accepted method for calculating evaporation, and visual hydrological observations were used to estimate infiltration volumes and rates through the depression bottoms and account for their spatial and temporal variability. Considering that groundwater recharge from the depressions' bottom is very area-focused and occurs episodically during the snowmelt, the process is not usually accounted for in the regional-scale evaluation of the groundwater resources in Tambov region. To include the impacts of spatial heterogeneity and dynamic fluctuation the depression-focused infiltration may be modeled numerically [8]. To examine the early hypothesis [21] on a critical role of depressions in ground-water recharge in a forest-steppe region through the simplified approach for estimating recharge, this paper aims: (1) to determine the variations of the pond recession and infiltration rate in time and between the depressions due to systematic (vertically-varying hydraulic conductivity) and random factors (presence of clogging or frozen layers, pond- surface drainage network connection); (2) to determine the role of such spatial variations through numerical analysis of shallow groundwater model for the simplified 2D case; and (3) to determine a method for calculating the volume of recharge through depression in other catchments both with and without the requirement of numerical modeling and data assimilation. The result allowed identification of the volume of intercepted water during snowmelt and calculation of the rate of water recession and infiltration rates in closed depressions for the first time for the study region. Use of the horizontal variation in parameters obtained along the studied transect substantially improved results of the shallow ground water model developed in the cited study [22].

2. Materials and Methods

Remote-sensing data were used to construct a mass balance and estimate volumes of ephemeral ponds by object-oriented supervised classification of high-resolution Planet Labs images of the Tambov area acquired from April to May 2021. The data acquired on dynamic changes in delineated ponds, in combination with a DEM, observations using an unmanned aerial vehicle (UAV), a widely accepted method for calculating evaporation, and visual hydrological observations were used to estimate infiltration volumes and rates through the depression bottoms and account for their spatial and temporal variability. The acquired time series of changes in the volume of nine temporary ponds enabled parametrization with a negative exponential curve. A time series of the infiltration rate, calculated from the water balance, was used to estimate the total amount accumulated during the event, and both the initial (maximum) and saturated (minimum) infiltration rates per unit area.

2.1. Study Area

The study area covers approximately 560 ha in the center of the Oka-Don lowland (52°37′ N, 40°2′ E) in the Petrovsky district of the Tambov region, Russia (Figure 1). The lowland is the largest in the forest-steppe biome. With elevation ranging from 120 to 180 m above sea level, on average it is 100 m lower than adjacent territories. The lowland has a semi-arid climate with long winters, pronounced spring snowmelt events and relatively dry summers with an annual precipitation to potential evapotranspiration ratio of 0.8. According to data recorded at a meteorological station 10 km north of the study site, during the period 2005–2020 the annual temperature was 6.9 °C, and average monthly temperatures in January and July were −8.6 °C and 21.0 °C, respectively [34]. Mean annual precipitation during this period amounted to 550 mm (of which 113 mm fell during periods with sub-zero temperatures), and the mean snow height before onset of snowmelt was 320 mm, very similar to the recorded historical climatic norm for 1961–1990 (290 mm).

The soils are mainly chernozems and the area is mainly used for cultivating crops (typically wheat, corn, sunflower, soy, sugar beet), despite hindrance by water shortages. Clay and loamy deposits, generally 5–15 m (but sometimes up to 40 m) thick, with boulders of glacial origin, underlie a layer of loess-like loam with thickness ranging from 2 m in the lower parts of slopes to 30 m in the interfluve. The upper layer is porous and can both accumulate and retain moisture, while the glacial clays and loams form a local aquiclude for infiltrated surface waters. Shallow groundwater above this aquiclude is permanent and forms a continuous layer in the focal catchment. Evidence of stagnic condition in topsoil is restricted to the presence of albic material in the lower part of the humus horizon in grey gleysols in the depression bottom. There is clear evidence of gleyic conditions in the soil morphology (Fe-Mn concretions, Fe masses, pore lining, reduced matrix) in the catchment and continuous presence of water saturation below 2–3 m depth in the poorly drained soils and 1 m depth in the waterlogged soils. The latter was confirmed by a few cases of drilling in different years and seasons (the WTD is consistently highest after the springflood) and automated measurements in 2019 (a year with extremely low snow accumulation). Physical properties of the surface loams have contributed to development of closed depressions of multifactorial genesis, which are widely spread throughout the Oka-Don lowland. These depressions delay the runoff of surface waters into rivers [21] and transfer surface runoff to groundwater, thereby replenishing the groundwater and moistening the surrounding soil. The closed depressions are filled with water in the spring when the snow melts. Snow located in the catchment area of each basin melts and replenishes it, usually in mid-March to early April. At the end of spring, surface water only remains in small parts of the depressions, and in summer they usually dry up completely, in contrast to the closed depressions of the Pothole Prairie Region. Monitoring the dynamics of water volume and its filtration enables estimation of amounts of valuable additional moisture entering the soil in this semi-arid region.

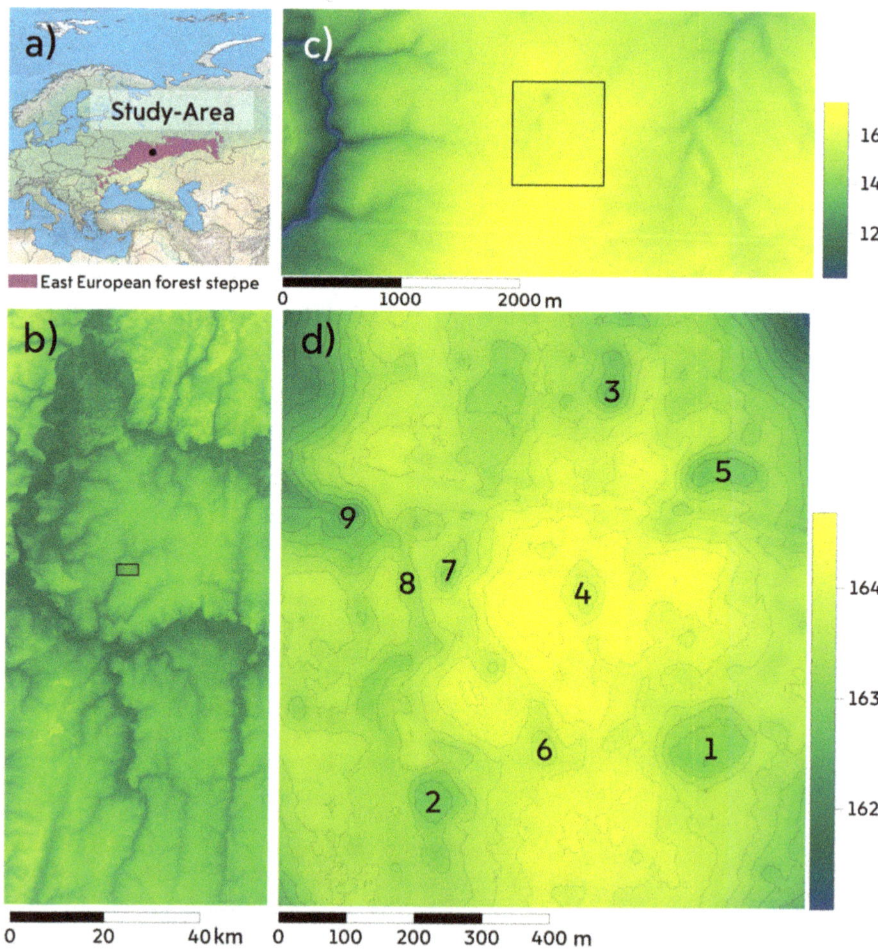

Figure 1. The study area. (**a**) Location of the study area in the forest-steppe biome of Eurasia. (**b**) Location of the study area in the catchment of the Matyr River-Oka-Don Lowland (SRTM). (**c**) Digital elevation model (DEM) of the interfluve of the Samovets brook; the studied territory of the depressions is marked with a rectangle. (**d**) Unmanned aerial vehicle (UAV) DEM of the study area, the numbers indicate numbers of closed depressions filled with pond water in the spring of 2021 (for details, see Table 2).

2.2. Input Data

Water dynamics in a closed depression in spring 2021 was tracked and modeled using the following three types of data (Figure 2): ultra-high resolution (25 cm) digital terrain and elevation models obtained using a small UAV—DJI Mavic 2 Pro, orthophotomaps of terrain in the visible range with a ultra-high resolution (25 cm) from UAV, orthophotomaps of high resolution (3 m) in the visible range from the sensors of the RapidEye and SkySat mini-satellites of the Planet Labs system [25]. Precipitation data were obtained from the nearest weather station with daily resolution. Evaporation data from the water surface was obtained using Penman's equation and meteorological input from the same station.

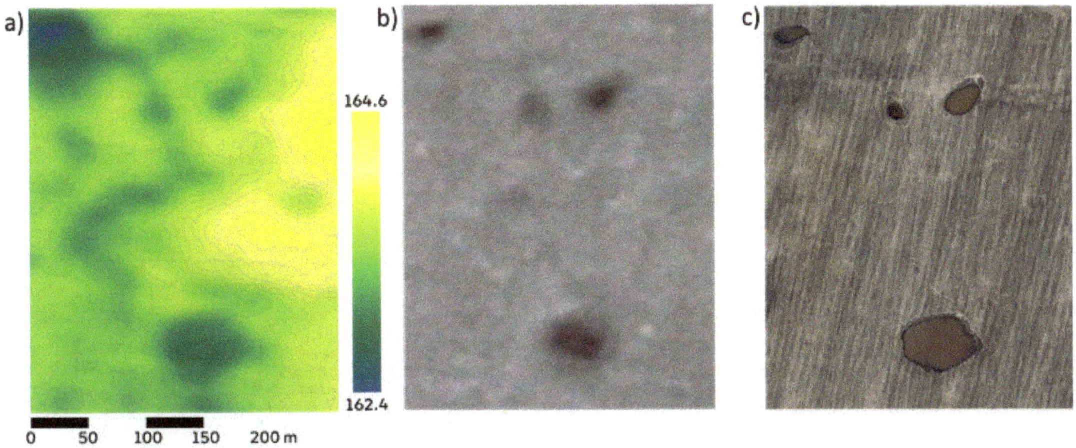

Figure 2. Sources of spatial data. (**a**) High-resolution DEM obtained photogrammetrically with colors indicating heights. (**b**) Planet Labs' digital images of terrain in the visible spectrum with 3 m resolution. (**c**) High-resolution digital image obtained using the UAV in the visible spectral range.

In this study we used stereophotogrammetry, i.e., estimation of three-dimensional coordinates of points on an object from two or more photographic images taken from different positions by the small UAV. In this article, we used a standard method for constructing a digital terrain model using a small UAV with an accuracy of 0.03 m (hereafter, the UAV DEM).

For this, we used the Mavic 2 Pro routing app (DroneDeploy.com). Geolocation markers were located on the ground, and their positions were determined using the STONEX GNSS system (flight altitude, 150 m; image overlap, 75%). We processed the data using Agisoft Metashape and created a dense point cloud to generate a digital terrain model. We manually filtered points associated with agroforestry areas within the fields in ArcGis Pro using the field mask and the vegetation mask. The masks were obtained by manual decoding the UAV materials. Orthophoto maps generated from free satellite photos obtained via Bing were used to identify trees. Points related to heights of the trees were removed. A digital model of the territory with 25 cm resolution was created from the remaining point cloud using the kriging interpolation tool in ArcMap. An orthomosaic was created in Agisoft Metashape and exported with 25 cm resolution.

A time series of high-resolution visible orthomosaics (with 1–3 m resolution) at times when there was no cloud cover, from the beginning of spring snowmelt to the drying up of temporary water bodies in early summer were downloaded from Planet Labs Inc. Images of the scenes were downloaded when there was no cloud cover, from the beginning of spring snowmelt to the drying up of temporary water bodies in early summer.

2.2.1. Delineation of Water Bodies

Orthophotomaps generated using an UAV allow correct interpretation of water surfaces, as they can be visually inspected to delineate water/dry surface boundaries accurately. Orthomosaic maps from Planet Labs have lower resolution and higher atmospheric noise. Therefore, we used the Interactive Supervised Classification tool in the ArcGIS Pro desktop application to delineate water bodies in them. For this we created a water feature training set from the ultra-high resolution orthomosaic, and used it to enable automatic recognition of water bodies in the Planet Labs orthomosaics via object-oriented supervised classification, as implemented in the ArcGis Pro raster classification tool [35]. It is well established that object-oriented classification is superior to pixel-based classification for high-resolution images [36], and it has been previously used to delineate similar depression-shaped natural systems [37,38].

The classification involved the following steps. First, the analyzed raster layer was constrained by a field cadastral border [39] buffered 15 m on each side to prevent inclusion of objects rather than a bare soil surface without water (e.g., an agricultural field with no vegetation in early spring) and surfaces that may be flooded with water. Masking was applied to avoid possible classification errors by excluding unnecessary objects (trees, roads, buildings, etc.). The second step was imaging segmentation, based on a mean shift procedure, by criteria of the minimum segment size expressed in pixels [40,41] implemented in ArcGIS Pro, to merge adjacent pixels of relative homogeneity—preferentially based on spectral (color) characteristics—into image objects. Unitless segmentation scale parameters determining the average size of objects governing the degree of homogeneity allowed for pixel merging was set to 10 on the RGB scale. The third step was creation of a training set. As summer approaches, ponds in the depressions always shrink (Figure 3). Thus, water surfaces present on the date of a UAV flight were always present on the preceding dates, and three training samples were created for groups of dates before each UAV survey (Figure 3). Each training sample contained two categories: water and soil surface. Finally, the random forest (RF) method [42,43] and support vector machine (SVM) [44] for supervised classification of segmented images was applied, yielding a binary (water–not water) raster. The testing set from the next UAV survey was used to validate the resulting binary models. This enabled identification of the water surface areas in each period. However, in the classified images, the boundary of the ponds does not have a constant height relative to the DEM of the UAV. To avoid this unnatural variation of height the classified raster was transformed into a vector containing only water polygons. Along the outer boundary of the water polygon, the DEM values of the UAV were sampled with a frequency of 25 centimeters. The median was calculated from the extracted values. The contour was then drawn for the second time, now in accordance with the average value on the UAV DEM, thus, the outer boundaries of the ponds were forced to have a constant height value. The described procedure allowed us to avoid misclassification of the Planet Labs mixed pixels due to relatively low resolution as we operated with the vector area, not the raster area. For the comparison purely pixel-based classification was also made. The DEM and water polygons vector were used to calculate the volume of water in each depression on the days the images were taken. We used the Surface-Volume tool from ArcGIS Pro to calculate the area and volume between the surface and the reference plane (Polygon Volume (3D Analyst). This provided the water content in each depression in cubic meters on each of the days.

The maximum surface water area of a depression corresponds to the volume of water up to its overflow point, defined here as the minimum value of the height in the UAV DEM along its drainage basin vector boundary [45]. The watershed boundary was defined by the Basin tool in ArcGis based on the raster of the flow direction, derived from the UAV DEM using the "Direction of flow" tool in ArcGis Pro. The water layer (mm) in the catchment area of each depression required for its maximum volume is equal to its total maximum volume of water divided by its entire catchment area.

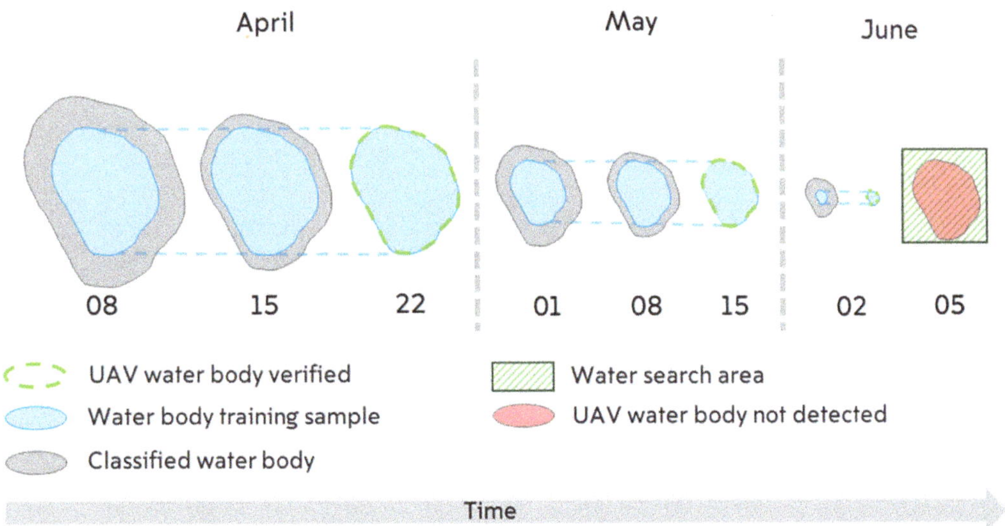

Figure 3. Schematic illustration of the water surface classification method. The arrow at the bottom indicates the general trend of depressions drying (from left to right) in spring. The green dashed lines around the blue areas show the extent of the water as photographed by the UAV on the control dates (22 April, 15 May and 2 June). The gray color indicates the water surface resulting from object-oriented image classification trained on a subset of points from the later control surveys by the UAV. Three time intervals were used for the training. The green shading at the end is the area searched for a water mirror surface by the UAV on a day when the pond had already disappeared (light red fill).

2.2.2. Evaporation

Results of a previous comparison suggest that all of three conventional methods for estimating evapotranspiration from water-filled and vegetated depressions have acceptable applicability for estimating evaporation from open water [46]. The most convenient of these methods, the classical form of the Penman equation [47,48], was used in this study to estimate potential evaporation:

$$E_{PEN} = \frac{\Delta}{\Delta+\gamma}\frac{R_n}{\lambda}\frac{\gamma}{\gamma+\Delta}\frac{6.43E_A}{\lambda}, \tag{1a}$$

Here: E_{PEN} is potential (open water) evaporation (mm/d); R_n is net radiation at the surface (MJ/m^2/d); Δ is the slope of the saturation vapor pressure curve (kPa/C); γ is a psychrometric coefficient (kPa/°C); λ is the latent heat of vaporization (MJ/kg); and E_A is the drying power of the air, which can be found using the following Dalton-type formulation:

$$E_A = f(U)D = (1+0.536U)(e_s - e_a) \tag{1b}$$

Here: $f(U)$ is a wind function with linear coefficients for the original Penman equation (1948, 1963); u is the wind speed at 2 m height (m/s), $D = (e_s - e_a)$ is vapor pressure deficit (kPa); e_S is saturation vapor pressure (kPa); and e_a is actual vapor pressure (kPa).

Open-water evaporation was computed from readily available data as previously described [49] and implemented in the Evaplib Python library [50]. Input data for this were air temperature (T, °C), solar radiation (RS, MJ/m^2/d), relative humidity (RH, %), and wind velocity (u, m/s).

In the absence of actinometric measurements of net radiation at the surface, this was calculated from amounts of cloud cover recorded at the weather station and a previous regional calibration [51].

$$R_a = \frac{24(60)}{\pi} G_{sc} d_r (\omega_s \sin(\phi) \sin(\delta) + \cos(\phi) \cos(\delta) \sin(\omega_s)) \tag{2a}$$

Here: R_a is extraterrestrial radiation (MJ/m^2/day**), G_{sc} is the solar constant (0.0820 MJ/m^2/min), d_r is the inverse of the relative distance between the Earth and Sun, ω_s is the sunset hour angle (rad), ϕ is latitude (rad), and δ is solar declination (rad).

$$d_r = 1 + 0.033 \cos\left(\frac{2\pi}{365} J\right) \tag{2b}$$

where J is the day of the year; $\delta = 0.409 \sin\left(\frac{2\pi}{365} J - 1.39\right)$;

$$\omega_s = \arccos(-\tan(\phi)\tan(\delta))\ ; N = \frac{24}{\pi}\omega_s$$

Solar radiation, R_s, can be calculated from the amount of cloud:

$$R_s = (a_s + b_s(1 - N))R_a \tag{3}$$

where N is the amount of cloud (ranging from 0 for clear sky to 1 for full cloud cover), while a_s and b_s are Angstrom values, and without regional calibration values of 0.25 and 0.50, respectively, are recommended [52].

Net longwave radiation (R_{nl}) can be estimated from the air temperature, actual vapor pressure, and solar radiation. Net longwave radiation is expressed by the Stefan–Boltzmann law:

$$R_{nl} = \sigma\left(\frac{T}{max^2_{4^4_{min}}}(0.34 - 0.14\sqrt{e_a})\left(1.35\frac{R_s}{R_{so}} - 0.35\right)\right) \tag{4}$$

where T_{max} is daily maximum air temperature (K), T_{min} is daily minimum air temperature (K), and R_{so} is clear-sky radiation (MJ/m^2/day) according to:

$$R_{so} = (a_s + b_s)R_a \tag{5}$$

We applied a constant albedo of 7% (0.07) for water surfaces in the calculations, based on the latitude and published mean reference values [51].

$$R_n = (1 - \alpha)R_s - R_{nl} \tag{6}$$

We calculated daily evaporation values. Input data for Equations (1a), (1b) and (4) and daily precipitation were obtained from the nearest meteorological station (at Lipetsk city).

2.2.3. Water Balance and Groundwater Model Recalibration

Infiltration rates (mm/day) were calculated from the daily water balance equation:

$$F = 1000 \cdot [-\Delta V - A_t(E_{PEN} - P)]/A_t \tag{7}$$

where $-\Delta V$ is the daily rate of reduction in pond volume (m^3), A_t is the current pond area (m^2), and P is the daily precipitation (mm/d).

The volume of a pond on a given Julian day (*dayT*) was derived from the volume on the first Julian day in a series (*dayF*) and the following negative exponential equation:

$$V = a \cdot e^{-c \cdot (dayT - dayF)} \tag{8}$$

The scaling coefficient *a* and the power of the exponent *c* (the pond's approximate initial volume and decay rate, respectively) were obtained by the least square method, which has given a fit with $R^2 > 0.9$ for each of studied lakes (Table 2, Figure 4d). The estimated volume on each day was used to calculate the rate of reduction in pond volume in Equation (7) with a daily time step.

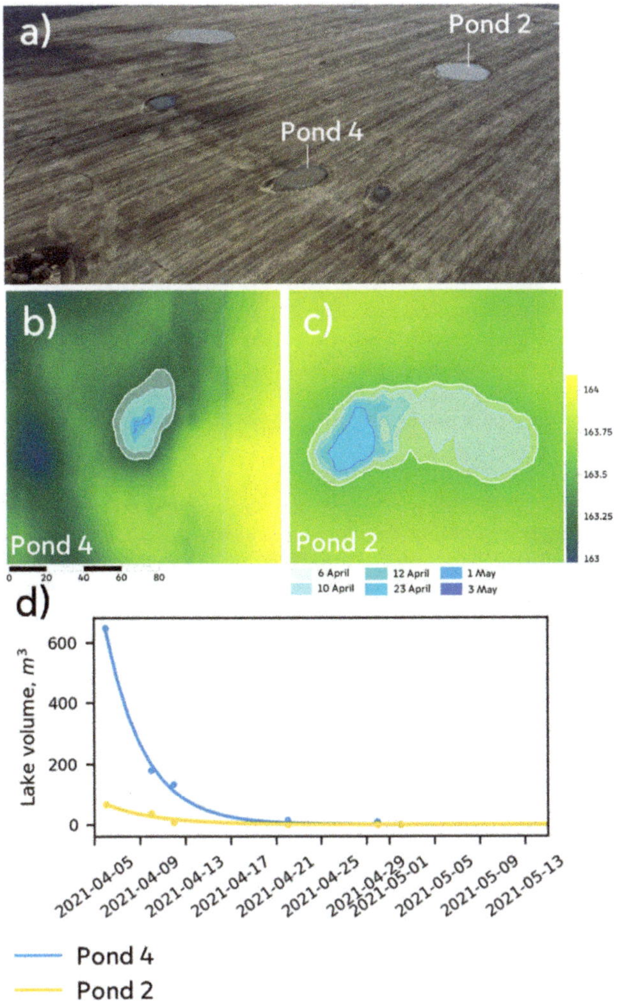

Figure 4. Temporal dynamics of selected ponds in the depressions. (**a**) Positions of Ponds 4 and 2 in a UAV photo image. (**b**,**c**) Groundwater levels in basins of the ponds on indicated dates based on the classification of images (highlighted in color by day). (**d**) Water volumes in the basins of ponds 4 and 2 on indicated dates (points) and negative exponential fits derived by Equation (8) (lines). Pond numbering as in Figure 1 and Table 2.

A limitation of the method lies in the choice of the first day of infiltration, because it is impossible to determine the water boundary in depressions when they are covered with snow. Water infiltrates the soil when the temperature is already above 0 degrees Celsius, but classification of images with partial snow cover is problematic. Thus, the first Planet Labs image that was subjected to classification was the first when there was no snow cover according to the nearest (Lipetsk) weather station.

A 2D profile of the steady-state shallow water table depth was obtained by the analytical form of continuity equation with calibration based on soil redoximorphic features. In [22] the hypothesized relationship between archived morphological properties (redoximorphic features as indicators of gleyic conditions) of soils and a current hydrological process indicator (WTD) were established based on the expert knowledge of soil types, WTD co-occurrence, then verified under a hillslope flow continuity constraint expressed mathematically as a steady-state solution with two free parameters: hydraulic conductivity and recharge rate. Here the input horizontal transect of groundwater recharge rate was taken as a time integral of Equation (7). Spatially, it varied along the transect according to positions of the depressions in the landscape. Infiltration into the soil is not equal to the groundwater recharge rate, so relative values in the [0, 1] interval were used to describe the variability along the transect while the formal calibration of the absolute values of recharge rate to hydraulic conductivity (N/k) was preserved in the method. Time-averaged infiltration was calculated based on the volume of water that infiltrated in closed depressions. We established 10 regular topographic profiles representing the generalized transect, 3 km long and crossing the interfluve along the main slope with regularly (5 m) spaced points. At each standard point, the value of the water layer (mm) was extracted, which was filtered out in a closed depression. The 10 topographic lines were combined into a single profile by averaging values corresponding to the order of the points of the water layer. The regular placement of topographic profiles and sampling points was intended to optimize the two-dimensional characterization of additional moisture infiltration along the studied transect.

3. Results

The proposed combination of object-oriented image classification based on a time series of Planet Labs images and an orthomosaic derived from UAV surveys to verify the satellite data enabled highly accurate identification of the water mirrors of closed depressions during their drying (Cohen's kappa = 0.99). Moreover, high-precision digital terrain models obtained using UAVs can be used to calculate volumes of water in closed depressions.

We compared different methods of pond extraction for the scene on 9 April 2021, when the reference UAV image was obtained. Two supervised pixel-based classification methods were compared: random forest (RF) and support vector machines (SVM) providing results as a raster. Then the ponds boundaries were brought to a constant median value of the DEM to obtain vector pond polygons (also both from RF and SVM classification). Root mean square error (RMSE) and mean absolute percentage error (MAPE) of pond volume and area were the lowest for the vector approach and notably higher for raster approach (Table 1). SVM and RF errors were almost the same within the vector approach (Table 1), and it was decided to use RF as the most common in such studies. Contrary to [52], the novel way to vectorize the polygons based on idea of flat pond mirror with constant height brought a very notable increase in the quality of area and volume estimate.

Table 1. Root mean square error (RMSE) and mean absolute percentage error (MAPE) of the area and volume with true value taken from ultra-high resolution UAV estimate of ponds' boundaries. Random forest (RF) and support vector machine (SVM) methods are compared for the pixel-based (raster) and median DEM height-based (vector) delineation of the image taken on 9 April 2021.

	Errors of Area Estimate		Errors of Volume Estimate	
	RMSE, m^2	**MAPE, %**	**RMSE, m^3**	**MAPE, %**
RF vector	7.8	1.4	2.5	5.9
RF raster	43.2	3.1	16.0	21.9
SVM vector	8.7	1.6	2.1	5.1
SVM raster	44.9	3.6	16.3	22.2

Results obtained using the described procedure show that the drainage process of the focal depressions follows an exponential equation (Figure 4, Table 2), with coefficients (Table 2) that presumably depend on various factors (e.g., the depressions' source rocks and filtration areas), but we found no systematic quantitative relationships between the coefficients and considered parameters.

Table 2. Derived characteristics of the ponds during the decreasing volume phase after snowmelt from the maximum (starting day) to zero (final day). Coefficients a and c are from Equation (8) and R^2 is the coefficient of determination for the negative exponential fit of the lake volume by the least square method.

Pond No. [1]	Maximum Volume, m^3	Maximum Area, m^2	Coefficient a	Coefficient c	R^2	Total Infiltration, m^3	Initial Infiltration Rate, m^3/d	Soil Refill Capacity, mm/dd [2]	ksat mm/d	Total Evaporation, m^3	Total Precipitation, m^3
1	640	6123	604	0.11	0.91	475	47	54	3	320	155
2	647	4275	645	0.29	1.00	594	153	163	3	113	59
3	173	1756	173	0.38	0.98	167	50	172	12	18	12
4	66	1283	67	0.23	0.95	56	10	53	4	24	14
5	416	3546	428	0.19	0.94	366	64	103	6	103	54
6	16	540	16	0.53	1.00	14	5	37	1	6	4
7	103	978	104	0.28	1.00	92	23	113	2	24	13
8	36	543	37	0.34	0.97	34	9	93	4	7	5
9	12	295	13	0.19	0.99	9	1	34	2	7	4

[1] The numbering follows Figure 1. [2] dd—melt water peak event duration (days).

During the initial phase the rate of pond recession is much higher than later in the season (Figure 5 top). Notably less water is evaporated than infiltrates (Figure 5, middle and bottom), so the depression-focused replenishment of the groundwater is consistent with the previously mentioned hypothesis that the major source of recharge for shallow groundwater in the study area (and similar areas) is depression-focused infiltration during snowmelt [21]. There are two phases of infiltration—fast and slow (Figure 5, bottom). Measurements during the fast phase enable estimation of the unsaturated soil's refill rate and capacity (Table 2). During the slow phase the change in infiltration rate from day to day is much smaller. The saturated hydraulic conductivity decreases strongly with depth under a depression [8], reflecting the effects of the decreasing frequency of fractures with depth, and the flow is presumably limited by the lowest layer with the smallest frequency. Thus, the infiltration rate estimated during the slow phase provides an approximation of the hydraulic conductivity (Table 2), corresponding to the maximum possible flux out of the soil column.

Overflow can occur from any closed depression (Figure 6). The probability of spillage depends on multiple factors, including elevations of the lowest point in the catchment area and the depression's overflow point. In 2018, water reached the overflow point in almost all the depressions considered here (Figure 6). Thus, the initial volumes (a coefficients) obtained for the nine studied ponds can be used in speculation regarding the effects of the landscape morphometry and meltwater input on initial volumes of ponds after snowmelt.

The results also indicate that the hypothesis of a quantitative linear relationship between the volume of water accumulated in a depression and the catchment area of the basin is only partly correct. The water volumes do not appear to be linearly related to the depressions' catchment, because the amount of water in a depression depends on the catchment area and maximum volume that can be stored in it. Excess water will flow through the overflow point without replenishing the water table. Limits of the possible volume and layer of water intercepted by the focal depressions, which limit their ability to converge surface runoff into underground flows, were identified. The maximal layer depends on the catchment area of the depression and height of its overflow (Figure 6, right).

The water layer filling closed depressions during snowmelt in the forest-steppe zone is highly dynamic. From 2005 to 2021, the snowmelt water layer (snow water equivalent, SWE), reconstructed from the statistically corrected snow height and snow density data series, varied from 50 to 300 mm. The derived snowmelt water layer during this period has a binomial distribution with two maxima, at 50 and 200 mm SWE. Analysis of the meteorological data showed that closed depressions did not overflow during snowmelt in 60% of cases, on average, from 2005 to 2021. This corroborates the finding that in most cases closed depressions intercept the surface runoff and transfer it to groundwater. Frequencies of overflow were lowest for Ponds 2 and 4 (around 10%) and highest for Pond 9 (90%).

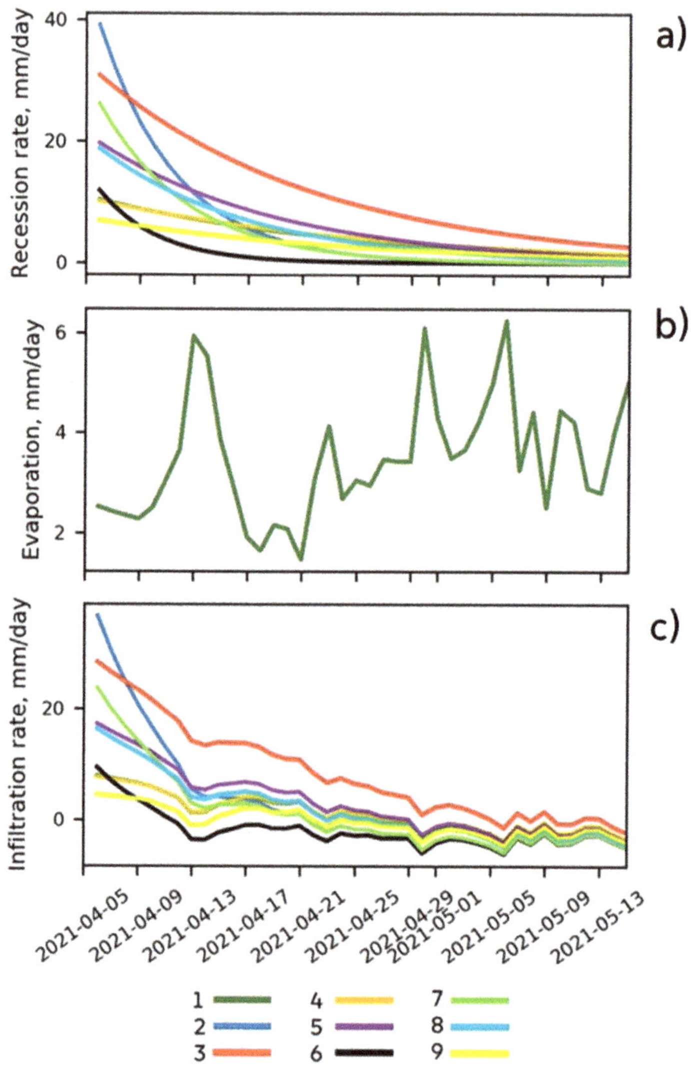

Figure 5. Rate of recession of the indicated ponds' water levels (**a**), daily evaporation rate (**b**) and infiltration rates of the ponds estimated from the mass balance (bottom panel) expressed in mm of the water layer (**c**). The numbering of the ponds in the color legend follows Figure 1 and Table 2.

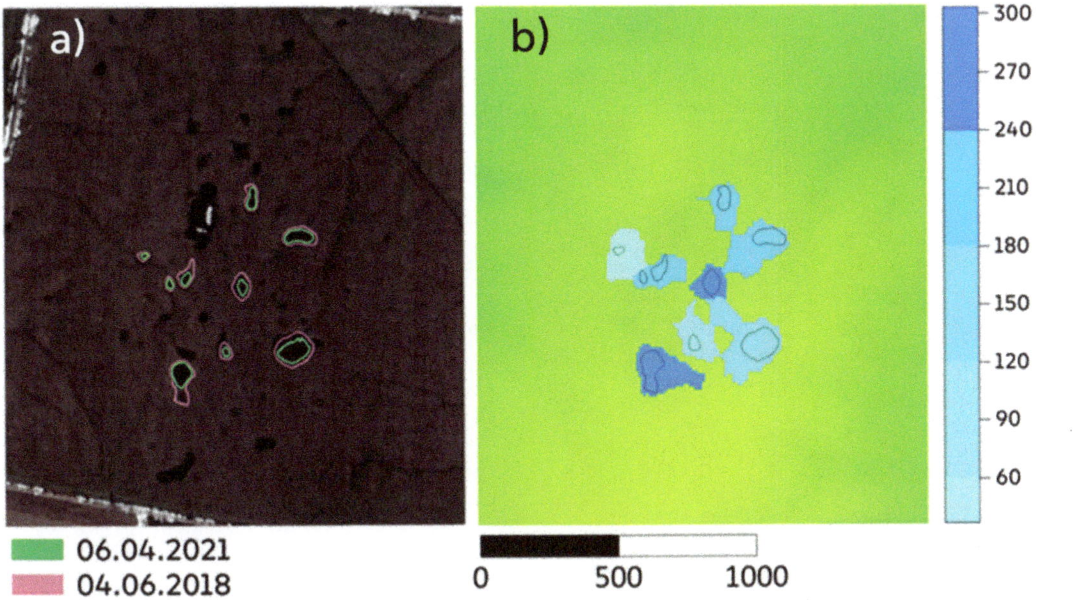

Figure 6. (**a**) Boundaries of the maximum potential filling of closed depressions (up to their overflow heights) shown in white in a Planet Labs image from 4 April 2018, with color shading and magenta contours, and the maximum pond boundaries in 2021 from 6 April 2021 (green contours). In 2018, the investigated closed depressions were overflowing. (**b**) Shades of blue indicating the layers of water (in mm) that must enter the depressions from their catchment areas to completely fill them.

Figure 7 illustrates the simulations of the shallow groundwater level for cases with the recharge rate either constant or spatially varied along the transect. The parameter N/k was restricted by the requirement for correspondence between the simulated WTD and range of WTD for soils of each type from expert knowledge (Table 2 in [22]) in distance intervals across the catena's whole toposequence. For example, if very poorly drained soils (under depression bed) are present in M unit intervals, those in which WTD > 3 m (too deep) were counted with 0 weight and the others with 1 weight. The same procedure was then applied for each of the intervals with the other soil types, then the sums were added for all groups and scaled to the total number of unit intervals in the catena toposequence to obtain the accuracy in percent. Simulation of WTD was successful for the generalized transect in terms of correspondence between the simulated WTD and ranges of WTD obtained from the indirect soil indicators (redoximorphic features) and expert knowledge both in the cases of constant and spatially variable recharge. However, the required accuracy threshold was set at 97%, and was met for the spatially variable recharge. A significantly lower threshold (80%) was satisfied for the constant recharge case. Therefore, the method to estimate depression-focused infiltration proposed here can make the shape of the water table profile more realistic.

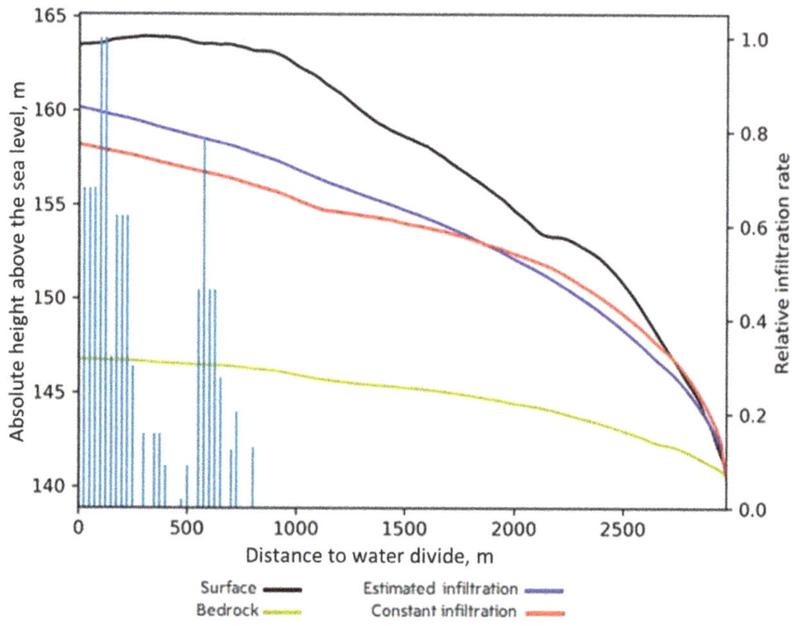

Figure 7. Cross-section of the catena with water table depth (WTD, blue and red lines) adjusted to correspond to soil group [22]. The yellow and black lines indicate the position of the bedrock and DEM profile, respectively. The blue and red lines respectively indicate WTD obtained with infiltration along the transects estimated from the depressions' positions and specific infiltration rates (as indicated by the blue bars and right y axis) and constant infiltration along the transect.

4. Discussion

The word steppe is usually associated with the Russian plains, but the northern part of this ecoregion has notable similarity to North American prairies. The lithological and geomorphological similarity of the Tambov region to the Saskatchewan and Alberta provinces in Canada enables direct comparison of the depression-focused infiltration into their soils through temporal ponds that are very similar in size distribution and shape. The recession rate of the ponds after snowmelt obtained in this study is similar to that derived from an artificial flooding experiment in the C24 depression, northwest of Calgary, Alberta, Canada, in 2004 [8]. As in the cited study [8] and another previous investigation [4], we found that evaporation accounts for a much smaller proportion of the pond water balance loss term than infiltration into the soil. The pre-event pore space available for filling with infiltration water was not directly measured in this study. However, data from a depression monitoring site in the study region in the years 2003–2005 show a spread of 30–400 mm of water deficit to saturation. An assumption underlying our two-stage infiltration conceptual model is that pores of the soils below the bottom of a pond are all filled to saturation during the first stage down to the shallow groundwater depth (approximately 2 m). Thus, the inflow is restricted by the bottleneck hydraulic conductivity below this point, which is an order of magnitude lower than in the upper soil layers [8] and also by the gradual rise of the water table when two fronts of water are jointing. The soil refill amount of 34 to 172 mm recorded in Table 2 fits well into this range. There is also similarity with the refill amount (148 mm) obtained in the cited Canadian study [8]. A strength of our study is that the infiltration rate was estimated for nine ponds, not just one pond such as the well-studied experimental pond C24. The variation (four-fold) in infiltration between those ponds (Table 2) could not be explained by the pond size or topographical settings. Thus, it is not sufficient to apply infiltration data from one pond to other ponds as this leads to large errors. The differences are likely due to diverse factors, inter alia physical properties of the soil associated with their lithological and textural characteristics, the thawing rate and ice content, abundance of root channels and other pathways for preferential flow. We conclude that there is no straightforward analytical way to characterize this spatial variability, but use of data obtained by the methods proposed here in conjunction with appropriate hydrological models and high-resolution satellite images is highly promising.

Here, we used the steady-state continuity equation in kinematic wave form parameterized using expert knowledge of the links between typical water table depth (WTD) and redoximorphic features of soils with different hydromorphy degrees [22]. In this simulation, we were able to account for variation in infiltration rates in the catena using real data on depressions' positions within the transect. However, calibration was still necessary because the infiltration and recharge are split in time by unsaturated zone processes. In future research, we plan to develop a model conceptually similar to the VSMB Depression-Upland System (VSMB-DUS) model [8] using data acquired in investigations of the surface water–groundwater interaction in individual depressions and their catchments. The planned model will be based on the watershed hydrological WASA-SED model [53], which already discretizes focal watersheds into hierarchical levels (sub basins, land units, terrestrial components, soil-vegetation components). Land units are representative catenas and terrestrial components can be easily supplemented with depressions and uplands providing surface flow to them by an already incorporated horizontal flow mechanism. For the terrestrial components prescribed as depressions, the temporally varying fluxes obtained by the method developed here will be used as upper boundary conditions. Collection of field data is planned to obtain saturated hydraulic conductivity values for the vertical levels besides the bottom soil layer. Groundwater depth measurements will provide calibration for the drainage rates from the deepest soil layer and validation for the dynamic version of the WASA-SED shallow groundwater flow sub-model. In this manner, groundwater recharge will fully account for the spatial variability of depression density, such as prevailing areas of numerous depressions at the water divide.

In this study, we derived the saturated hydraulic conductivity, not for a single point, but aggregated for the area of depressions. Most grid data used represent points, but landscape-level data are essential inputs for a hydrological model. A hypothesis under test is that soil hydraulic properties are related to landscape position and topography [54]. If so, elucidation of these relationships could greatly enhance pedotransfer functions for estimating saturated hydraulic conductivities at the level of land units and terrestrial components, not just points. Our study, based on remote sensing, provides an example of such derivation because the hydraulic conductivity is based on the depressions' water balance accounting for their positions in the landscape.

A limitation of this study lies in the assumption that all snowmelt runoff from the upland was routed to the depressions before the initial day of the study, and water volume within each depression exceeding its maximum storage capacity overflowed directly into surface runoff with no contribution to infiltration into the soil. However, it is widely acknowledged that depressions tend to form fill-spill networks, where overflow from one depression feeds an adjacent depression [5,6]. This process can be modeled [6,10,15–17], but studies of fill-spill processes have primarily focused on effects of depression storage on surface flow to streams rather than depression-focused groundwater recharge. Visual observations during the hydrological phase after the most active snowmelt showed no signs of connectivity between depressions at our study site, but that was typical for the active snowmelt phase itself of about a week duration. We justify the restriction of our approach with the hypothesis that non-stationary volumes of the depression ponds when snow is still present contribute little to total infiltration, partly due to the frozen state of the soil.

5. Conclusions

Estimation of infiltration through ponds is an important step toward the challenging goal to estimate depression-focused recharge of groundwater, and thus evaluate this important resource, in the forest-steppe zone of Russia. Using high-resolution Planet Labs images and widely evaluated tools for object-based image recognition, we have developed a relatively simple method to reconstruct a time series of infiltration into the soil under ponds and estimate landscape-scale saturated hydraulic conductivity. The simulation of the steady-state groundwater profile for the topographical transect fed with data on relative water supplies through depressions along the transect was more consistent with observations (based on soil redoximorphic indicators of water level) than the simulation fed with a uniform recharge function. Further development is needed to assimilate the data generated with consideration of the spatial variability of pond infiltration into a process-based model of groundwater recharge that accounts for interactions between depressions and their catchments.

Author Contributions: Conceptualization, A.Y.Y. and P.P.F.; methodology, P.P.F. and A.Y.Y.; software, P.P.F., A.D. and A.Y.Y.; validation, P.P.F. and A.D., formal analysis, A.Y.Y.; investigation, P.P.F. and A.Y.Y.; resources, P.P.F.; data curation, P.P.F.; writing—original draft preparation, P.P.F., A.D. and A.Y.Y.; writing—review and editing, A.Y.Y. and D.K.; visualization, P.P.F.; supervision, D.K.; project administration, D.K.; funding acquisition, D.K. All authors have read and agreed to the published version of the manuscript.

Funding: This research was funded by the RFBR, grant number 19-29-05277.

Institutional Review Board Statement: Not applicable.

Informed Consent Statement: Not applicable.

Data Availability Statement: The data presented in this study are openly available in Fil, Pavel; Yurova, Alla (2021), "Water bodies recognition for depression-focused recharge. Tambov region, Russia", Mendeley Data, V1, doi:10.17632/pn3gdzhdy4.1.

Acknowledgments: We are grateful to three anonymous reviewers for the useful comments that have helped to improve the manuscript.

Conflicts of Interest: The authors declare no conflict of interest.

References

1. de Vries, J.J.; Simmers, I. Groundwater recharge: An overview of process and challenges. *Hydrogeol. J.* **2002**, *10*, 5–17. [CrossRef]
2. Hayashi, M.; Van Der Kamp, G.; Schmidt, R. Focused infiltration of snowmelt water in partially frozen soil under small depressions. *J. Hydrol.* **2003**, *270*, 214–229. [CrossRef]
3. Bam, E.K.P.; Ireson, A.M.; van der Kamp, G.; Hendry, J.M. Ephemeral Ponds: Are They the Dominant Source of Depression-Focused Groundwater Recharge? *Water Resour. Res.* **2020**, *56*, e2019WR026640. [CrossRef]
4. Pavlovskii, I. *Groundwater Recharge in the Canadian Prairies: Mechanisms, Constraints, and Rates*; University of Calgary: Calgary, AB, Canada, 2019.
5. Shook, K.; Pomeroy, J.W.; Spence, C.; Boychuk, L. Storage dynamics simulations in prairie wetland hydrology models: Evaluation and parameterization. *Hydrol. Process.* **2013**, *27*, 1875–1889. [CrossRef]
6. Shook, K.; Papalexiou, S.; Pomeroy, J.W. Quantifying the effects of Prairie depressional storage complexes on drainage basin connectivity. *J. Hydrol.* **2021**, *593*, 125846. [CrossRef]
7. Ahmed, M.I.; Elshorbagy, A.; Pietroniro, A. A novel model for storage dynamics simulation and inundation mapping in the prairies. *Environ. Model. Softw.* **2020**, *133*, 104850. [CrossRef]
8. Noorduijn, S.L.; Hayashi, M.; Mohammed, G.A.; Mohammed, A.A. A coupled soil water balance model for simulating depression-focused groundwater recharge. *Vadose Zo. J.* **2018**, *17*, 170176. [CrossRef]
9. Muhammad, A.; Evenson, G.R.; Stadnyk, T.A.; Boluwade, A.; Jha, S.K.; Coulibaly, P. Impact of model structure on the accuracy of hydrological modeling of a Canadian Prairie watershed. *J. Hydrol. Reg. Stud.* **2019**, *21*, 40–56. [CrossRef]
10. Kiss, J. *Predictive Mapping of Wetland Types and Associated Soils through Digital Elevation Model Analyses in the Canadian Prairie Pothole Region*; University of Saskatchewan: Saskatoon, SK, Canada, 2018.
11. Shaw, D.A.; Pietroniro, A.; Martz, L.W. Topographic analysis for the prairie pothole region of Western Canada. *Hydrol. Process.* **2013**, *27*, 3105–3114. [CrossRef]
12. Wang, N.; Chu, X. A new algorithm for delineation of surface depressions and channels. *Water* **2020**, *12*, 7. [CrossRef]
13. Wu, Q.; Lane, C.R. Delineating wetland catchments and modeling hydrologic connectivity using lidar data and aerial imagery. *Hydrol. Earth Syst. Sci.* **2017**, *21*, 3579–3595. [CrossRef]

14. Wu, Q.; Lane, C.R.; Wang, L.; Vanderhoof, M.K.; Christensen, J.R.; Liu, H. Efficient Delineation of Nested Depression Hierarchy in Digital Elevation Models for Hydrological Analysis Using Level-Set Method. *J. Am. Water Resour. Assoc.* **2019**, *55*, 354–368. [CrossRef]
15. Brooks, J.R.; Mushet, D.M.; Vanderhoof, M.K.; Leibowitz, S.G.; Christensen, J.R.; Neff, B.P.; Rosenberry, D.O.; Rugh, W.D.; Alexander, L.C. Estimating Wetland Connectivity to Streams in the Prairie Pothole Region: An Isotopic and Remote Sensing Approach. *Water Resour. Res.* **2018**, *54*, 955–977. [CrossRef] [PubMed]
16. McDonough, O.T.; Lang, M.W.; Hosen, J.D.; Palmer, M.A. Surface Hydrologic Connectivity Between Delmarva Bay Wetlands and Nearby Streams Along a Gradient of Agricultural Alteration. *Wetlands* **2015**, *35*, 41–53. [CrossRef]
17. McLaughlin, D.L.; Kaplan, D.A.; Cohen, M.J. A significant nexus: Geographically isolated wetlands influence landscape hydrology. *Water Resour. Res.* **2014**, *50*, 7153–7166. [CrossRef]
18. Rover, J.; Wright, C.K.; Euliss, N.H.; Mushet, D.M.; Wylie, B.K. Classifying the hydrologic function of prairie potholes with remote sensing and GIS. *Wetlands* **2011**, *31*, 319–327. [CrossRef]
19. Vanderhoof, M.K.; Lane, C.R. The potential role of very high-resolution imagery to characterise lake, wetland and stream systems across the Prairie Pothole Region, United States. *Int. J. Remote Sens.* **2019**, *40*, 5768–5798. [CrossRef]
20. Tiner, R.W. Geographically isolated wetlands of the United States. *Wetlands* **2003**, *23*, 494–516. [CrossRef]
21. Rode, A.A. *The Water Regime of Soils and Its Regulation*; Academy of USSR Publishing: Moscow, Russia, 1963.
22. Yurova, A.Y.; Smirnova, M.A.; Lozbenev, N.I.; Fil, P.P.; Kozlov, D.N. Using soil hydromorphy degree for adjusting steady-state water table simulations along catenas in semiarid Russia. *Catena* **2021**, *199*, 105109. [CrossRef]
23. Jawak, S.D.; Kulkarni, K.; Luis, A.J. A Review on Extraction of Lakes from Remotely Sensed Optical Satellite Data with a Special Focus on Cryospheric Lakes. *Adv. Remote Sens.* **2015**, *4*, 196. [CrossRef]
24. Montgomery, J.; Mahoney, C.; Brisco, B.; Boychuk, L.; Cobbaert, D.; Hopkinson, C. Remote Sensing of Wetlands in the Prairie Pothole Region of North America. *Remote Sens.* **2021**, *13*, 3878. [CrossRef]
25. Yuan, C.; Gong, P.; Bai, Y. Performance Assessment of ICESat-2 Laser Altimeter Data for Water-Level Measurement over Lakes and Reservoirs in China. *Remote Sens.* **2020**, *12*, 770. [CrossRef]
26. Zhang, Y.; Pulliainen, J.T.; Koponen, S.S.; Hallikainen, M.T. Water quality retrievals from combined Landsat TM data and ERS-2 SAR data in the Gulf of Finland. *IEEE Trans. Geosci. Remote Sens.* **2003**, *41*, 622–629. [CrossRef]
27. Verpoorter, C.; Kutser, T.; Tranvik, L. Automated mapping of water bodies using Landsat multispectral data. *Limnol. Oceanogr. Methods* **2012**, *10*, 1037–1050. [CrossRef]
28. Huang, W.; DeVries, B.; Huang, C.; Lang, M.; Jones, J.; Creed, I.; Carroll, M. Automated Extraction of Surface Water Extent from Sentinel-1 Data. *Remote Sens.* **2018**, *10*, 797. [CrossRef]
29. Martin, R.; Brabyn, L.; Beard, C. Effects of class granularity and cofactors on the performance of unsupervised classification of wetlands using multi-spectral aerial photography. *J. Spat. Sci.* **2014**, *59*, 269–282. [CrossRef]
30. Talukdar, S.; Mankotia, S.; Shamimuzzaman, M.; Mahato, S. Wetland-Inundated Area Modeling and Monitoring Using Supervised and Machine Learning Classifiers. In *Advances in Remote Sensing for Natural Resource Monitoring*; Wiley: Hoboken, NJ, USA, 2021.
31. Tulbure, M.G.; Broich, M. Spatiotemporal dynamic of surface water bodies using Landsat time-series data from 1999 to 2011. *ISPRS J. Photogramm. Remote Sens.* **2013**, *79*, 44–52. [CrossRef]
32. Mahdianpari, M.; Rezaee, M.; Zhang, Y.; Salehi, B. Wetland Classification Using Deep Convolutional Neural Network. In Proceedings of the IGARSS 2018—2018 IEEE International Geoscience and Remote Sensing Symposium, Valencia, Spain, 22–27 July 2018.
33. Zhaohui, Z.; Prinet, V.S.M. Water body extraction from multi-source satellite images. In Proceedings of the IGARSS 2003–2003 IEEE International Geoscience and Remote Sensing Symposium, Toulouse, France, 21–25 July 2003.
34. RIHMI-WDC (Roshydromet) The Database of the Main Meteorological Parameter at the Russian Weather Stations. Available online: http://aisori-m.meteo.ru/ (accessed on 10 September 2021).
35. ArcGis Pro The Image Classification Wizard. Available online: https://pro.arcgis.com/en/pro-app/latest/help/analysis/image-analyst/the-image-classification-wizard.htm (accessed on 10 September 2021).
36. Li, M.; Zang, S.; Zhang, B.; Li, S.; Wu, C. A review of remote sensing image classification techniques: The role of Spatio-contextual information. *Eur. J. Remote Sens.* **2014**, *47*, 389–411. [CrossRef]
37. Waz, A.; Creed, I.F. Automated Techniques to Identify Lost and Restorable Wetlands in the Prairie Pothole Region. *Wetlands* **2017**, *37*, 1079–1091. [CrossRef]
38. Gadal, S.; Zakharov, M.; Kamicaityte, J.; Danilov, Y. Alas landscape modeling by remote sensing image analysis and geographic ontology: Study case of central yakutia (Russia). In Proceedings of the GISTAM 2020—6th International Conference on Geographical Information Systems Theory, Applications and Management, Prague, Czechia, 7–9 May 2020; pp. 112–118. [CrossRef]
39. Public Cadastral Map of Russia. Available online: https://pkk.rosreestr.ru (accessed on 10 September 2021).
40. Fukunaga, K.; Hostetler, L.D. The estimation of the gradient of a density function, with applications in pattern recognition. *IEEE Trans. Inf. Theory* **1975**, *21*, 32–40. [CrossRef]
41. Comaniciu, D.; Meer, P. Mean shift: A robust approach toward feature space analysis. *IEEE Trans. Pattern Anal. Mach. Intell.* **2002**, *24*, 603–619. [CrossRef]
42. Breiman, L. Random forest. *Mach. Learn.* **2001**, *45*, 5–32. [CrossRef]

43. Belgiu, M.; Drăgu, L. Random forest in remote sensing: A review of applications and future directions. *ISPRS J. Photogramm. Remote Sens.* **2016**, *114*, 24–31. [CrossRef]
44. Vapnik, V.N. *The Nature of Statistical Learning*; Springer: New York, NY, USA, 1995.
45. ArcGIS Pro Polygon Volume (3D Analyst). Available online: https://pro.arcgis.com/en/pro-app/latest/tool-reference/3d-analyst/polygon-volume.htm (accessed on 10 September 2021).
46. Armstrong, R.N.; Pomeroy, J.W.; Martz, L.W. Evaluation of three evaporation estimation methods in a Canadian prairie landscape. *Hydrol. Process. Int. J.* **2008**, *22*, 2801–2815. [CrossRef]
47. Penman, H.L. Natural evaporation from open water, hare soil and grass. *Proc. R. Soc. Lond. A. Math. Phys. Sci.* **1948**, *193*, 120–145. [CrossRef] [PubMed]
48. Penman, H.L. Evaporation: An introductory survey. *Netherlands J. Agric. Sci.* **1956**, *4*, 9–29. [CrossRef]
49. Valiantzas, J.D. Simplified versions for the Penman evaporation equation using routine weather data. *J. Hydrol.* **2006**, *331*, 690–702. [CrossRef]
50. Evaplib Python Library. Available online: http://python.hydrology-amsterdam.nl/moduledoc/_modules/evaplib.html (accessed on 10 September 2021).
51. Allen, R.G. FAO Irrigation and Drainage. *Irrig. Drain.* **1998**, *300*, 300.
52. Wade, T.G.; Wickham, J.D.; Nash, M.S.; Neale, A.C.; Riitters, K.H.; Jones, K.B. A Comparison of Vector and Raster GIS Methods for Calculating Landscape Metrics Used in Environmental Assessments. *Photogramm. Eng. Remote Sensing* **2003**, *69*, 1399–1405. [CrossRef]
53. Bronstert, A.; de Araújo, J.C.; Batalla, R.J.; Costa, A.C.; Delgado, J.M.; Francke, T.; Foerster, S.; Guentner, A.; López-Tarazón, J.A.; Mamede, G.L.; et al. Process-based modelling of erosion, sediment transport and reservoir siltation in mesoscale semi-arid catchments. *J. Soils Sediments* **2014**, *14*, 2001–2018. [CrossRef]
54. Van Looy, K.; Bouma, J.; Herbst, M.; Koestel, J.; Minasny, B.; Mishra, U.; Montzka, C.; Nemes, A.; Pachepsky, Y.A.; Padarian, J.; et al. Pedotransfer Functions in Earth System Science: Challenges and Perspectives. *Rev. Geophys.* **2017**, *55*, 1199–1256. [CrossRef]

sensors

Article

A Wireless Sensor Network Deployment for Soil Moisture Monitoring in Precision Agriculture

Jaime Lloret *, Sandra Sendra, Laura Garcia and Jose M. Jimenez

Instituto de Investigación Para la Gestión Integrada de Zonas Costeras (IGIC), Universitat Politècnica de València, Paraninf 1, 46730 Valencia, Spain; sansenco@upv.es (S.S.); laugarg2@teleco.upv.es (L.G.); jojiher@dcom.upv.es (J.M.J.)
* Correspondence: jlloret@dcom.upv.es; Tel.: +34-6-0954-9043

Abstract: The use of precision agriculture is becoming more and more necessary to provide food for the world's growing population, as well as to reduce environmental impact and enhance the usage of limited natural resources. One of the main drawbacks that hinder the use of precision agriculture is the cost of technological immersion in the sector. For farmers, it is necessary to provide low-cost and robust systems as well as reliability. Toward this end, this paper presents a wireless sensor network of low-cost sensor nodes for soil moisture that can help farmers optimize the irrigation processes in precision agriculture. Each wireless node is composed of four soil moisture sensors that are able to measure the moisture at different depths. Each sensor is composed of two coils wound onto a plastic pipe. The sensor operation is based on mutual induction between coils that allow monitoring the percentage of water content in the soil. Several prototypes with different features have been tested. The prototype that has offered better results has a winding ratio of 1:2 with 15 and 30 spires working at 93 kHz. We also have developed a specific communication protocol to improve the performance of the whole system. Finally, the wireless network was tested, in a real, cultivated plot of citrus trees, in terms of coverage and received signal strength indicator (RSSI) to check losses due to vegetation.

Keywords: electromagnetic induction; soil moisture; precision agriculture; low cost; water management; Internet of Things (IoT); wireless sensor network

Citation: Lloret, J.; Sendra, S.; Garcia, L.; Jimenez, J.M. A Wireless Sensor Network Deployment for Soil Moisture Monitoring in Precision Agriculture. *Sensors* **2021**, *21*, 7243. https://doi.org/10.3390/s21217243

Academic Editors: Vicenç Puig and Miquel À. Cugueró-Escofet

Received: 26 September 2021
Accepted: 27 October 2021
Published: 30 October 2021

Publisher's Note: MDPI stays neutral with regard to jurisdictional claims in published maps and institutional affiliations.

1. Introduction

Given the basic need to provide food to the world's population, it is necessary to introduce technology to the agriculture sector to reduce the environmental impact caused by the crops and to increase the conservation of natural resources, among others [1]. Efficient Irrigation is one of the essential factors to increase the development of sustainable agriculture, especially in arid and semi-arid regions where there are the greatest limitations. Irrigation methods can be classified into three generic categories; these are (1) gravity irrigation, (2) sprinkler irrigation, and (3) drip irrigation. The gravity irrigation system is the oldest method and the least efficient for the conservation of natural resources. However, in order to determine the specific irrigation needs of crops, sensing devices must be deployed to obtain data such as soil moisture.

Precision agriculture is a concept that appeared in the USA in the 1980s. It is a management strategy that allows making decisions to improve farming productivity and to achieve more sustainable activity. It is based on the management of crops by observing, measuring, and acting against the variability of the many factors that affect them. Using Internet of Things (IoT) solutions, the soil where the crops are planted can be monitored to make decisions and perform more effective irrigation. These solutions may include not only the electronic devices deployed in the fields but also the use of vehicles such as drones to support the network [2] and to manage the use of pesticides on the crops [3,4]. However, in crop monitoring tasks, especially in those where fruit trees are grown, it is important to be able to control soil moisture levels accurately. For the correct progress of a fruit tree, it is necessary to ensure that the roots have the right levels of moisture. High humidity levels can facilitate the proliferation of fungi in the roots and leaves, thus affecting production. However, an extremely low soil moisture level can provoke the soil to crack. causing broken roots and the tree to die. This fact negatively affects the growth of plants and consequently their production.

One of the main drawbacks that hinder the use of precision agriculture is the cost of the sensors and the utilized technology. For farmers that want to use technology on a massive scale, it is necessary to provide low-cost systems to make easier deployments.

The available commercial sensors for soil monitoring use different methods to assess the water content of the soil. The most relevant existing methods for obtaining moisture values from the soil [5] are the gravimetric method, tensiometric method, neutronic method, gamma-ray attenuation method, dielectric method, Wenner or resistive method, and light method infrared [6]. Generally, when low-cost sensors are used to measure soil moisture, conductivity-based sensors are based on the use of two electrodes [1]. These types of sensors have two fundamental disadvantages, lack of reliability, and durability. On the one hand, depending on the type of soil and its salt content, the conductivity measurement can vary even when the amount of water in the soil is maintained. On the other hand, the electrodes must be in contact with the ground, and consequently, they can suffer rapid deterioration. Inductive sensors are also employed to measure soil moisture. However, they do not integrate the system into a sensor node to be able to read the parameters.

The network design is an important aspect to consider as well. Usually, fields are located in remote areas. These areas may not have access to the internet infrastructure and the power grid. Therefore, PA systems should include a form of energy harvesting such as solar panels, and some characteristics of these networks should be considered when designing the deployment of sensing devices [7]. Wireless communications are a good solution because it eliminates the cost and hindrance of deploying cabled networks on extensive areas where machinery is utilized. However, the foliage of the crops affects the quality of the signal, resulting in reduced coverage between the devices. It is therefore necessary to determine the optimal deployment design for the area of interest according to the type of crop, and size of the field. Furthermore, the available protocols may not provide all the functionalities desired for a particular crop and the resources available for the area.

In this paper, we present a group-based wireless sensor network to efficiently irrigate cultivated lands. The network is composed of both actuators and sensor nodes that will collect data from the soil and will activate different irrigation systems as a function of the plot needs. Additionally, we design a new soil moisture sensor able to measure the amount of water content in the root ball of a tree. The design includes the sensor and the power circuit required to generate the bi-phase signals to power the coils. The paper presents the design of the operation algorithm and the message exchange for efficient use of water. Finally, the entire system is tested in a real environment to check the correct operation in terms of soil moisture measurements and network performance.

The rest of the paper is structured as follows. Section 2 presents some previous and related works where soil moisture systems are developed. Section 3 presents an overall description of our proposed sensor as well as the features of the different coils used to develop our soil moisture sensor and the experimental tests performed with the coils. This section also includes the power circuit in charge of generating the required signals as well as the integration of both the sensor and the power circuit with an ESP32 module. Section 4 explains the network operation algorithm and message exchange between nodes. In Section 5, the tests performed in a real environment are shown. Section 6 explains the conclusion and future works.

2. Related Work

In this section, we summarize some previous works related to our proposal. The gap in current solutions for soil moisture monitoring is also identified.

Authors such as Ojha et al. [8] present a study where they analyze the wireless sensor network (WSN) implementations for various agricultural applications. We will look at surveys such as the one presented by Garcia et al. [1], aimed at summarizing the current state of the art regarding smart irrigation systems and schemes for Internet of Things (IoT) irrigation monitoring. This survey includes the review of more than 100 scientific works. Other authors, such as Susha Lekshmi et al. [9], present a review of techniques employed for soil moisture measurement. The authors highlight the limitations of the techniques and the influence of soil parameters. Tumanski [10] describes the use of a coil to develop sensors. The work compares, summarizes, and analyzes coil design methods and frequency properties of the coil as well as the use of coil sensor applications such as magnetic antennas. Jawad et al. [11] describe applications of WSNs in agricultural research, and classify and compare wireless communication protocols, the taxonomy of energy efficiency, and energy harvesting techniques for WSNs used in agricultural monitoring systems. They also explore the challenges and limitations of WSNs in agriculture, highlighting energy reduction and agricultural management techniques for long-term monitoring. Hamami et al. [12] present a review of the application of WSNs in the field of irrigation. Mekonnen et al. [13] present a review of the application of different machine learning algorithms in the analysis of sensor data observed using WSNs in agriculture. In addition, they analyze a case study on a smart farm prototype, based on IoT data, as an integrated food, energy, and water (FEW) system. Nabi et al. [14] present a comparative study of different studies to provide a deeper insight into these implemented systems. They also present a study of apple disease prognostic systems, highlighting their key characteristics and drawbacks. The result of their study can be used to select appropriate technologies to build a WSN-based system, optimized for precision apple cultivation, which will help farmers avoid the ravages caused by disease outbreaks.

Kabashi et al. [15] present a framework to design WSNs for agricultural monitoring in developing regions, taking into account the particularities of said environments. They propose new solutions and research ideas for sensor network design, including zone-based joint topology control and power scheduling mechanism, multi-sink architecture with complementary routing associated with backlink/storage, and a task scheduling approach with parameter, energy, and environment recognition. Authors such as Kassim et al. [16] present WSNs as the best way to solve agricultural problems related to optimization of agricultural resources, decision support, and land monitoring in order to perform those functions in real time. They explain in detail the hardware architecture, network architecture, and software process control of the precision irrigation system. García et al. [7] study different WSN deployment configurations for a soil monitoring PA system, to identify the effects of the rural environment on the signal and to identify the key aspects to consider when designing a PA wireless network. The PA system is described, providing the architecture, the node design, and the algorithm that determines the irrigation requirements. The results of their testbed show high variability in densely vegetated areas. These results are analyzed to determine the theoretical maximum coverage for acceptable signal quality for each of the studied configurations. Furthermore, there are aspects of the rural environment and the deployment that affect the signal. Zervopoulos et al. [17] present the design and deployment of a WSN capable of facilitating the sensing aspects of smart and precision agriculture applications. They describe a simple synchronization scheme, which was installed in an olive grove, to provide time-correlated measurements using the receiving node's clock as a reference. The obtained results indicate the general effectiveness of the system, although they appreciate a difference in the time correlation of the acquired measurements. Bayrakdar [18] investigated an intelligent insect pest detection technique with underground wireless sensor nodes for precision agriculture using a mathematical simulation model. To evaluate performance, he examined the received signal strength and path loss parameters. He observed the need for transmission of signals with different transmission powers for depth-based communication in wireless underground sensor networks.

Other authors study the application of WSNs to monitor specific crops. Khedo et al. [19] describe the implementation of the PotatoSense application, for precision agriculture with WSNs, to monitor a potato plantation field in Mauritius. They employ different energy efficiency algorithms, to ensure that the life of the system is prolonged. Additionally, they have developed a monitoring application to process the data obtained from the simulated WSN. Rasooli et al. [20] propose using WSNs and IoT to help increase wheat and saffron production in Afghanistan in the future. Using both techniques, they predict the control of the condition and growth of the crop as well as the ability to check soil, temperature, humidity, and other environmental parameters.

Some authors propose the observation of parameters utilizing WSNs in greenhouses. Chaudhary et al. [21] propose and discuss the use of the programmable system on chip technology (PSoC) as part of the WSN to monitor and control various greenhouse parameters. Srbinovska et al. [22] propose a WSN architecture for vegetable greenhouses, in order to achieve scientific cultivation and reduce management costs from the aspect of environmental monitoring. They have designed a practical and low-cost greenhouse monitoring system based on wireless sensor network technology to monitor key environmental parameters such as temperature, humidity, and lighting.

There are authors also studying energy savings in WSNs used in monitoring agriculture. Hamouda et al. [23] study the problem of selecting the sampling interval, for precision agriculture using WSNs, due to the energy limitation that appears when deploying sensors in WSNs. They propose a Variable Sampling Interval Precision Agriculture (VSI-PA) system to measure and monitor agricultural parameters for appropriate agricultural activities, such as water irrigation. Compared to other fixed sampling interval schemes, the proposed VSI-PA system provides a significant improvement in energy consumption, while maintaining a small variation in soil moisture, regardless of soil temperature values. Qureshi et al. [24] propose Gateway Clustering Energy-Efficient Centroid (GCEEC)-based routing protocol, where a cluster head is selected from the centroid position and gateway nodes are selected from each cluster. The results obtained, after evaluating the proposed protocol in comparison to last-generation protocols, indicated a better performance of the proposed protocol, and provided a more feasible WSN-based monitoring for temperature, humidity, and lighting in the agricultural sector.

Table 1 summarizes different previous studies, carried out by other authors, regarding the use of WSNs in soil monitoring for agriculture.

Table 1. Previous studies regarding the use of WSNs in soil monitoring for agriculture.

Types	References
Surveys of WSN Implementations for Agriculture.	[1,8–14]
Frameworks, Studies, Designs and Deployments for WSN.	[7,15–18]
WSNs for the Monitoring of Specific Crops.	[19,20]
WSNs in Greenhouses.	[21,22]
Energy Savings Studies.	[23,24]

Regarding the available sensors for soil monitoring, there are works, such as [25], that study farmed podzolic soils since these types of soils are under-represented in the relevant literature. In the study, the authors established the relationship between apparent electrical conductivity (ECa) and soil moisture content (SMC). The authors also evaluated the estimated SMC with ECa measurements obtained with two electromagnetic (EMI) induction sensors. The authors concluded that ECa measurements obtained through multi-coil or multi-frequency sensors had the potential to be successfully used for field-scale SMC mapping. Others, such as [26], designed and manufactured an integrated passive wireless sensor to monitor the moisture in the sand. The sensor was made of a printed spiral inductor embedded within the sand and it contained an inductive-capacitive (LC) resonant circuit. The authors measured the level of internal moisture by monitoring the resonance frequency using a sensing coil. Kizito et al. [27] presented a study where ECH20 sensors were used to measure soil moisture content, bulk electrical conductivity, and temperature for a range of soils, across a range of measurement frequencies between 5 and 150 MHz. The authors affirmed that the measurements carried out on soil were accurate enough to work at 70 MHz. Finally, Nor et al. [28] discussed the development of a low-cost sensor array based on planar electromagnetic sensors to determine the contamination levels of nitrate and sulfate in water sources. The authors proposed three types of sensors: parallel, star, and delta. According to their experiments, the star sensor array was the one with the highest sensitivity.

After analyzing the exhibited works and many others not included in this paper, we can conclude that our work improves the existing systems. In either very few or no cases in the other works reviewed do the authors present complete or easily integrable systems in commercial nodes, such as Arduino or similar, and many of them use working frequencies that are too high (on the MHz scale). This fact makes it difficult to develop a simple and inexpensive signal generator circuit. Our proposal aims to take a step beyond the current state of the art, proposing a complete system, consisting of a sensor based on coils whose working frequency is around 93 kHz, and a power circuit that can be easily integrated into commercial modules for the development of a more complex wireless sensor network to monitor a large-scale crop.

3. Network Nodes Description

This section describes the proposed system and the different parts that comprise our proposed system. Additionally, it presents the features of the different coils used to develop our soil moisture sensor as well as the experimental tests performed to determine the best prototype.

3.1. Overall System Description

When trying to develop complete monitoring systems for precision agriculture, it is important to take into account different aspects. On the one hand, agriculture is an essential activity for the survival and development of society; this fact is evidenced by the amount of global- and regional-scale agricultural monitoring systems [29] to assess the crop growing conditions, crop status, and agro-climatic conditions that may have an impact on global production of any type of crop. Some examples are Group on Earth Observations Global Agricultural Monitoring Initiative (GEOGLAM) [30] or CropWatch [31], among others.

On the other hand, it is necessary to know the kind of crop wanted to be developed in order to design adapted methods for monitoring activity. Considering the crop to monitor and the location of the plot, the network should use a specific wireless communication technology. Currently, it is possible to use cellular technologies by paying for subscriptions to a service or by using low-power technologies such as ZigBee, LoRa, LoRaWAN, Bluetooth BLE, or Sixfox, among others; most of these services do not require payment for using their communication network infrastructure [32]. However, the wireless technology par excellence for developing wireless sensors networks continues to be Wi-Fi. Although its energy consumption is still high, it allows transmitting any type of content without the bandwidth limitations that other technologies present. In addition, it is a widely studied standard so it is easy to develop new optimized protocols. Therefore, by making a good design of a power system based on renewable energies, it is possible to use Wi-Fi to develop a Wi-Fi-based agriculture monitoring network with very interesting properties.

In the end, the completion of the design of the system led to precisely defining the type of parameter to be monitored since this fact will indicate the type of sampling, and analysis we should do. After that, the data interpretation and the scoring curves will help us to define the correct operation of our actuator network system. Lastly, the correct processing of collected data will help us to know the soil health and its characteristics for determining if these are optimal for our crop.

Therefore, considering these previous issues, we propose the development of a group-based wireless sensor network for soil moisture monitoring in precision agriculture. The network is composed of a set of nodes with different roles and functions. Some nodes are able to collect data from the environment, particularly data from soil moisture and other parameters required to ensure the correct progress of a tree (See Figure 1). The rest of the nodes have actuators to control the activity of ditch gates and drip irrigation elements. So, we will have 3 different sets of nodes that will communicate between them. Additionally, sensor nodes will provide data to the actuator nodes performing the required computation and decision making in the edge. Edge computing is recommended in scenarios where nodes present in the network are able to analyze the data and take decisions. Edge computing enables data produced by Internet of Things (IoT) devices to be processed closer to where it was created rather than being sent over long journeys to reach data centers and computing clouds. One of the fundamental advantages of this type of computing is that it allows analyzing important data in near real time [33]. In citrus groves, it is common to distribute them by forming rows of trees separated at a distance of approximately 6 m, being able to opt for a denser plantation, with a minimum separation of 4.5 m. The minimum depth that a citrus tree usually reaches is 45 cm. Considering these facts and taking into account that a field can have different extensions, different topologies of sensor nodes can be created. An important aspect is to ensure complete coverage between nodes to guarantee stable communication. A distributed ad-hoc network is optimal for this kind of scenario.

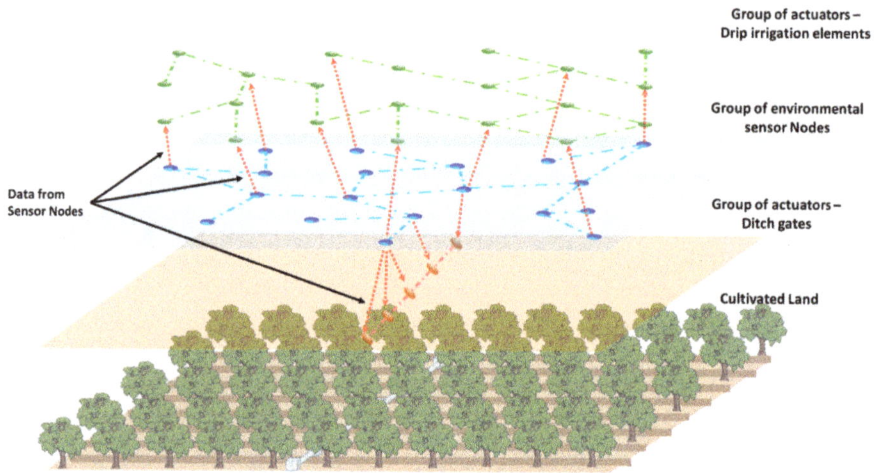

Figure 1. Proposed group-based network.

One of the main characteristic aspects of this proposal is its hierarchical structure by layers where each layer has a series of nodes that, if necessary, could change their role. That is, all sensor nodes and actuator nodes are wireless devices with the ability to act as a packet relay. In a hypothetical situation where a node falls, communications can be rerouted by other nodes of the same layer. If there is a fall of several nodes and one of them is isolated but active, it could use nodes of the upper or lower layer as an alternative way to carry out communications. However, these nodes would only forward the packet to nodes of the isolated node's layer.

To deal with the failure of a sensor or actuator node, it is convenient to establish an alarm system, based on keep-alive messages. It is a task periodically scheduled, once per day. It is possible to work with a large periodicity because the irrigation tasks of a field are not considered a critical task. If there is a node or several nodes not responding to these requests, the system will consider a node is down.

Additionally, developing a low-cost system was required to measure the moisture in the soil depths. This system consists of four coil-based sensor elements equally distributed along 60 cm. The coils are connected to a processor module in charge of collecting the data and wirelessly share them with the rest of the nodes of its group. Finally, and considering the values of moistures collected by the sensor nodes, the actuator nodes will enable/disable the ditch gates or the drip irrigation.

When talking about moisture or soil humidity, we refer to the amount of water the soil contains. A gravimetric analysis method gives the relative comparison between the mass of dry soil and the mass of watered soil (which will always be higher). The moisture given in percentage is the result of dividing the difference between these two values by the mass of dry soil. If there is no difference, moisture will be 0%. In the opposite case, when the watered soil mass doubles the dry one, the moisture level will be 100%.

The development of our coil-based soil moisture sensor is based on the principle of electromagnetic induction of the coils and how it varies as a function of the type of core the coil has inside [34–36].

The soil moisture sensor is composed of two solenoid coils wound on the same PVC pipe support. Coil 1 receives the sinusoidal signal generated by the power circuit based on the integrated ICM7555. Coil 1 induces a current on Coil 2 which is largely affected by the content of the coil core since the magnetic field is affected by the type of soil and water content inside it. Finally, this current is measured, collected, and stored with an electronic module. In our case, a module ESP32 DevKIT [37] with an integrated Wi-Fi interface has been chosen. Figure 2 shows the diagram of the proposed soil moisture sensor.

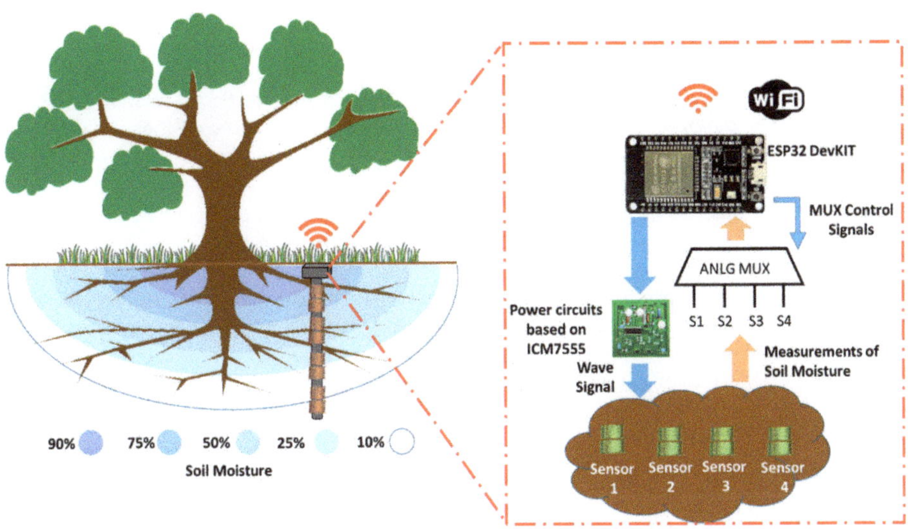

Figure 2. Diagram of proposed soil moisture sensor base on coils.

Since this kind of module usually presents one or two analog inputs to collect data, we also propose the use of an analog multiplexor of four inputs which can be controlled by using two digital outputs. With this, our system will be able to take measurements from the four soil moisture sensors.

3.2. Soil Moisture Sensor Based on Coils

As we mentioned before, it is possible to develop soil moisture sensors based on several principles and chemical processes. However, we want to use a method based on physical principles such as the variation of electromagnetic flow as a function of the nature of the coil core.

In a coil distribution such as the one shown in Figure 3, coil 1 generates a magnetic field that affects coil 2. This effect is known as mutual inductance and refers to the electromotive force (EMF) in a coil due to the change of current in another coil attached. The induced EMF in a coil is described by Faraday's law and its direction is always opposite to the change in the magnetic field produced in it by the coupled coil (Lenz's law). The EMF in coil 1 (left) is due to its own inductance L.

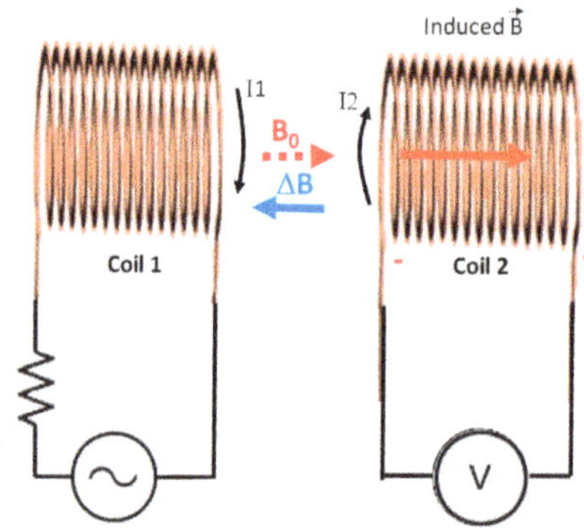

Figure 3. Principle of operation for our developed sensor.

The induced EMF in coil 2, generated by the changes of current I1, can be expressed as (see Equation (1)):

$$emf_2 = -N_2 A \frac{\Delta B}{\Delta t} = -M \frac{\Delta I_1}{\Delta t} \quad (1)$$

where N_2 is the number of spires of coil 2, M the coefficient of mutual self-induction, A is the cross-sectional area of the coil, $\frac{\Delta B}{\Delta t}$ the variation of the magnetic field as a function of the time, and $\frac{\Delta I_1}{\Delta t}$ the variation of current in coil 1 as a function of time. Mutual inductance (M) can be defined as the ratio between the electromagnetic force (EMF) generated in coil 2, and the changes in current in coil 1 that causes that EMF. Likewise, M is highly affected by the characteristics of the medium that surrounds the coils, usually expressed by its magnetic permeability.

Since it is difficult to measure the value of the magnetic permeability of the earth core as a function of the moisture level, two theoretical approximations of the air core are introduced [38]. Based on Equation (2) (which presents the coil inductance), we can state Equation (3) where l is the length of the coil and r is the radius to the center of the coil of the innermost layer of the conductor while R is the radio for the outermost layer.

$$L = \frac{\Phi N}{I} = \frac{\mu N^2 A}{l} = \frac{\mu_0 \mu_r N^2 \pi r^2}{l} \ (\text{H}) \quad (2)$$

$$L_{layer} = \frac{N^2 r^2}{2.54(9r + 10l)} \ (\mu\text{H}) \quad (3)$$

where L is the inductance of our coil (in H), Φ is the magnetic flow (in Wb), N represents the number of turns (dimensionless), l is the length of the coil (in m), r expresses the radius of the inner coil's layer (in m), R is the radius of the outer coil's layer (in m), A is the area of the coil's surface (in m^2), μ_0 is the magnetic permeability (free space) (H/m) and finally, μ_r is the relative magnetic permeability (medium) (dimensionless).

This approximation allows estimating the components of the circuit for an air core, which would be similar to those obtained with a large amount of pure water; so this value will vary depending on the type of soil, its composition, and the level of soil moisture presented by the soil that contains coil. The resonance peak of our coils can be calculated by Equation (4).

$$f_r = \frac{1}{2\pi} \sqrt{\frac{1}{LC_d} - \frac{R_S^2}{L^2}} \approx \frac{1}{2\pi\sqrt{LC_d}} \ (\text{Hz}) \quad (4)$$

where f_r is resonance frequency (in Hz), C_d is coil's parasite capacity (in F), L is the coil's inductance (in H) and, R_s is the coil's resistance (in Ω).

We should take into account that the primary coil and secondary coil will have different resonance frequencies because the secondary coil has a different number of coils. However, our sensor only intends to detect changes in the induced current due to the presence of a changeable medium and, finally, we want to relate this value of current with the amount of water content in the soil.

Equations from (1) to (3) are theoretical approaches to explain how important it is to know the relationship between the physical and electrical characteristics of the coil. Equations (1)–(3) explain how the coil inductance, and hence mutual inductance, depend on its geometry (length, radius, and the number of turns) for single-layer coils. Equation (4) helps us to design a resonant circuit to obtain the maximum power transfer. It is highly important to consider the appearance of a possible parasite capacity due to coil geometry and working frequency.

We previously performed several tests with different combinations of coils, varying the number of spires, the ratio of spires between coils, and the diameter [39]. In these previous works, we performed many experiments with combinations of spires and the best results were determined for a ratio of 1:2 with a medium value of spires and larger diameter. For a fixed diameter, if we reduced the number of spires, the working frequency increased for a fixed number of spires; if we increased the coil diameter, the working frequency decreased. Additionally, developing a simple and cheap electronic system to generate the signals was required. For such a system, it is highly recommended to have a sensor that requires the lowest working frequency. Therefore, we chose to set some parameters such as the type of copper and number of spires, we only varied the diameter of coils.

In developing our coils, 0.6 mm enameled copper wire was used. The process entails winding copper wire along a cylinder, forming two solenoids. The distance between the primary coil and the secondary coil is five mm. Figure 4 shows the developed coils with a single layer of spires. Table 2 shows the physical features of each prototype.

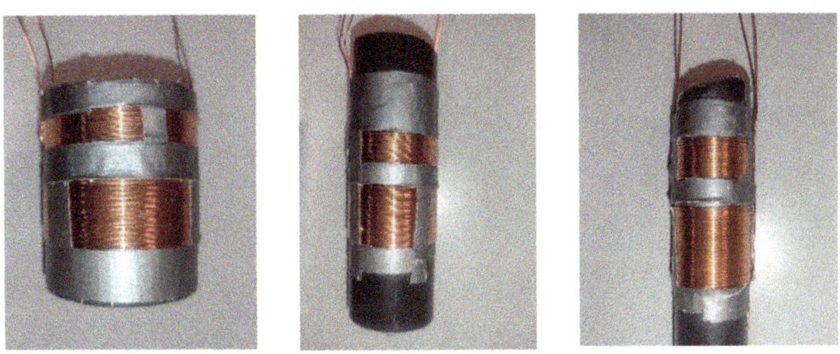

Figure 4. Coils used in our developed sensor: (**a**) P1, coil of 50 mm;(**b**) P2, coil of 32 mm; (**c**) P3, coil of 20 mm.

Table 2. Prototypes to measure soil moisture.

Prototype	Caliber	N° Layers	N° Spires 1st Coil	N° Spires 2nd Coil	N:n	Diameter
P1 (Figure 4a)	0.6 mm	1	15	30	1:2	50 mm
P2 (Figure 4b)	0.6 mm	1	15	30	1:2	32 mm
P3 (Figure 4c)	0.6 mm	1	15	30	1:2	20 mm

The procedure to perform these tests with the coils consists of introducing each model into a container filled with dry and compacted soil to observe the behavior of the output voltage as a function of the amount of water. Therefore, for the same moisture level, a frequency sweep will be carried out to find the frequency that shows a peak in the induced voltage. This value will be the sought-out resonance frequency. After that, the linearity of each model will be analyzed. For each test, 4000 g of soil is used, with increments of 250 mL of water, for each moisture level up to 1000 mL of water. When water is added to the soil, the sample is reposed for an hour to obtain a homogeneous sample inside the coil. Specifically, five levels of content of water in soil will be measured: 0%, 6.25%, 15.5%, 18.75%, and 25% (see Table 3). For this type of soil, 25% of the content of water in soil implies a land completely flooded. Measurements have been taken at 25 °C.

Table 3. Samples used during the tests.

Sample	Mass of Dry Soil (g)	Mass of Wet Soil (g)	Mass of Water (g)	% Volumetric Water Content
1	4000	4000	0	0
2	4000	4250	250	6.25
3	4000	4500	500	12.5
4	4000	4750	750	18.75
5	4000	5000	1000	25

In order to determine in which type of soil our sensor can be used, we endeavored to determine which one presents the biggest linearity. The idea of this concept is that an increase in the percentage of moisture is equivalent to an increase in output voltage without instabilities.

3.3. Experimental Results with the Developed Coils

This subsection presents the test performed to determine the most suitable prototype selected to develop the system, followed by the testing of the selected coil with different types of soils and different levels of moisture.

Considering the types of soil, we can conclude that soils are usually made up of different proportions of sand, silt, and clay. Each of them has morphological characteristics:

- Sandy soils, coarse texture (sand and clay sand).
- Silty soils, moderately coarse texture (sandy clay and fine sandy clay), medium texture (very fine sandy clay, silt, silt loam, and sediments), and moderately fine texture (clay silt, clay sand loam, and sand loam soil silty).
- Clay soils, fine texture (sandy clay, silty clay, and clay).

In order to perform our tests, we have selected three different types of soils, i.e., sand from the beach, soil from cultivated land, and commercial universal substrate.

The sand on the beaches is formed by sediments from rocks and other marine debris such as shells, corals, animals, algae, and even sand that travels through the rivers until flowing into the sea. Due to the erosion of water and wind, due to rain and waves, or temperature differences, the grain size of the sand tends to be reduced.

The soil from cultivated land is usually made up of an organic fraction, organic matter more or less degraded into humus and humic and fulvic acids. These elements provide the fertile part of the earth. The rest of the soil is considered as physical support. Some farmlands have a high degree of clay which also intervenes in ion exchange and water retention, facilitating the release of fertile elements according to the needs of the plants.

The raw materials used in the manufacture of a commercial universal substrate are usually blonde peat from sphagnum moss, coconut fiber, compost, perlite, organic fertilizer, mineral fertilizer, algae extract, etc. In addition, this type of soil usually contains a high level of aeration.

Performing the measurements entails three identically constructed sensors simultaneously placed in three samples of each soil. The results shown in our graphs are the average value of the three measurements collected, which in all cases were identical.

In order to perform the test, the primary coil is powered by using a wave of 7 Vpp with positive and negative values. For example, it is possible to use sine or square waves, such as the one shown in Figure 5. In Figure 5, we can see, in blue, the signal used to power the primary coil while the result of the induced current is shown in yellow.

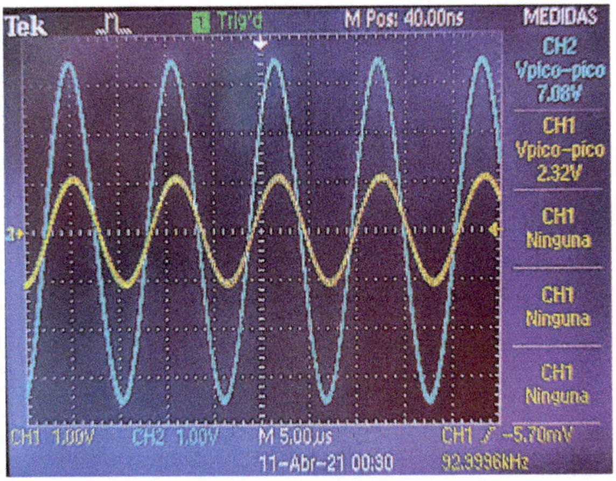

Figure 5. Example of generated and obtained signals.

Figure 6 shows the preliminary results obtained with the 3 coils. Figure 6 shows the value of the resonance frequency and the maximum voltage value.

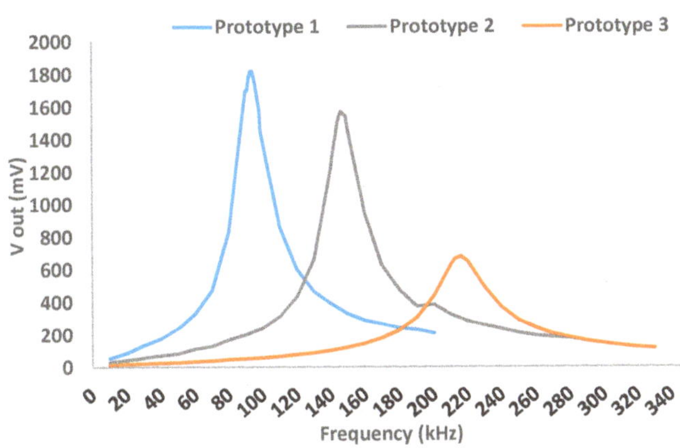

Figure 6. Resonance frequency obtained for each prototype.

Table 4 shows the resonance frequency values (in kHz) of the developed prototypes and the maximum voltage value (in mV) obtained in the induced coil.

Table 4. Prototypes to measure soil moisture.

Prototype	Working Frequency (kHz)	Maximum Voltage (mV)
1	93	1820
2	146	1570
3	216	680

After analyzing the results obtained in Figure 7, we can conclude that the prototype that gives the best results is Prototype 1, with a working frequency of 93 kHz and a maximum output voltage of 1.82 V.

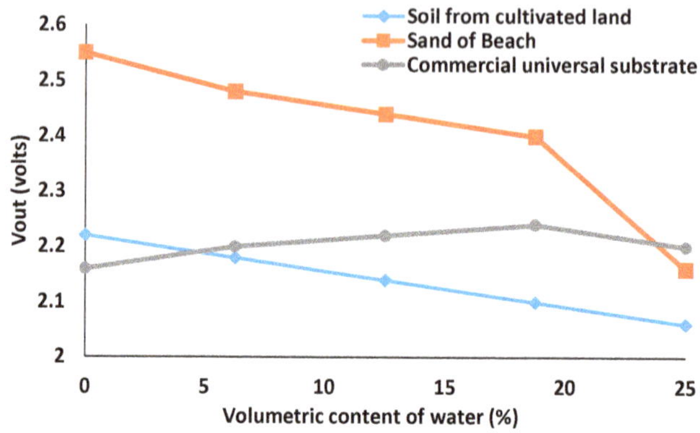

Figure 7. Behavior for prototype 1 in several types of soils.

Once the most suitable sensor for further development has been determined, this sensor is tested on different types of soil to determine its versatility. In all cases, we will look for the maximum linearity in the sensor response.

As we can see in the previous figure, the selected model has a linear behavior for all three cases up to a humidity degree of 18.75%, i.e., a total volume of water of 750 mL for 4000 g of sand.

Another important aspect to highlight is that the behavior of the sensor for a universal substrate is inverse to the behavior shown in the case of beach sand or cultivated soil. This aspect should be considered when the results are processed in a real environment.

3.4. *Power Circuit Design*

Locating the resonance frequencies for the selected prototype of coil systems requires a power supply and excitation circuit able to generate an alternating signal. To do this, a 555 series oscillator integrated circuit [40] will be used. A series of components will be used to obtain the desired output signal with a resonance frequency of 93 kHz. Our circuit is based on the ICM7555 [41]. According to the manufacturers' specifications, this integrated circuit can generate signals up to 3 MHz. Figure 8 shows the schema of our entire circuit. This kind of integrated circuit has been conceived to be customized regarding the duty cycle of signal and the required frequency. In this case, we can use R2 and C1 to change the working frequency while C3, C4, and R1 are used to control the ripple signal and its form. Modifying C4 and R1, it is possible to obtain both a sine wave and square signal, such as the one shown in Figure 9.

To directly read a value of voltage proportional to the amount of water content in the soil, we include a Graetz bridge or double wave rectifier bridge followed by an RC filter that has been connected to the terminals of the secondary coil.

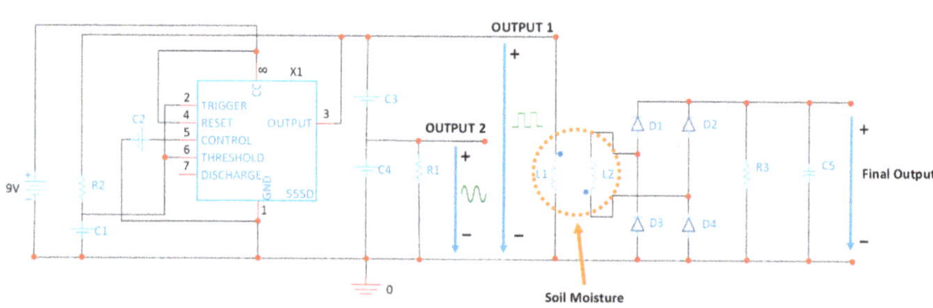

Figure 8. Enhanced power circuit schematic.

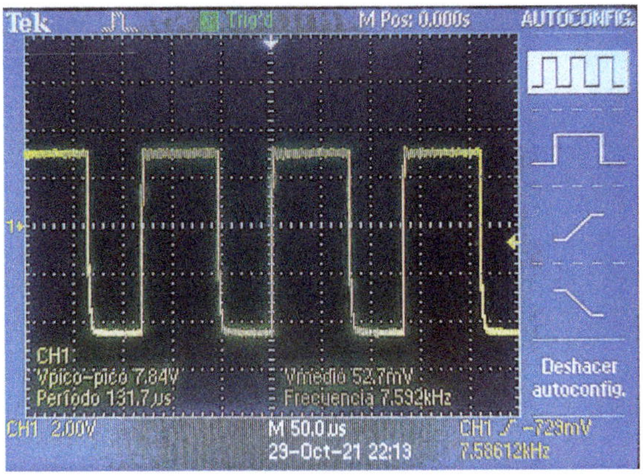

Figure 9. Example of signal obtained in Output 1.

3.5. Comparision and Discussion with Existing Published Systems

We have compared our sensor model with existing and commercial soil moisture sensors. Table 5 shows this analysis. It is important to consider that this table only contains the price of the sensor, with exception of references [42–45] which include an electronic module. In the rest of the cases, a microprocessor module must be included similar to the one used in this paper that can cost approximately $10–$15.

Table 5. Comparison of soil moisture sensors with our proposal.

Ref.	Model	Sensitivity	Power	Size	Cost
[42]	RK520-02 Soil Moisture Sensor, Temperature Probe & EC Sensor	0–100%	5 VDC, 12–24 VDC	136 × 45 mm	$58.00–$72.00
[43]	S-Soil MT-02A	±3% (0–53%) ±5% (53–100%)	3.6–30 VDC	149 × 45 mm	$79.00
[44]	S-Temp&VWC&EC-02A	±2% (0–50%) ±3% (50–100%)	3.6–30 VDC	149 × 45 mm	$99.00
[45]	SenseCAP Wireless Soil Moisture & Temperature Sensor	±2% (0–50%); ±3% (50–100%)	3.6 V	149 × 45 mm	$219.00
[46]	Sensor YL-69	0–95%	3.3–5 VCD	60 × 30 mm	$2.65
[47]	Keyes Brick Soil Moisture Sensor Module	-	3.3–5V VDC	63 × 22 × 8 mm	$1.34
[48]	KeeYees	-	3.3–5 VDC	3.858 × 0.906 mm	$7.99
[49]	Seeed Studio Grove—Capacitive Moisture Sensor—101020614	-	3.3–5 V	-	$5.95
[50]	Grove—Soil Moisture Sensor	0–95%	3.3–5 V	60 × 20 × 6.35 mm	$2.99
[51]	Seeed Studio Moisture Sensor—101020008	-	3.3 V CC y 5 VDC	-	$3.99
[52]	SEN0308	0–57%	3.3–5.5 VDC	175 × 30 mm	$15.51
[52]	MSE020SMS	-	3.3V–12 VDC	-	$5.80
-	Our proposal	-	9 V	50 m × 30 m	$2.15

It is evident that our proposed sensor, based on 2 coils, is one of the models that presents the lowest prices. The price includes the pipe and wire, because it can be added to any electronic platform to gather the data.

When this type of system is designed and developed, it is extremely important to consider practical implementation problems and challenges.

One of the main problems in the outdoors is how to protect the electronics from adverse conditions. A waterproof protection is highly recommended to protect the different devices since the places where the devices are deployed can be highly changeable. Additionally, the time during which the sensor nodes should work and the exposure to environmental temperature and humidity can cause some variations in the measurements. This issue should be controlled since a wrong reading would cause anomalous values and consequently wrong behavior. To shore up this problem, it is possible to use artificial intelligence and redundancy mechanisms.

Furthermore, there is another important issue regarding manufacturing techniques of certain sensors and probes. In several cases, they are manufactured with copper. An improvement in the system implementation could be the replacement of these sensors for ones protected with the process of gold plating which help to combat the corrosion of probes.

Coverage estimations do not usually match practical experimentation because the emulation of environmental conditions is difficult. For this reason, we highly recommend performing practical experiments and test benching, as presented in this paper.

4. Network Protocol Design and System Procedure

This section presents the network protocol used in our topology. In addition, it also presents the algorithm designed to collect data from sensors and control the different actuators as well as the messages exchanged between devices and the algorithm designed for the system procedure.

As we presented before, our network is composed of three different types of nodes, which can be classified as sensor nodes and actuator nodes. Additionally, we consider an additional node that is placed in the engine to provide water to the plot. This node will be in charge of starting the process of monitoring the entire network. Sensor nodes collect data from the soil and provide the required warning alarms to the actuators for enabling or disabling the irrigation systems. Figure 10 shows the diagram of our entire network deployed in the plot.

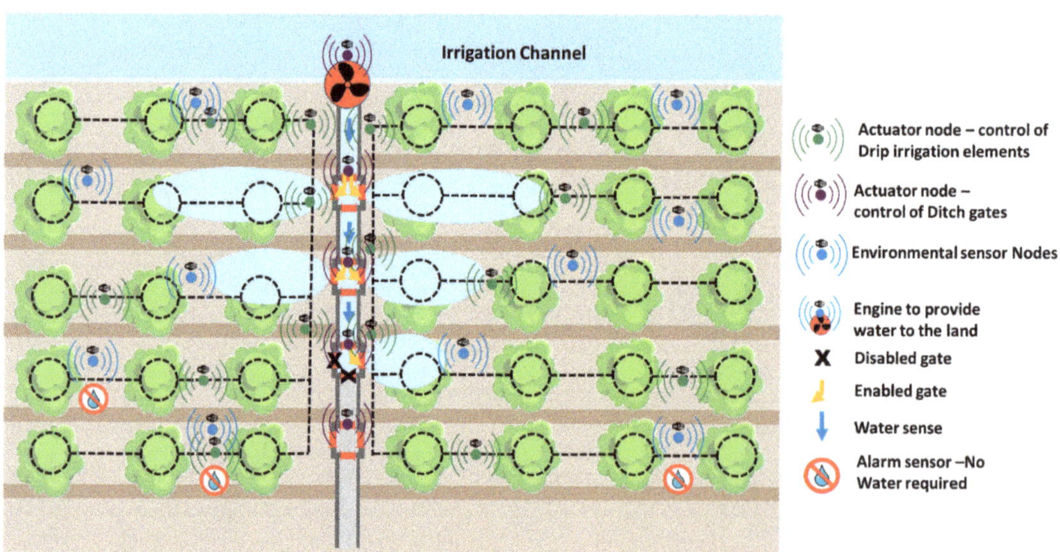

Figure 10. Diagram of our entire network in the plot.

In order to obtain high performance, we have developed a specific network protocol. This section presents the message exchange between devices, the fields of the messages exchanged between devices, and the algorithm designed for the system procedure.

The designed network is a distributed network made of sensor nodes (each one has one or several physical moisture sensors), and one or several actuators which activate the engine, for the drip irrigation system and/or the ditch gates, depending on the case.

Then the number of nodes (see Equation (5)) of the whole system (N) is:

$$N = n_s + n_a \tag{5}$$

where n_s is the number of sensor nodes and n_a is the number of actuator nodes.

Our network will use Ad Hoc On-Demand Distance Vector Routing (AODV) since it is one of the ad hoc routing protocols that presents the best performance [53].

4.1. Algorithm of the System

In order to determine when the irrigation process should be carried out, we need to collect the data from the different n_s which are placed and identified by zones (i). For each zone, we defined the maximum number of nodes comprised in the zone as counter. Carrying out this automation process of irrigation requires the design of an operation algorithm. Figure 11 shows the operation algorithm of our soil moisture monitoring system.

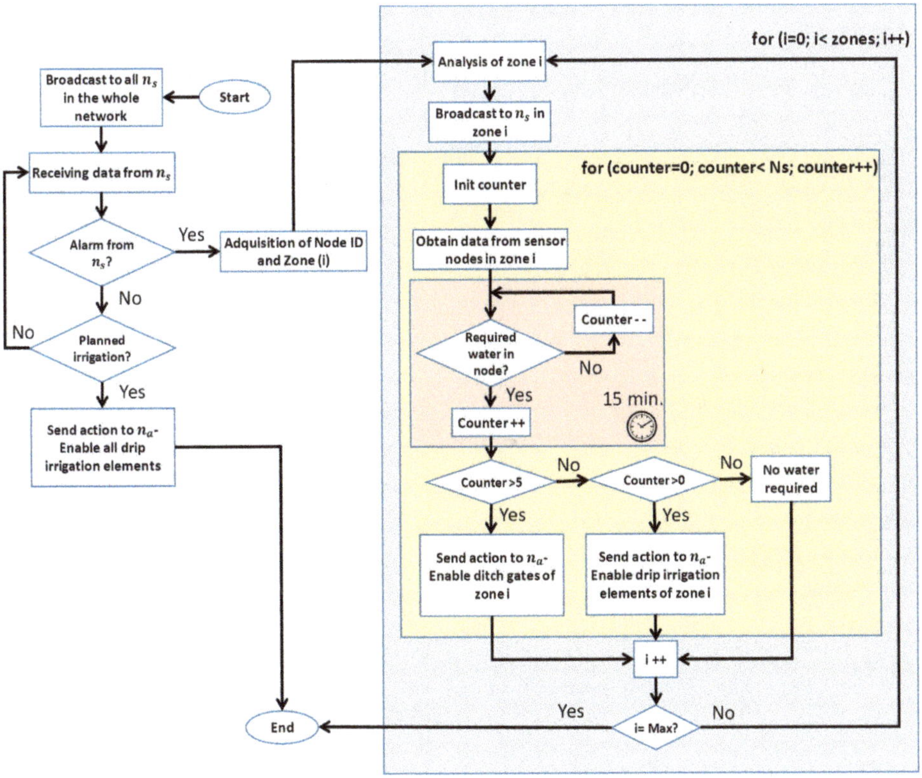

Figure 11. Operation algorithm.

As commonly done in agriculture, there are periodic planned irrigations that should be performed. In this case, the system of drip irrigation elements is enabled and it covers the entire extension of trees. If an alarm is registered from a sensor, the system will request the data from all nodes of that zone. If the number of nodes that register the need for water is higher than 5, the system will enable the ditch gate of this zone. Even if only some sensors warn about the need for water, the system will enable the drip irrigation elements of this zone. The rest of the zones in the plot will be analyzed to check if it is required to proceed with irrigation. The different orders will be sent to the sink node by the nearest node of that area to the sink node which will be in charge of enabling/disabling the irrigations systems.

Finally, if the plot does not require any action, the system will remain in idle mode waiting for new information.

4.2. Message Flow between Nodes

Finally, in order to send the required actions to the correct actuator nodes, it is important to design the message exchange between nodes. In this sense, we should consider three different situations (see Figure 12). Firstly, the most frequent situation is the one in which the plot does not require any type of irrigation. In this case, if the sensor nodes do not send any message in the next 30 min, the system will consider that no irrigation is required (1).

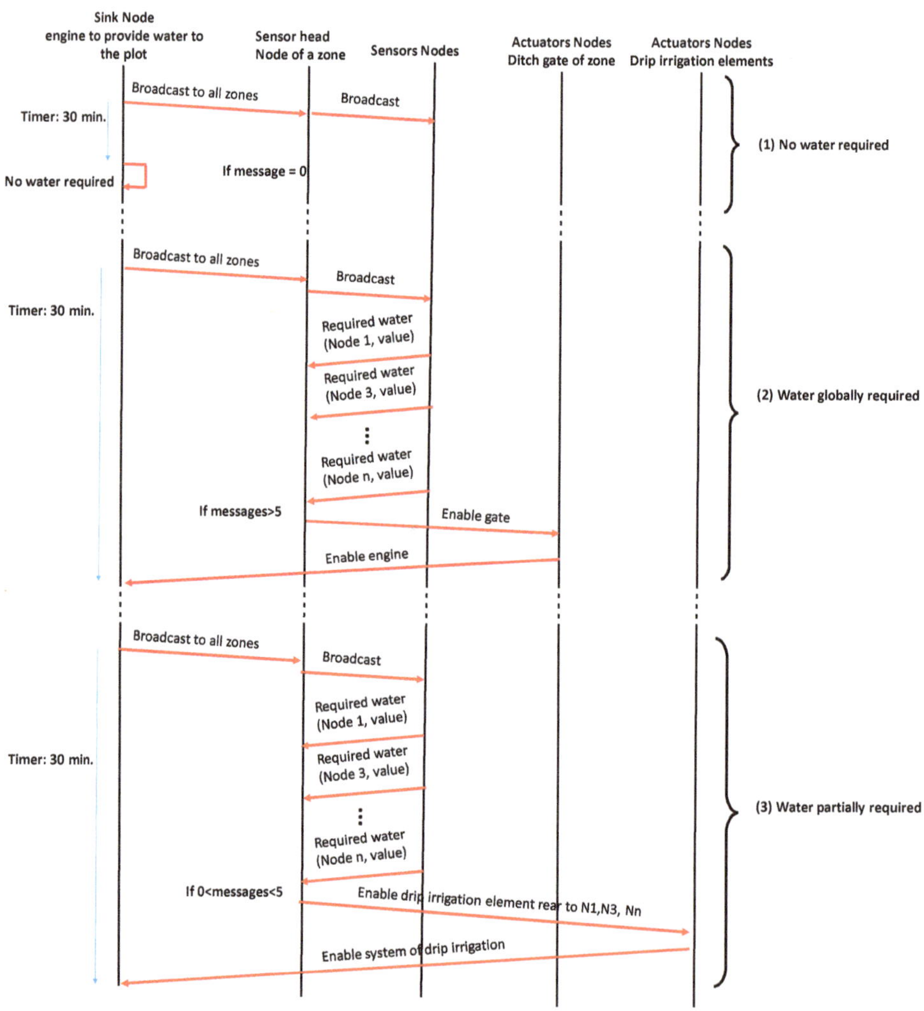

Figure 12. Message exchange between nodes.

The second situation (2) is when there is a global need for water in a zone of the plot. In this case, the sensor head node will wait 30 min for messages from sensor nodes. If more than 5 messages are received, the system will consider that global irrigation for this zone is required. Then, the sensor head node will send a message to the actuator node in charge of enabling the gates. After that, this node will inform the sink node to enable the engine to provide water to the ditch.

The third situation (3) will be done when there is a partial need for water in a zone of the plot. In this case, the sensor head node will wait 30 min for messages from sensor nodes. If less than 5 messages are received, the system will consider that partial irrigation for this zone is required. Then, the sensor head node will send a message to the actuator node in charge of enabling the drip irrigation element of the affected trees. After that, this actuator node will signal the sink node to enable the system of drip irrigation.

To make easier the process of forwarding messages from sensor nodes to actuator nodes or sink nodes, it is possible to use any node present in the network. In this sense, a node can receive several packets but if it is not the destination of this message, the node will relay the message without processing it. When the sensor nodes of a zone are communicating, an intragroup routing protocol will be utilized. When the message exchange is performed between sensor nodes and actuator nodes or between actuator nodes for ditch gates and actuator nodes for drip irrigation elements, these nodes will use an intergroup routing protocol [54].

5. Experimental Results in a Practical Deployment

In this section, the results obtained in the deployments on orange groves are presented. In order to perform the test, we have used several ESP32 DevKit nodes placed at different heights. This will allow us to study the coverage of the nodes at different heights, so it will be kept as a recommendation for practical deployments. The different deployment strategies that were tested are presented in Figure 13. As can be seen, different configurations of emitter height and receiver height were tested. The emitters were deployed at heights of 0.5 m, 1 m, 1.5 m, and 2 m. The receivers were placed at 0 m for the on-ground deployment, 0.5 m for the near-ground deployment, and 1.5 m for the above-ground deployment. The emitter and receiver were separated for each test. The trees are spaced in four-meter intervals and the field is located in an area with a Mediterranean climate. The foliage of the trees affects the wireless communication among the devices. Testing different configurations of transmitter and receiver provides us with the knowledge to design the best deployment for optimal communication with this type of crop. The tests were performed with sunny weather and temperatures of 20 °C. The measurement carried out is the received signal strength indicator (RSSI) at different measuring points. The Esp32 DevKit nodes were encapsulated on a protective box.

The results for the emitter at a height of 0.5 m and the receiver at different deployment configurations are presented in Figure 14. The positions of the trees are indicated by the bold orange numbers on the X-axis. As can be seen, the overall higher RSSI values considering multiple trees along the tested distance were obtained for the near-ground position of the receiver. This configuration of the receiver is also the most stable. Moreover, some small fluctuations occurred for tree number 1 and tree number 4. However, the foliage in the space between trees 2 to 4 presented a higher density, which lead to higher fluctuations. One of the reasons for these fluctuations may be the multipath effect. Thus, avoiding node deployments in areas of high foliage density is best to obtain more stable signals.

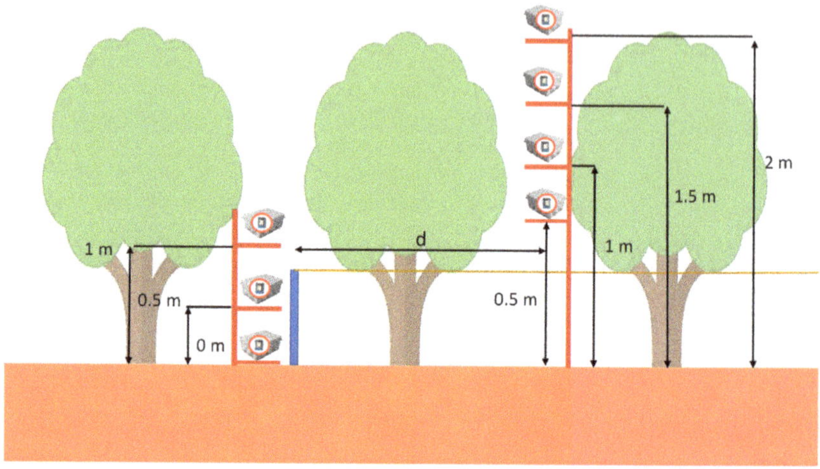

Figure 13. Testbed.

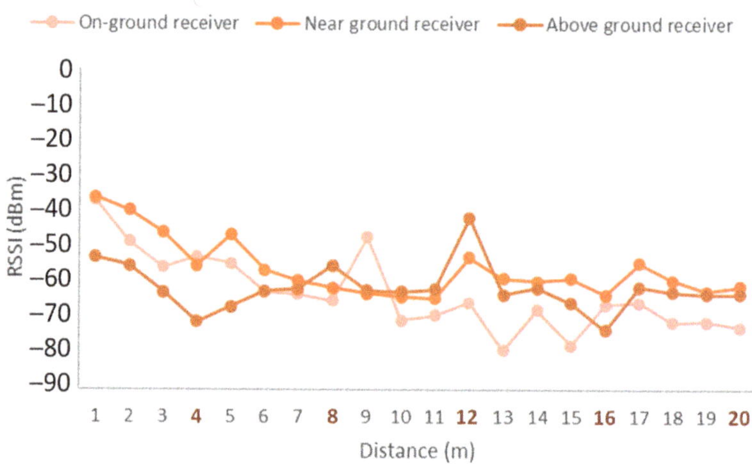

Figure 14. Emitter at height of 0.5 m.

For the case of the emitter deployed at a height of 1 m, the results are presented in Figure 15. As it can be seen, the near-ground receiver is the one with the best results. In this case, the signal quality is reduced between trees 2 and 4. However, for the near-ground receiver, the signal presents some recovery after the area with high foliage density. The above-ground deployment presents similar results for the area with high foliage density but worse signal quality for the rest of the measurement points. Lastly, the on-ground receiver deployment presents the worst results.

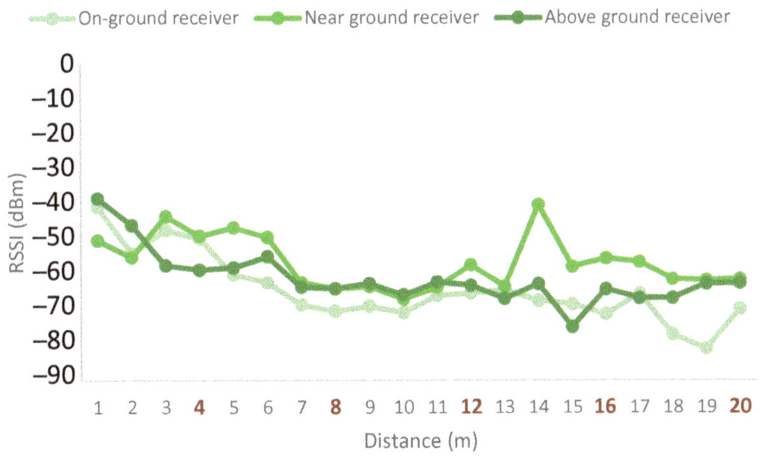

Figure 15. Emitter at height of 1 m.

Figure 16 presents the results for the emitter height of 1.5 m. The near-ground deployment has the highest signal quality values at almost all measuring points. As can be seen, it experiences some fluctuations between trees 2 and 3. However, even with the fluctuations, the signal quality is better than that of the other configurations. The next best option is the above-ground receiver. In this case, the signal is more stable while remaining below the quality levels of the near-ground receiver. Lastly, the on-ground deployment presented the worst results and the highest fluctuations. Another final aspect to consider is that the average signal quality for this emitter height was lower than the signal quality obtained for lower emitter heights.

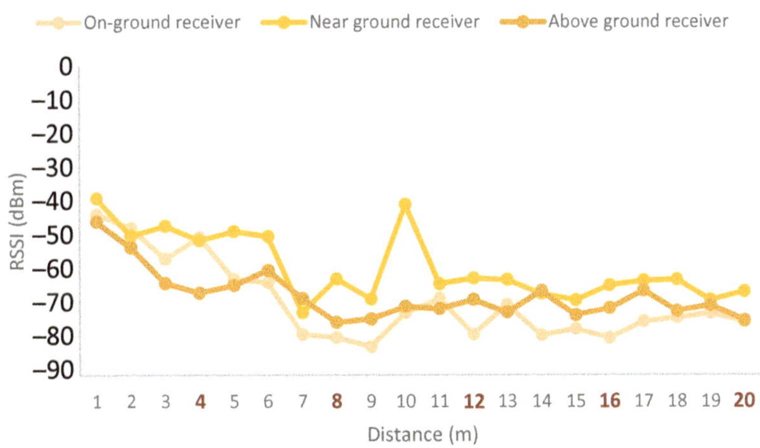

Figure 16. Emitter at height of 1.5 m.

Lastly, the results for the emitter height of 2 m are presented in Figure 17. This emitter height obtains the worst signal quality results compared to all the emitter heights. Regarding the receiver height, in this case as well, the near-ground deployment obtained the best results. However, as shown in the figure, all receiver configurations present similar results, while the results for this emitter height present the least fluctuations. As in the other cases, the on-ground configuration was the worst option.

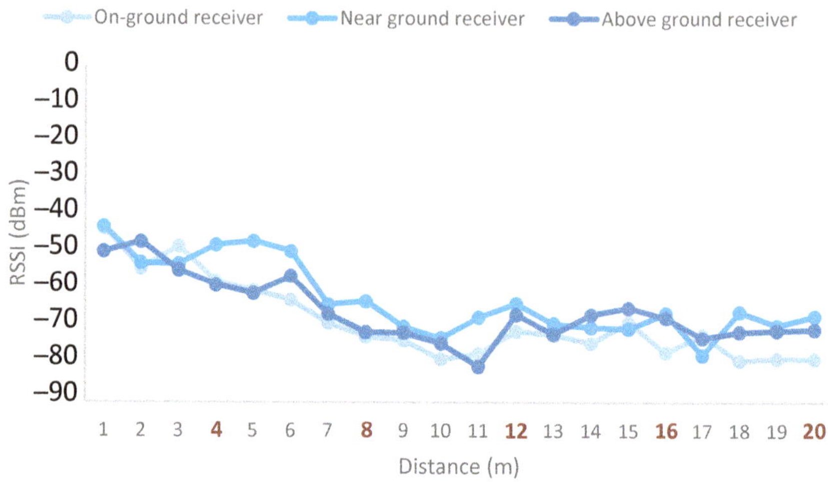

Figure 17. Emitter at height of 2 m.

Considering the results for all the emitter heights, we can conclude that in the case of orange groves, emitter heights of 0.5 and 1 m present the best signal quality and the near-ground receiver deployment is the best option for all emitter heights. Therefore, near-ground configurations are the optimal deployment style for both emitters and receivers.

The coverage results obtained from the tests performed on the orange groves have been utilized to obtain a heuristic signal attenuation model for all emitter heights as specified in [7]. The outlier values were discarded to perform this model. Equations (6)–(9) show the model for emitter heights of 0.5 m, 1 m, 1.5 m, and 2 m respectively.

$$P_{0,5\ m} = -7.182 \ln d(m) - 45.276 \tag{6}$$

$$P_{1\ m} = -7.69 \ln d(m) - 44.194 \tag{7}$$

$$P_{1,5\ m} = -9.545 \ln d(m) - 44.475 \tag{8}$$

$$P_{2\ m} = -10.34 \ln d(m) - 43.493 \tag{9}$$

Furthermore, the model, confidence intervals, and prediction intervals are presented in Figure 15, where the dots represent the values obtained from the tests on the fields. As can be seen, the model reflects that the configurations of emitter heights of 0.5 m and 1 m (See Figure 18a,b) present better signal quality. Lastly, Figure 18c shows the graphic representation for the case of the emitter height at 1.5 m and Figure 18d presents the results for the emitter height of 2 m.

Considering the results for all the emitter heights, we can conclude that for the case of orange groves, emitter heights of 0.5 and 1 m present the best signal quality and the near-ground receiver deployment was the best option for all emitter heights. Therefore, near-ground configurations are the optimal deployment style for both emitters and receivers.

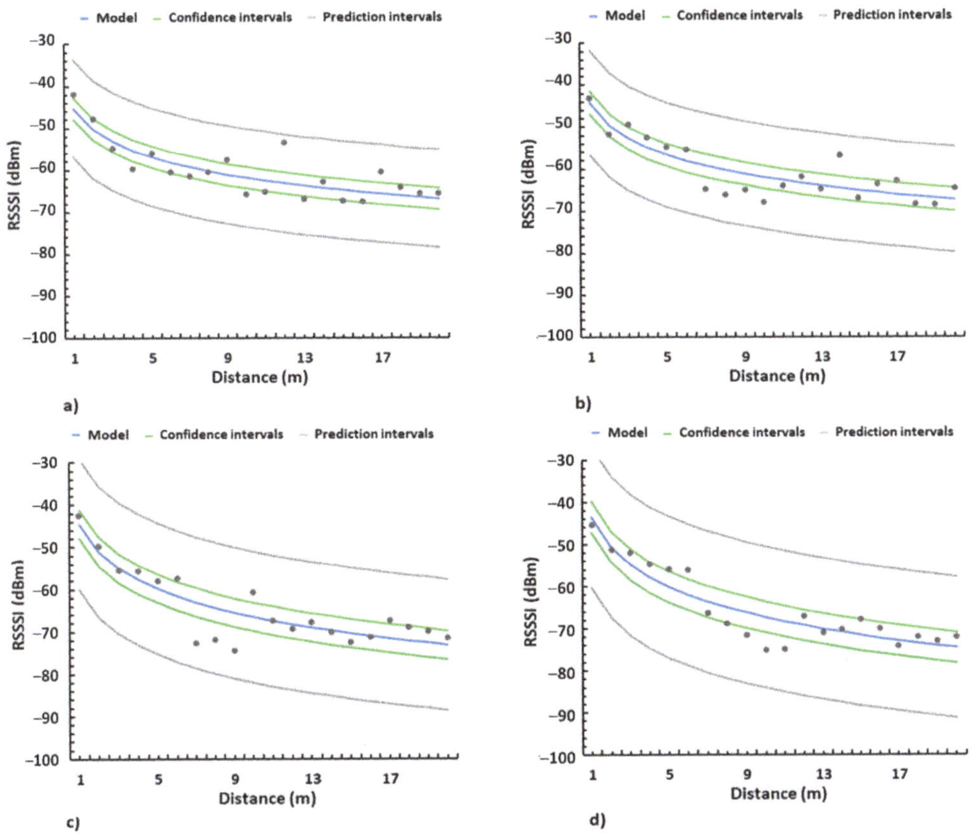

Figure 18. Heuristic model for (**a**) emitter at height of 0.5 m, (**b**) emitter at height of 1 m, (**c**) emitter at height of 1.5 m, and (**d**) emitter at height of 2 m.

6. Conclusions and Future Work

Estimating the amount of water needed to irrigate a crop is essential to carry out efficient use of a scarce resources such as water. The introduction of technology in the agricultural sector is also important to improve the sustainability and competitiveness of the sector. For this reason, this paper has presented the prototype of a low-cost sensor based on coils for measuring soil moisture. For this, three prototypes composed of two coils with different characteristics have been presented. These coils have been tested to analyze their behavior based on the humidity level of the soil. After the observed results, it has been concluded that the sensor that has had the best performance is prototype 1 working at 93 kHz. Additionally, a power circuit based on the ICM7555 has been designed to be able to generate the biphase signal to power the soil moisture sensor. This sensor is able to measure the percentage of water content in the soil at the desired depth. This fact helps us to ensure the correct irrigation of the root ball. The sensor and power supply circuit is connected to an ESP32 module for reading and storing humidity measurements. The entire system has been tested with real samples for the extraction of its mathematical behavior model. The results show that our sensor demonstrates that by using these models we can achieve accuracies close to 95%.

Additionally, the network performance has been tested in a real, cultivated plot. According to the results, and after modeling mathematically the results of the network coverage, we can conclude that for the case of orange groves, the best results are obtained when the emitter is placed at 0.5 and 1 m and the receiver is placed near the ground.

So, near-ground configurations are the optimal deployment style for both emitters and receivers.

In future work, we would like to perform more practical experiments with more models of coils and different kinds of soils to design a more versatile sensor capable of working with several sorts of soils without changing the sensor. It will also study the possibility of including a system to automatically adapt the working frequency to the type of soil. Because in our practical experiments we have included only the measurements of signal amplitude, it could be interesting to measure the quadrature component and phase of the obtained signal and trying to relate these parameters with changes of pH of water. We also want to include other sensors in a multi-parametric node to place in the crop field [55,56] to enhance the efficiency of water management in precision agriculture [57]. In this sense, we want to check if soil temperature has some effect over the soil moisture measurements and, if required, over obtaining the soil moisture values compensated with temperature. Finally, as the last step, we will study the most appropriate enclosures to protect our entire system.

Author Contributions: Conceptualization, S.S. and J.L.; methodology, S.S. and J.L.; validation, L.G. and J.M.J.; formal analysis, S.S. and J.L.; writing—review and editing, S.S., J.L., L.G. and J.M.J.; All authors have read and agreed to the published version of the manuscript.

Funding: This work has been partially supported by the European Union through the ERANETMED (Euromediterranean Cooperation through ERANET joint activities and beyond) project ERANETMED3-227 SMARTWATIR, by the "Programa Estatal de I+D+i Orientada a los Retos de la Sociedad, en el marco del Plan Estatal de Investigación Científica y Técnica y de Innovación 2017–2020" (Project code: PID2020-114467RR-C33) and by "proyectos de innovación de interés general por grupos operativos de la Asociación Europea para la Innovación en materia de productividad y sostenibilidad agrícolas (AEI-Agri)" in the framework "Programa Nacional de Desarrollo Rural 2014–2020", GO TECNOGAR. This work has also been partially funded by the Universitat Politècnica de València through the post-doctoral PAID-10-20 program.

Institutional Review Board Statement: Not applicable.

Informed Consent Statement: Not applicable.

Data Availability Statement: No dataset has been created.

Conflicts of Interest: The authors declare no conflict of interest. The funders had no role in the design of the study, in the collection, analyses, or interpretation of data, in the writing of the manuscript, or in the decision to publish the results.

References

1. García, L.; Parra, L.; Jimenez, J.M.; Lloret, J.; Lorenz, P. IoT-Based Smart Irrigation Systems: An Overview on the Recent Trends on Sensors and IoT Systems for Irrigation in Precision Agriculture. *Sensors* **2020**, *20*, 1042. [CrossRef]
2. Ueyama, J.; Freitas, H.; Faical, B.S.; Filho, G.P.R.; Fini, P.; Pessing, G.; Gomes, P.H.; Villas, L.A. Exploiting the use of unmanned aerial vehicles to provide resilience in wireless sensor networks. *IEEE Commun. Mag.* **2014**, *52*, 81–87. [CrossRef]
3. Faiçal, B.S.; Pessin, G.; Filho, G.P.R.; Carvalho AC PL, F.; Gomes, P.H.; Ueyama, J. Fine-Tuning of UAV Control Rules for Spraying Pesticides on Crop Fields: An Approach for Dynamic Environments. *Int. J. Artif. Intell. Tools* **2016**, *25*, 1660003. [CrossRef]
4. Faiçal, B.S.; Pessin, G.; Filho, G.P.R.; Furquim, G.; Carvalho, A.C.P.L.F.; Ueyama, J. Exploiting Evolution on UAV Control Rules for Spraying Pesticides on Crop Fields. In Proceedings of the International Conference on Engineering Applications of Neural Networks (EANN 2014), Sofia, Bulgaria, 5–7 September 2014; pp. 49–58.
5. Instituto de Hidrología Meteorología y Estudios Ambientales. Métodos de Medición de Humedad del Suelo: Del Campo al Laboratorio. In PortalFruticola Web Site. 2018. Available online: https://www.portalfruticola.com/noticias/2018/08/30/metodos-de-medicion-de-humedad-del-suelo-del-campo-al-laboratorio/#:~{}:text=El%20m%C3%A9todo%20gravim%C3%A9trico%20es%20el,una%20temperatura%20de%20105\T1\textquoterightC (accessed on 20 September 2021).
6. Martini, E.; Werban, U.; Zacharias, S.; Pohle, M.; Dietrich, P.; Wollschläger, U. Repeated electromagnetic induction measurements for mapping soil moisture at the field scale: Validation with data from a wireless soil moisture monitoring network. *Hydrol. Earth Syst. Sci.* **2017**, *21*, 495–513. [CrossRef]
7. Garcia, L.; Parra, L.; Jimenez, J.M.; Parra, M.; Lloret, J.; Mauri, P.V.; Lorenz, P. Deployment Strategies of Soil Monitoring WSN for Precision Agriculture Irrigation Scheduling in Rural Areas. *Sensors* **2021**, *21*, 1693. [CrossRef] [PubMed]

8. Ojha, T.; Misra, S.; Raghuwanshi, N.S. Wireless sensor networks for agriculture: The state-of-the-art in practice and future challenges. *Comput. Electron. Agric.* **2015**, *118*, 66–84. [CrossRef]
9. Susha Lekshmi, S.U.; Singh, D.N.; Baghini, M.S. A critical review of soil moisture measurement. *Measurement* **2014**, *54*, 105.
10. Tumanski, S. Induction coil sensors—A review. *Meas. Sci. Technol.* **2007**, *18*, R31. [CrossRef]
11. Jawad, H.M.; Nordin, R.; Gharghan, S.K.; Jawad, A.M.; Ismail, M. Energy-Efficient Wireless Sensor Networks for Precision Agriculture: A Review. *Sensors* **2017**, *17*, 1781. [CrossRef] [PubMed]
12. Hamami, L.; Nassereddine, B. Application of wireless sensor networks in the field of irrigation: A review. *Comput. Electron. Agric.* **2020**, *179*, 105782. [CrossRef]
13. Mekonnen, Y.; Namuduri, S.; Burton, L.; Sarwat, A.; Bhansali, S. Review—Machine Learning Techniques inWireless Sensor Network Based Precision Agriculture. *J. Electrochem. Soc.* **2020**, *167*, 037522. [CrossRef]
14. Nabi, F.; Jamwal, S.; Padmanbh, K. Wireless sensor network in precision farming for forecasting and monitoring of apple disease: A survey. *Int. J. Inf. Technol.* **2020**. [CrossRef]
15. Kabashi, A.H.; Elmirghani, J.M.H. Technical Framework for Designing Wireless Sensor Networks for Agricultural Monitoring in Developing Regions. In Proceedings of the Second International Conference on Next Generation Mobile Applications, Services, and Technologies, Cardiff, UK, 16–19 September 2008.
16. Kassim, M.R.M.; Mat, I.; Harun, A.N. Wireless Sensor Network in precision agriculture application. In Proceedings of the 2014 International Conference on Computer, Information and Telecommunication Systems (CITS), Jeju, Korea, 7–9 July 2014.
17. Zervopoulos, A.; Tsipis, A.; Alvanou, A.G.; Bezas, K.; Papamichail, A.; Vergis, S.; Stylidou, A.; Tsoumanis, G.; Komianos, G.; Koufoudakis, G.; et al. Wireless Sensor Network Synchronization for Precision Agriculture Applications. *Agriculture* **2020**, *10*, 89. [CrossRef]
18. Bayrakdar, M.E. A Smart Insect Pest Detection Technique with Qualified Underground Wireless Sensor Nodes for Precision Agriculture. *IEEE Sens. J.* **2019**, *19*, 10892–10897. [CrossRef]
19. Khedo, K.K.; Hosseny, M.R.; Toonah, M.Z. PotatoSense: A Wireless Sensor Network System for Precision Agriculture. In Proceedings of the 2014 IST-Africa Conference Proceedings, Pointe aux Piments, Mauritius, 7–9 May 2014.
20. Rasooli, M.W.; Bhushan, B.; Kumar, N. Applicability of Wireless Sensor Networks & IoT in Saffron & Wheat Crops: A Smart Agriculture Perspective. *Int. J. Sci. Technol. Res.* **2020**, *9*, 2456–2461.
21. Chaudhary, D.D.; Nayse, S.P.; Waghmare, L.M. Application of wireless sensor networks for greenhouse parameter control in precision agriculture. *Int. J. Wirel. Mob. Netw. IJWMN* **2011**, *3*, 140–149. [CrossRef]
22. Srbinovska, M.; Gavrovski, C.; Dimcev, V.; Krkoleva, A.; Borozan, V. Environmental parameters monitoring in precision agriculture using wireless sensor networks. *J. Clean. Prod.* **2015**, *88*, 297–307. [CrossRef]
23. Hamouda, Y.; Msallam, M. Variable sampling interval for energy-efficient heterogeneous precision agriculture using Wireless Sensor Networks. *J. King Saud Univ.—Comput. Inf. Sci.* **2020**, *32*, 88–98. [CrossRef]
24. Qureshi, K.N.; Bashir, M.U.; Lloret, J.; Leon, A. Optimized Cluster-Based Dynamic Energy-Aware Routing Protocol for Wireless Sensor Networks in Agriculture Precision. *J. Sens.* **2020**, *2020*, 9040395. [CrossRef]
25. Badewa, E.; Unc, A.; Cheema, M.; Kavanagh, V.; Galagedara, L. Soil moisture mapping using multi-frequency and multi-coil electromagnetic induction sensors on managed Podzols. *Agronomy* **2018**, *8*, 224. [CrossRef]
26. DeRouin, A.J.; You, Z.; Hansen, M.; Diab, A.; Ong, K.G. Development and application of the single-spiral inductive-capacitive resonant circuit sensor for wireless, real-time characterization of moisture in sand. *J. Sens.* **2013**, *2013*, 894512. [CrossRef]
27. Kizito, F.; Campbell, C.S.; Campbell, G.S.; Cobos, D.R.; Teare, B.L.; Carter, B.; Hopmans, J.W. Frequency, electrical conductivity and temperature analysis of a low-cost capacitance soil moisture sensor. *J. Hydrol.* **2008**, *352*, 367–378. [CrossRef]
28. Nor, A.S.M.; Yunus, M.A.M.; Nawawi, S.W.; Ibrahim, S. Low-cost sensor array design optimization based on planar electromagnetic sensor design for detecting nitrate and sulphate. In Proceedings of the 2013 Seventh International Conference on Sensing Technology (ICST), Wellington, New Zealand, 3–5 December 2013; pp. 693–698.
29. Fritz, S.; See, L.; Bayas, J.C.L.; Waldner, F.; Jacques, D.; Becker-Reshef, I.; Whitcraft, A.; Baruth, B.; Bonifacio, R.; Crutchfield, J.; et al. A comparison of global agricultural monitoring systems and current gaps. *Agric. Syst.* **2019**, *168*, 258–272. [CrossRef]
30. Group on Earth Observations Global Agricultural Monitoring Initiative (GEOGLAM). In Earth Observations Website. Available online: https://earthobservations.org/geoglam.php (accessed on 20 September 2021).
31. CropWatch. Available online: http://www.cropwatch.com.cn/ (accessed on 20 September 2021).
32. Shafi, U.; Mumtaz, R.; García-Nieto, J.; Hassan, S.A.; Zaidi, S.A.R.; Iqbal, N. Precision agriculture techniques and practices: From considerations to applications. *Sensors* **2019**, *19*, 3796. [CrossRef]
33. Khan, W.Z.; Ahmed, E.; Hakak, S.; Yaqoob, I.; Ahmed, A. Edge computing: A survey. *Future Gener. Comput. Syst.* **2019**, *97*, 219–235. [CrossRef]
34. Trumper, R.; Gelbman, M. Investigating electromagnetic induction through a microcomputer-based laboratory. *Phys. Educ.* **2000**, *35*, 90. [CrossRef]
35. Kholmetskii, A.L. The Faraday induction law and field transformations in special relativity. *Apeiron* **2003**, *10*, 118.
36. Monk, P. *Finite Element Methods for Maxwell's Equations*; Oxford University Press: Oxford, UK, 2003.
37. ESP32 DevKit Features. In Espressif Website. Available online: https://www.espressif.com/en/products/devkits (accessed on 20 September 2021).

38. The national association for amateur Radio. *The ARRL Handbook for Radio Communications 2017*, 94th ed.; Amer Radio Relay League: Newington, CT, USA, 2017.
39. Parra, L.; Sendra, S.; Lloret, J.; Bosch, I. Development of a conductivity sensor for monitoring groundwater resources to optimize water management in smart city environments. *Sensors* **2015**, *15*, 20990–21015. [CrossRef]
40. Texas Instruments. 555 IC Datasheet. Available online: http://www.ti.com/lit/ds/symlink/ne555.pdf (accessed on 20 September 2021).
41. ICM7555 Datasheet. Available online: https://www.nxp.com/docs/en/data-sheet/ICM7555.pdf (accessed on 20 September 2021).
42. RK520-02 Soil Moisture Sensor, Temperature Probe& EC Sensor. Available online: https://www.rikasensor.com/rk520-02-soil-moisture-sensor-temperature-probe-ec-sensor.html (accessed on 20 September 2021).
43. RS485 Soil Moisture & Temperature & EC Sensor. Available online: https://files.seeedstudio.com/products/101990667/res/RS485%20Soil%20Moisture%20&%20Temperature%20&%20EC%20Sensor%20(S-Temp&VWC&EC-02A).pdf (accessed on 20 September 2021).
44. SenseCAP LoRaWAN Sensor User Manual. Available online: https://sensecap-docs.seeed.cc/pdf/SenseCAP%20LoRaWAN%20Sensor%20User%20Manual-V1.1.pdf (accessed on 20 September 2021).
45. Waterproof_Capacitive_Soil_Moisture_Sensor_SKU_SEN0308-DFRobot. Available online: https://docs.rs-online.com/8389/A700000007238462.pdf (accessed on 20 September 2021).
46. YL-69. Available online: https://www.electronicoscaldas.com/datasheet/OBSoil-01_ElecFreaks.pdf (accessed on 20 September 2021).
47. Keyes Brick Soil Moisture Sensor Module. Available online: https://www.fasttech.com/product/9720737-keyes-brick-soil-moisture-sensor-module (accessed on 20 September 2021).
48. KeeYees moisture sensor. Available online: https://www.amazon.com/KeeYees-Capacitive-Moisture-Sensor-Module/dp/B07R174TM1 (accessed on 20 September 2021).
49. Grove—Capacitive Soil Moisture Sensor (Corrosion Resistant). Available online: https://www.seeedstudio.com/Grove-Capacitive-Moisture-Sensor-Corrosion-Resistant.html (accessed on 20 September 2021).
50. Grove—Soil Moisture Sensor. Available online: https://www.seeedstudio.com/Grove-Moisture-Sensor.html (accessed on 20 September 2021).
51. RS485 Soil Moisture & Temperature Sensor (S-Soil MT-02A). Available online: https://files.seeedstudio.com/products/101990668/res/RS485%20Soil%20Moisture%20&%20Temperature%20Sensor%20(S-Soil%20MT-02)-Datasheet.pdf (accessed on 20 September 2021).
52. Soil Moisture Sensor. Available online: https://www.makerfabs.com/soil-moisture-sensor.html (accessed on 20 September 2021).
53. Xin, H.M.; Yang, K. Routing protocols analysis for Internet of Things. In Proceedings of the 2nd International Conference on Information Science and Control Engineering (CISCE 2015), Shanghai, China, 24–26 April 2015; pp. 447–450.
54. Garcia, M.; Sendra, S.; Lloret, J.; Canovas, A. Saving energy and improving communications using cooperative group-based wireless sensor networks. *Telecommun. Syst.* **2013**, *52*, 2489–2502. [CrossRef]
55. Sendra, S.; García, L.; Lloret, J.; Bosch, I.; Vega-Rodríguez, R. LoRaWAN Network for Fire Monitoring in Rural Environments. *Electronics* **2020**, *9*, 531. [CrossRef]
56. Nikolidakis, S.A.; Kandris, D.; Vergados, D.D.; Douligeris, C. Energy efficient automated control of irrigation in agriculture by using wireless sensor networks. *Comput. Electron. Agric.* **2015**, *113*, 154–163. [CrossRef]
57. Navarro-Hellín, H.; Torres-Sánchez, R.; Soto-Valles, F.; Albaladejo-Pérez, C.; López-Riquelme, J.A.; Domingo-Miguel, R. A wireless sensors architecture for efficient irrigation water management. *Agric. Water Manag.* **2015**, *151*, 64–74. [CrossRef]

sensors

MDPI

Review

Towards Synoptic Water Monitoring Systems: A Review of AI Methods for Automating Water Body Detection and Water Quality Monitoring Using Remote Sensing

Liping Yang [1,2,3,*], Joshua Driscol [1,2], Sarigai Sarigai [1,2], Qiusheng Wu [4], Christopher D. Lippitt [1,2] and Melinda Morgan [1]

1 Department of Geography and Environmental Studies, University of New Mexico, Albuquerque, NM 87131, USA; joshuadr@unm.edu (J.D.); rsa@unm.edu (S.S.); clippitt@unm.edu (C.D.L.); mhbenson@unm.edu (M.M.)
2 Center for the Advancement of Spatial Informatics Research and Education (ASPIRE), University of New Mexico, Albuquerque, NM 87131, USA
3 Department of Computer Science, University of New Mexico, Albuquerque, NM 87106, USA
4 Department of Geography, University of Tennessee, Knoxville, TN 37996, USA; qwu18@utk.edu
* Correspondence: lipingyang@unm.edu

Abstract: Water features (e.g., water quantity and water quality) are one of the most important environmental factors essential to improving climate-change resilience. Remote sensing (RS) technologies empowered by artificial intelligence (AI) have become one of the most demanded strategies to automating water information extraction and thus intelligent monitoring. In this article, we provide a systematic review of the literature that incorporates artificial intelligence and computer vision methods in the water resources sector with a focus on intelligent water body extraction and water quality detection and monitoring through remote sensing. Based on this review, the main challenges of leveraging AI and RS for intelligent water information extraction are discussed, and research priorities are identified. An interactive web application designed to allow readers to intuitively and dynamically review the relevant literature was also developed.

Keywords: surface water; water body detection; surface water extraction; water quality monitoring; remote sensing; artificial intelligence; computer vision; machine learning; deep learning; convolutional neural networks

Citation: Yang, L.; Driscol, J.; Sarigai, S.; Wu, Q.; Lippitt, C.D.; Morgan, M. Towards Synoptic Water Monitoring Systems: A Review of AI Methods for Automating Water Body Detection and Water Quality Monitoring Using Remote Sensing. *Sensors* **2022**, *22*, 2416. https://doi.org/10.3390/s22062416

Academic Editors: Miquel À. Cugueró-Escofet and Vicenç Puig

Received: 2 January 2022
Accepted: 15 March 2022
Published: 21 March 2022

1. Introduction and Motivation

Water is fundamentally necessary to all forms of life, and it is also the primary medium through which climate change impacts Earth's ecosystem and thus the livelihood and wellbeing of societies [1]. While water covers about 71% of the Earth's surface, only approximately 3% of the Earth's water bodies are freshwater [2]. Climate change will bring unique challenges to these water bodies. Many rivers and streams are heavily dependent on winter snowpack, which is declining with rising temperatures and changing precipitation patterns [3]. Sea level rise is also impacting the continued quality and quantity of water supplies [4]. Both the quantity and the quality of freshwater systems are critical environmental features essential to increasing resilience in the face of climate change [5,6]. Resilience is defined here as the capacity of a system to absorb disturbance and still retain its basic function and structure [7]. Climate change will bring new disturbances in many forms, including increased pollution from wildfires, saltwater intrusion, and deteriorated water quantity resulting from prolonged drought [1,8]. It is critical that we gather, ideally automatically, as much information as possible about freshwater bodies and how they function in order to increase our capacity to respond to a changing climate. Rockström [5,6] and his colleagues conceptualize freshwater use and biogeochemical flows that threaten the integrity of freshwater (via pollution) as two of seven variables key to overall Earth system function. Each of these variables, they argue, can be thought of as having "planetary boundary", a threshold that should not be crossed if we are to maintain the Earth in its current system state [5]. In this sense, the integrity and functioning of freshwater systems are essential not only in the local scale in which they provide critical ecosystem services; they also create a "safe operating space" for humanity as a whole, as we seek to achieve global solutions to the larger environmental challenges we face with climate change and associated stressors [6].

Responding to climate change challenge impacts on water resources requires adaptation strategies at the local, regional, national, and global scales. Countries are urged to improve their water resources management systems and to identify and implement "no regrets" strategies in order to be resilient to climate change [1]. The changing spatial and temporal patterns of surface water are important, in both practical and scientific terms, for water resources management, biodiversity, emergency response, and climate change [9]. More specifically, automated monitoring of water bodies is critical for adapting to climate change, water resources, ecosystem services, and the hydrological cycle, as well as for urban hydrology, which can facilitate timely flood protection planning and water quality control for public safety and health [10–12]. Accurate water quality monitoring is essential for developing sustainable water resource management strategies and ensuring the health of communities, ecosystems, and economies [13]. However, current knowledge of water quality is often disconnected in time and space across different measurement techniques and platforms that may fail to capture dynamic ecosystem changes. This disconnection indicates an inefficiency and redundancy in research and monitoring activities. A major challenge for water resource management is how to integrate multiple sources of water quality data and indices into usable and actionable information of environmental, social, economic, and infrastructural value [13,14].

Geospatial big data are leading to transformative changes in science (with the advent of data-driven and community science) and in society (with the potential to support the economy, public health, and other advances). Artificial intelligence (AI), especially its branches machine learning (ML), deep learning (DL), and computer vision (CV), are central to leveraging geospatial big data for applications in both domains. Remote sensing (RS) is the single largest source of geospatial big data and has increased dramatically in terms of both spatial and temporal resolution. This poses serious challenges for effective and efficient processing and analysis [15]. Meanwhile, recent advances in DL and CV have significantly improved research in RS and geosciences [16–18]. These advances, if integrated in creative and appropriate ways, host potential to enable the automated identification and monitoring of large-scale water bodies and water quality effectively and efficiently.

In this article, we argue specifically that bridging research into extracting important water information (e.g., water body extent, water quality) from RS imagery will provide an important computational foundation for the development of smart, RS-enabled water resource management systems. We review a range of recent developments in the relevant fields that can be leveraged to support intelligent automation of water body extraction and water quality detection and monitoring through RS imagery. An accompanying interactive web application allows our readers to intuitively track scholars and publications covered in this review (the web app tool URL and its brief demo video link are provided in Appendix A).

1.1. Selection Criterion for Reviewed Papers and Brief Graphic Summary

In the literature review process, we performed a systematic search on Google Scholar with the keywords and search strategy detailed in Table 1. In addition, our search was restricted to research articles published in English and in peer-reviewed journals or conference proceedings. For water body detection, we combined the water body keywords with some combination of the general keywords. The process for finding publications related to water quality was the same, only with the water quality keywords list. Beyond the keywords listed in this table, references (those cited in the papers we reviewed) cited by the keyword-identified papers were also retained. A total of 90 papers relevant to the topic of water body and/or water quality from RS imagery using AI/ML/DL/CV algorithms were identified. A total number of 56 highly relevant articles were identified by applying the following exclusion criteria: (1) papers related to plastic pollution and sewage/water treatment plants, (2) precipitation forecasting or groundwater detection (as it is not intuitive to detect groundwater from RS imagery), and (3) general land use classification. Figure 1 shows the spatial distribution and a simple statistics summary of the papers covered in this review, where (d) shows the number of published papers by year in the reviewed topics from 2011 to early 2022.

Table 1. Keywords used for article search.

Keyword Category	Search Strategy
General keywords [1]	"remote sensing" OR "satellite data" OR "UAV" AND "computer vision" OR "machine learning" OR "deep learning" OR "neural networks" OR "AI"
Water body	"water body" AND "detection" OR "extraction"
Water quality	"water quality" AND "sensing" OR "monitoring"

[1] A list of general keywords were combined with either the category of water body or water quality, respectively, to perform our search.

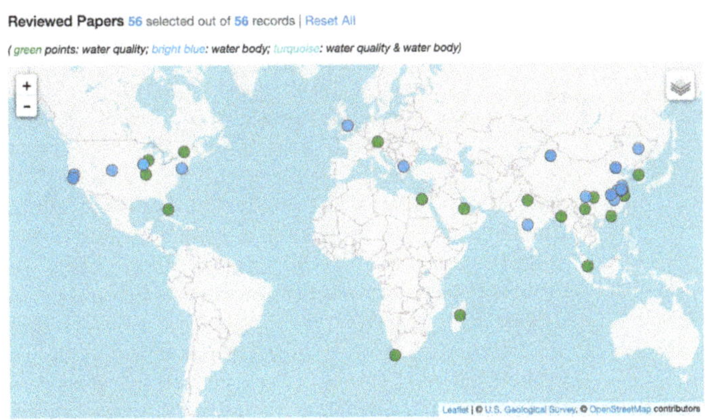

(**a**)

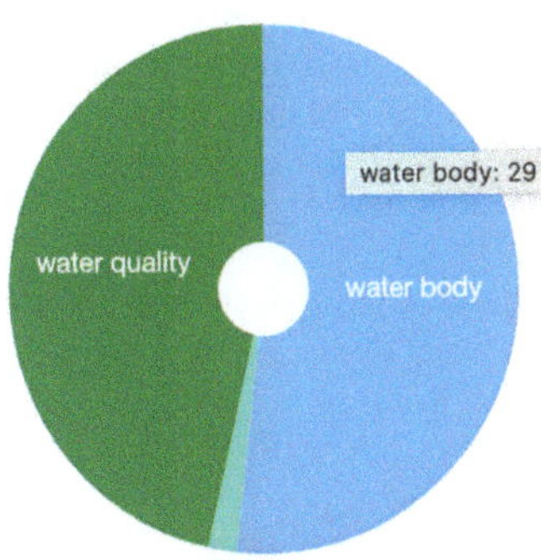

(**b**)

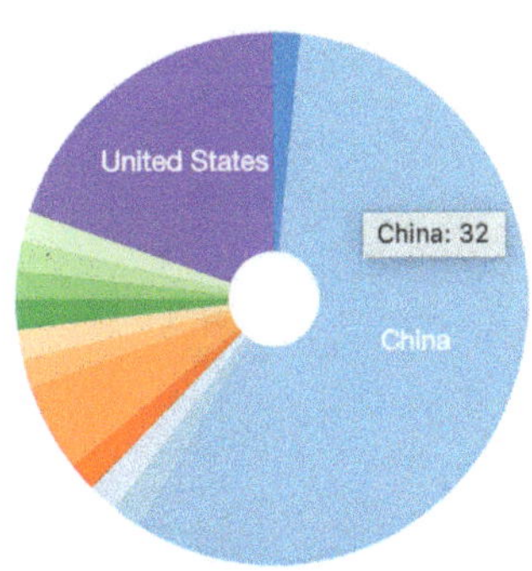

(**c**)

Figure 1. *Cont.*

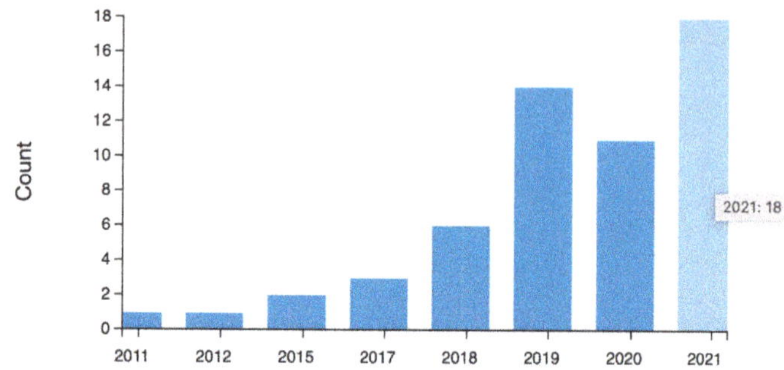

(**d**)

Figure 1. Geospatial distribution and simple statistics of the reviewed papers. Note that a freely accessible interactive version of the charts can be accessed via our web app tool (the web app tool URL and its brief demo video are provided in Appendix A). We can easily see that the major countries are China and the United States and that the number of published papers by year (2011 to 2021) has dramatically increased since 2018 and 2019. (**a**) Spatial distribution of reviewed papers based on the first author's institution location. (**b**) Topic distribution (water body, water quality, both). (**c**) Country distribution. (**d**) Number of published papers by year from 2011 to 2021 on the relevant topics.

1.2. Roadmap

Here, we provide a roadmap for the rest of the paper. Section 2 outlines the scope of this review and our intended audience. Section 3 is the core of the paper, focused on identifying important and recent developments and their implications to water body detection and water quality monitoring from RS imagery through the leverage of AI/ML/DL/CV. Here, we highlight recent advances in several subfields of AI that water domains and RS can leverage. Specifically, we provide general characteristics of the reviewed studies using word clouds (Section 3.1). We then examine and appraise key components of influential work in water body detection (Section 3.2) and water quality monitoring (Section 3.3). Section 4 starts with a brief summary (Section 4.1), followed with a discussion of key technical challenges (Section 4.2) and opportunities (Section 4.3). The paper concludes in Section 5.

To allow our readers to intuitively and dynamically review the relevant literature, we have developed a free-of-charge interactive web app tool (the web app URL and its brief demo video are provided in Appendix A). To provide background for readers (particularly those from water resources and RS) who are new to AI/ML/DL/CV, we introduce essential ML terms in Appendix B. As evaluation metrics are essential for measuring the performance of AI/ML/DL/CV models, we also provide an introduction to a set of commonly used evaluation metrics in Appendix C. In addition, as there are plenty of acronyms in this paper, we provide a full list of abbreviations right before the appendices.

2. Audience and Scope

It is important to know where water is and how its extent and quality are changing over time in a quick and accurate manner. Water quality is a key issue in water supply, agriculture, human and animal health, and many other areas [19]. Impaired water quality can be caused by natural disasters, but the most common cause is anthropogenic pollution. Pollutants, excessive nutrients from fertilizers, and sediment (e.g., from soil erosion) are carried into local lakes and rivers via runoff from urban or agricultural areas [19,20]. The quality of water varies from places and from time to time [19]. Affected surface waters are present in RS imagery and can be identified with the help of computational techniques such as ML. *To make near real-time intelligent water body detection and water quality monitoring possible, we need to first detect extent of water bodies from RS imagery, from which volume can be computed, and then recognize their corresponding water quality, eventually linking the two to allow water quality monitoring.*

Environmental nonprofits, government agencies, and water managers need access to this type of integrated spatial–time series of water body and water quality information to see how local water resources are changing and plan for future drought conditions. Collective detection and monitoring of water bodies and their associated water quality has applications for human health, as well as to private-sector industries including timber, agriculture, recreation, and tourism. Public policy planners need to be better informed as they make environmental preservation and restoration decisions based on changing water availability, and with this data we can be better equipped to monitor water quality that can quickly change due to floods, hurricanes, or human-caused pollution, *and yet, to date, water body detection and water quality monitoring research has been historically separate and does not focus enough on producing intuitive, operational products.*

Building on the long-term interest in ML and CV within the RS community, the main goals of this review paper are to (1) survey recent advances in water body detection and water quality monitoring from RS data using AI to identify commonly cited challenges in order to provide suggestions for new research directions, and (2) move towards automated, synoptic water quantity and quality monitoring to inform more robust water resource management.

This systematic review is relevant to multiple research domains, including, but not limited to RS, geographic information science, computer science, data science, information science, geoscience, hydrology, and water resource management. This paper does not attempt to review the application of RS to water resources and hydrology more generally; for recent reviews of these topics, see [13,21–24]. A survey of DL applications in hydrology and water resources can be found in [25]; a survey of AI in the water domain can be found in [26]; and a survey of water quality applications using satellite data solely focused on ML can be found in [27]. This review focuses on investigating recent AI methods, including its branches ML, DL, and CV, for water information extraction (specifically water body detection and/or water quality monitoring) from RS imagery. Our review has a narrowed scope in water resources and hydrological research, but a wider and deeper scope in terms of AI methods and metrics used to assess models in both water body detection and water quality research. *By integrating both domains, we hope to develop a basis for effective computational frameworks for intelligent water monitoring systems using RS and AI.*

3. The State of the Art: Advances in Intelligent Waterbody Information Extraction

3.1. General Characteristics of the Reviewed Studies

Note that we only included and reviewed the papers that use both RS and AI/ML/DL/CV for water body and/or water quality detection (that is, the number of papers cited in our reference section is much larger than the number of papers we review in this Section 3). A word cloud visualization of the titles, abstracts, and keywords of the reviewed 56 papers are provided in Figure 2, where the top figure indicates the word cloud for all reviewed papers. The bottom left word cloud is for reviewed water body papers, and the bottom right for reviewed water quality papers.

As we can see from the word cloud for both water body extraction and water quality (see the top word cloud in Figure 2), "remote sensing", "deep learning", "prediction", "classification", "extraction", "machine learning", "water body", "water quality", and "convolutional neural network" are prominent concepts and words captured by the word cloud. Our focus is on studies that use RS for water body extraction and water quality monitoring, so many of the keywords are to be expected. However, it is perhaps surprising to see DL featured so prominently given that the shift from ML to DL models is a relatively recent phenomenon.

Figure 2. Word cloud visualization of all the reviewed papers (**top**), water body papers (**bottom left**), and water quality papers (**bottom right**). Note that the word clouds are generated from paper titles, abstracts, and keywords. The word clouds provide an informative (general and specific) focus of each set of the papers. For example, both water body and water quality papers share the focus on RS, DL, and neural networks (NN). We can also see that water body extraction tasks tend to focus on the use of convolutional neural networks (CNN), whereas for water quality modeling the use of long short-term memory (LSTM) networks is more prevalent. We can also see that there are specific, unique keywords for water quality, such as "turbidity", "chl", and "algal bloom".

When we separate the keyword word clouds (see the bottom two word clouds in Figure 2), this trend becomes clearer. Deep learning is much more common in water body extraction, whereas in the word cloud for water quality monitoring, "neural network" and "machine learning" are about the same size. Additionally, in the water body extraction word cloud, "remote sensing" is featured much more heavily than it is in the water quality extraction literature. In our review, the water quality papers often involved other types of data, including in situ sensors or smaller RS devices (not satellites), whereas the water body extraction literature is dominated by RS imagery. This is related to the scale of projects in the two domains: water body extraction is usually undertaken across large spatial scales, whereas the water quality monitoring literature is still focused on smaller, often individual, bodies of water. This points to a future research direction in the water quality literature that we touch on in our review paper: we need to scale up water quality estimation using RS imagery by matching it with ground-truth water quality measurements.

Tables 2 and 3 provide a brief summary of the methods used for water body detection and water quality monitoring, elaborated in Sections 3.2 and 3.3, respectively. The general characteristics summarized by machines (i.e., the word clouds in Figure 2) align with the literature; convolutional neural network (CNN) models are indeed applied much more frequently for water body detection, and long short-term memory (LSTM) models are often used for water quality monitoring. The evaluation metrics used in the reviewed articles were also summarized and are provided in Tables 2 and 3 (a brief explanation of each metric is in Appendix C).

Table 2. Studies targeting water body detection from RS imagery using AI (note that it is ordered chronologically to show trends in data type and model usage; see the Abbreviations for a list of the acronyms).

Reference	Method	Model Comparison	RS Data Type	Evaluation Metrics
Li et al. (2011) [28]	DNN	NDWI	Landsat TM	coherence
Yang et al. (2015) [29]	AE	DNN, SVM	Landsat ETM+	accuracy
Huang et al. (2015) [30]	ELM	DT, LORSAL, RF, SVM, TB	GeoEye-1, WorldView-2	Kappa, F-score
Isikdogan et al. (2017) [31]	CNN	MDWI, MLP	Landsat	F1-score, CE, OE, precision, recall
Yu et al. (2017) [32]	CNN–LR hybrid	ANN, CNN, SVM	Landsat ETM+	accuracy
Jiang et al. (2018) [10]	MLP	MLC, NDWI	Landsat-8 OLI	Kappa, OA
Chen et al. (2018) [33]	CNN	CNN, NDWI, SVM	GaoFen-2, Zi Yuan-3	ECE, EOA, EOE, OA, PA, UA
Miao et al. (2018) [34]	CNN	DNN	Google Earth imagery	OA
Acharya et al. (2019) [35]	SVM	ANN, DT, GMB, NB, NDWI, RF, RPART	Landsat-8 OLI	Kappa, OA
Feng et al. (2019) [36]	CNN–CRF hybrid	CNN, CV-method	GaoFen-2, WorldView-2	Kappa, PCC, precision
Li et al. (2019) [37]	CNN	CNN, NDWI, SVM	GaoFen-2	F1-score
Li et al. (2019) [38]	CNN–CRF hybrid	CNN, NDWI	GaoFen-1	IoU, pixel accuracy, recall
Meng et al. (2019) [39]	CNN–SVM hybrid	CNN, SVM	GaoFen-2	accuracy, MA, UA
Isikdogan et al. (2020) [40]	CNN	CNN, MLP, MNDWI	Landsat-8	F1-score, precision, recall
Song et al. (2020) [41]	CNN	CART, KNN, RF, SVM	GaoFen-2, WorldView-3	IoU, precision, recall
Yang et al. (2020) [42]	CNN	CNN	GaoFen-2	IoU
Wang et al. (2020) [43]	CNN	CNN, NDWI	GaoFen-1	F1-score, mIoU, precision, recall
O'Neil et al. (2020) [44]	CNN	DEM, NDVI, RF	LiDAR DEMs, NAIP	precision, recall
Chen et al. (2020) [45]	CNN	NDWI, SVM	GaoFen-1, GaoFen-2, Zi Yuan-3	BOA, Kappa, OA
Dang and Li (2021) [46]	CNN	CNN	GaoFen-2, GID	mIoU, FWIoU, OA
Yuan et al. (2021) [47]	CNN	CNN, MNDWI, NDMI, NDWI	Sentinel-2	accuracy, mIoU
Tambe et al. (2021) [48]	CNN	CNN	Landsat-8 OLI	CA, F1-score, GA, IoU, precision, recall
Yu et al. (2021) [49]	CNN	CNN	GaoFen-2, Landsat-7	F1-score, OA, precision, recall
Li et al. (2021) [50]	CNN	CNN, CV-method, SVM	UAV	Kappa, F-score, OA, precision

Table 2. *Cont.*

Reference	Method	Model Comparison	RS Data Type	Evaluation Metrics
Zhang et al. (2021) [51]	CNN	CNN, MLC, NDWI, SVM	GaoFen-2	IoU, Kappa, pixel accuracy
Li et al. (2021) [52]	CNN	CNN, NDWI, SVM	GaoFen-2, GaoFen-6, Sentinel-2, Zi Yuan-3	F1-score, IoU, OA
Su et al. (2021) [53]	CNN	CNN	Landsat-8, Sentinel-2A	IoU, pixel accuracy, recall
Ovakoglou et al. (2021) [54]	KMeans	fuzzy-rules classification, Haralick's textural features of dissimilarity, Otsu valley-emphasis	Sentinel-1	Kappa, OA, precision, recall

Table 3. Studies targeting water quality monitoring from RS imagery using AI (where "/" means none. Note that it is ordered chronologically to show trends in data type and model usage) (See the Abbreviations for a full list of the acronyms).

Reference	Method	Model Comparison	RS Data Type	Evaluation Metrics
Chebud et al. (2012) [55]	DNN	/	Landsat TM	RMSE, R^2
Wang et al. (2017) [56]	SVR	index methods	spectroradiometer, water samples	RMSE, RPD, R^2
Lee and Lee (2018) [57]	LSTM	DNN, RNN	water quality time series	RMSE
Wang et al. (2019) [58]	LSTM	/	water quality time series	accuracy, cross-correlation
Pu et al. (2019) [59]	CNN	RF, SVM	Landsat-8	accuracy
Liu et al. (2019) [60]	LSTM	ARIMA, SVM	IoT data	MSE
Chowdury et al. (2019) [61]	MLP	/	IoT data	threshold value
Hafeez et al. (2019) [62]	DNN	CB, RF, SVR	Landsat	accuracy, relative variable importance
Li et al. (2019) [63]	RNN–DS hybrid	GRU, LSTM, SRN, SVR	water quality time series	MAE, MAPE, RMSE
Randrianiaina et al. (2019) [64]	DNN	/	Landsat-8	RMSE, R^2
Yu et al. (2020) [65]	LSTM	/	water quality time series	MAE, RMSE
Zou et al. (2020) [66]	LSTM	DNN, GRU, LSTM	meteorological time series, water quality time series	MAE
Peterson et al. (2020) [67]	ELR	MLR, SVR	Landsat-8, Sentinel-2	MAPE, RMSE, R^2
Hanson et al. (2020) [68]	LSTM	/	water quality time series	auto-correlation, MK statistics, RMSE
Barzegar et al. (2020) [69]	CNN–LSTM hybrid	CNN, LSTM	water quality data from multiprobe sensor	MAE, NSEC, Percentage of Bias, RMSE, Wilmott's index
Aldhyani et al. (2020) [70]	LSTM	ANN, DNN, KNN, NB, SVM	water quality time series	accuracy, F-score, MSE, precision, R, sensitivity, specificity
Li et al. (2021) [71]	RF	SVM	Sentinel-2 MSI	RMSE, RPD, R^2, Z-score
Sharma et al. (2021) [72]	CNN	CNN	UAV camera	precision, recall
Cui et al. (2021) [73]	CNN	KNN, index method, RF, SVM	Landsat-8, Sentinel-2	RPD, RMSE, R^2
Zhao et al. (2021) [74]	DNN	RBFNN	Landsat-8, water quality time series	MAE, MSE, R^2
Arias-Rodriguez et al. (2021) [75]	ELM	LR, SVR	Landsat-8, Sentinel-2 MSI, Sentinel-3 OLI	MAE, MSE, RMSE, R^2
Kravitz et al. (2021) [76]	DNN	KNN, RF, XGBoost	Landsat 8 OLI, Sentinel-2 MSI	MAPE, RMSE, RMSLE [1]
Sun et al. (2021) [77]	DNN	GPR, RF	proximal hyperspectral imager, water samples	accuracy, MRE, RMSE, R^2

[1] The authors use the abbreviation RMSELE for RMSLE in their paper (this might be a typographical error).

3.2. Recent Advances in Water Body Detection Using AI

From our systematic review (including Table 2), we provide a brief summary here about the recent advances in water body detection based on AI. (1) The most common satellite platforms were Landsat, GaoFen, Zi Yuan, WorldView, and Sentinel, although there were some manually annotated datasets. The use of UAVs and DEMs were noted but were not as common. (2) Precision, recall, overall accuracy (OA), F1-score, kappa, and intersection over union (IoU) are the most popular evaluation metrics for water body detection since it is mainly a classification task. (3) Convolutional neural networks (CNNs) are normally compared to normalized difference water index (NDWI) or another index-based method, some form of "shallow" ML model (e.g., random forest (RF), support vector machine (SVM)), or other CNN architectures). Below, we provide a more detailed review of the methods used for water body detection. As Table 2 and word clouds (see Figure 2) indicate, the dominant methods used in water body detection with AI are CNNs (Section 3.2.1). Beyond CNN-based methods, there are other methods including CNN hybrids (Section 3.2.2), artificial neural networks (ANN), multilayer perceptrons (MLP), dense neural networks (DNN), other DL methods (Section 3.2.3), and "shallow" ML based methods (Section 3.2.4).

3.2.1. CNN-Based Water Body Detection

CNN-based models are the dominant methods for water body detection, but each of them have addressed different challenges posed in water body detection from RS imagery. Based on our review, we identify the following five groups of use cases: (1) Addressing limitations of index-based methods; (2) sharpening blurred boundaries caused by CNNs; (3) Addressing spatial and spectral resolution challenges, which covers those methods that are able to recognize water body across scales, at multiple resolutions, from very high-resolution imagery, and/or integrating bands beyond RGB channels to use for CNN model training; (4) Robust detection of small/slender/irregular-shaped water bodies; (5) Others.

1. Addressing limitations of index-based methods:

Index methods (e.g., NDWI) are rule-based and fail to take advantage of context information. CNNs can overcome this, although they often blur boundaries in segmentation tasks because of the convolution operation [34]. A DenseNet was used in [43] for water feature extraction and the authors compared its performance with NDWI and several popular CNN architectures. While NDWI methods are quick, they are not as accurate as CNNs. The authors showed that DenseNet performed the best at distinguishing water from shadows and clouds. However, the authors argue that clouds often occlude optical imagery, so one way to improve their method is to combine it with microwave RS imagery.

The authors in [31] pointed out that index methods require careful calibration and that indices differ from place to place. They also suffer from false positives (from snow, ice, rock, shadows, etc.) and vary in different weather conditions (e.g., clouds). To overcome those limitations of index-based methods, the authors of [31] developed DeepWaterMap, which can classify water with high accuracy, even distinguishing it from snow, ice, shadow, and clouds. DeepWaterMap is able to classify land classes that are often misclassified as water (or vice versa); thus, it minimizes false positives during the classification process. Most importantly, the DeepWaterMap model also works across different terrains and in different weather conditions, although it is still affected by clouds. The same authors released a second version of the model, DeepWaterMap v2, in [40]. The major improvement from v1 is that the new version allows users to input large RS scenes without the need for tiling, and the authors made their network run efficiently with constant memory at inference time. This model should theoretically work across different sensor platforms as long as they have the visible, near-infrared, and shortwave infrared 1 and 2 bands, but will still sometimes classify clouds as water.

2. Sharpening blurred boundaries caused by CNNs:

CNN-based methods can overcome the limitations of index-based methods, as reported above in group (1) [34], but they often blur boundaries in segmentation tasks because of the convolution operation. To sharpen water body detection boundaries, in [34], a restricted receptive field deconvolution network (RRF DeconvNet) and a new loss function called edges weighting loss were proposed. However, the authors needed to retrain the entire network (which is very computationally expensive) instead of using transfer learning (TL).

Apart from blurring pixel boundaries, CNNs generally require many training parameters and very large training datasets to be successful. A novel convolution–inception block in a network, called W-Net, was proposed in [48], to extract water bodies from RS imagery. W-Net is able to train on fewer images compared with other CNN models and still extract water bodies accurately, and the authors pointed out that less computations are necessary due to use of inception layers. W-Net outperformed other CNN architectures, although the authors still needed to go through the time- and labor-intensive process of creating a dataset of manually annotating images.

3. Addressing resolution and band related challenges

High-resolution optical RS imagery allows for much finer detail in surface water body extraction. However, clouds and their shadows are often present in optical RS images [78]. The shadows (e.g., cloud shadows and building shadows) and water bodies share a very similar appearance in optical RS images. Therefore, water body extraction is not an easy task in the optical high-resolution RS images due to the limited spectral ranges (including blue, green, red, and near-infrared bands) and the complexity of low-albedo objects (cloud shadows, vegetation, and building shadows). Higher spatial resolution imagery often comes at the cost of less spectral channels and thus makes it difficult to extract features from complex scenes. To address this problem, a dense local feature compression (DLFC) was proposed [52] to extract bodies of water from RS imagery, and their DLFC outperformed other state-of-the-art (SOTA) CNNs, as well as an SVM and NDWI thresholding. Their results demonstrated that the DLFC is good at extracting slender water bodies and distinguishing water bodies from building shadows using multisensor data from multiple RS platforms.

TL and data augmentation (see Appendix B) are used in [37] to extract water bodies from satellite imagery. The authors showed that a CNN can outperform NDWI and an SVM in water body detection when the input data is very high resolution. There are tradeoffs, however, and the authors reported that the difficulty of hyperparameter tuning is one downside to using a CNN. A water body extraction NN, named WBE-NN, was proposed in [45] to extract water bodies from multispectral imagery at multiple resolutions while distinguishing water from shadows, and performed much better than NDWI, an SVM, and several CNN architectures. A self-attention capsule feature pyramid network (SA-CapsFPN) was proposed in [49] to extract water bodies from satellite imagery of different resolutions. SA-CapsFPN is able to recognize bodies of water across scales and different shapes and colors, as well as in varying surface and environmental conditions, although it is still entirely dependent on optical imagery as input to the CNN.

The novel MSResNet proposed in [46] learned from a large dataset of unlabeled RS imagery. MSResNet, in addition to being able to extract water bodies in an unsupervised manner, is able to recognize water bodies at multiple resolutions and of varying shapes. However, their network cannot distinguish water bodies from farms and barren areas. In addition, the CNN-based model name FYOLOv3, proposed in [51], is able to detect tidal flats at different resolutions. However, it does depend on a manually selected similarity threshold that introduces some subjectivity.

RGB band imagery is the primary focus in substantial research for water body extraction, but many more bands are available in RS imagery. A multichannel water body detection network (MC-WBDN) was created in [47], which fused the infrared and RGB channels and used them as input data for their CNN architecture. They demonstrated that when multispectral data is used, model performance for water body detection is increased and the model is more robust to lighting conditions. The proposed model MC-WBDN is much more accurate than index-based methods such as NDWI, modified NDWI (MNDWI), and normalized difference moisture index (NDMI). MC-WBDN also outperforms other SOTA architectures such as U-Net and DeepLabV3+ for water body detection tasks. However, this method still relies on preprocessing data to make sure each input image is the same shape and free of clouds.

4. Robust detection of small/slender/irregular-shaped water bodies

Small water bodies are hard to extract from RS imagery. In [33], the authors designed a CNN (named SAPCNN), which is able to extract high-level features of water bodies from input data in a complex urban background. NDWI and SVMs cannot distinguish between water and shadows and their architecture's performance partly relies on visual inspection. Ref. [53] utilized a modified DeepLabv3+ architecture to extract bodies of water at different scales. Their focus is on extracting water bodies in urban RS images. Their network performed well on small bodies of water, but the model has problems identifying many of them because they were not properly annotated.

Mask-region-based CNNs (R-CNNs) have demonstrated success in detecting small and irregular shape water bodies. Song et al. (2020) [41] employed an R-CNN for water body detection from RS imagery, and their model outperforms many traditional ML models in identifying small water bodies and bodies of water with differing shapes. However, it is still difficult to deploy a trained NN model into a usable, production-ready form for water mapping applications. The authors reported that using NN output to create and update a vector map of water resources for stakeholders is challenging.

Yang et al. (2020) [42] also used a mask R-CNN to automate water body extraction. The authors argued that this allows them to avoid manual feature extraction in complex RS imagery. They segmented small water bodies and bodies of water with irregular shapes, although their methods suffer from poor IoU accuracy. This is primarily due to a small training set, for which DL models are ill-suited, and resulted in their models having problems identifying multiple bodies of water in RS images.

A self-attention capsule feature pyramid network (SA-CapsFPN) was proposed in [49] to extract water bodies from satellite imagery. SA-CapsFPN is able to recognize bodies of water across scales and different shapes and colors, as well as utilizing different information channels. The novel MSResNet proposed in [46], learnt from unlabeled large RS imagery, is also able to recognize water bodies at multiple resolutions and of varying shapes; however, their network cannot distinguish water bodies from farms and barren areas.

A dense local feature compression (DLFC) was proposed in [52] to extract bodies of water from RS imagery, and their DLFC outperformed other SOTA CNNs, as well as an SVM and an NDWI. Their results demonstrated that the DLFC is good at extracting slender water bodies and distinguishing water bodies from building shadows using multisensor data from multiple RS platforms.

5. Others

Extracting water bodies from RS imagery quickly and reliably is still a difficult task. Based on U-Net, [50] developed a new model called SU-Net to distinguish between water bodies, shadows, and mixed scenes. However, the authors only focused on water body extraction in urban areas and only used RGB information during the extraction process. While SU-Net performed better than an SVM and classic U-Net, it suffered when extracting water bodies from RS imagery with high reflectivity or that contained aquatic plants.

Wetlands are important ecosystems because they can keep flooding at bay and store carbon; however, they are threatened by development, climate change, and pollution. For the task of identifying wetlands, [44] combined RS imagery with hydrological properties derived from digital elevation models (DEMs) to identify wetlands. They showed that an RF performs as well as a CNN, although both models had issues distinguishing roads and trees from wetlands. This is perhaps due to their small training set. To improve performance, the authors argued that larger datasets with finer labels should be created for wetland detection.

Substantial water body detection work has focused on water bodies in urban and inland settings. Very few focus on tidal flat extraction, where sediment levels are high and the boundary of the water body itself is blurry. A CNN model called FYOLOv3 was proposed in [51], where the authors compared their model to NDWI, an SVM, a maximum likelihood classifier, U-Net, and YOLOv3. FYOLOv3 performed the best and is able to detect tidal flats at different resolutions; however, it depends on a manually-selected similarity threshold during the training process, which is a source of subjectivity.

Large sets of unlabeled water body data are available and easy to acquire, and semantic segmentation networks cannot recognize different water body shapes. A recent, very novel encoder–decoder CNN architecture named MSResNet, proposed in [46], is able to overcome those limitations. MSResNet is able to learn from unlabeled data and can also recognize water bodies of varying shapes and at multiple resolutions. However, even though their network outperforms other SOTA architectures without supervised training, their network has some issues categorizing water bodies, farms, and barren areas.

3.2.2. CNN Hybrid-Based Water Body Detection

CNNs are the SOTA models in water body extraction tasks (detailed in Section 3.2.1 above); however, their output and decisions for why they make the predictions that they do are largely a black box. Recent studies have integrated CNNs with some ML models. Interpretability was improved by using a CNN and SVM in parallel to classify wetland water bodies [39]. Wetlands are difficult/complex to identify in high-resolution satellite imagery with any single ML model. Hybrid models have shown promise in a process called decision fusion. Here, the authors pick a decision fusion threshold value by performing cross-validation on the CNN to see when it is sure or not. They then use this threshold value for the decision fusion predictions (e.g., when the CNN is not that sure, they defer to the SVM). However, the authors did not explain why they used an SVM and not some other ML model. The classifier used in [32] combines a CNN with a logistic regression (LR) model to extract water bodies. The authors emphasized that traditional ML methods for water body extraction need multispectral data and rely on lots of prior knowledge. Thus, those ML-based methods would not generalize well to different tasks. The authors also argue that single-band threshold methods are subjective. Their results demonstrated that the hybrid CNN-LR model works better than an SVM, an ANN, and other CNNs. However, their method requires segmented RS images as input.

How to accurately extract water bodies from RS images, while continuously updating the surface water maps, is an active research question. Index methods and active contour models are popular methods for water body detection tasks but are sensitive to subjective threshold values and starting conditions. Deep U-Net model was proposed to be used with a conditional random field (CRF) and regional restriction to categorize water versus non-water in satellite images [36], while reducing the blurring of edges that often occurs from CNNs for image segmentation. Although this network is highly accurate, it takes a lot of data and computation power to train. Training ML models at a single scale in single channels can cause errors when generalizing to other scales or types of RS data. Multiscale RS imagery was used with DeepLabV3+ and a CRF for water body segmentation [38]. This approach works well for training models on data from different scales, and they concluded that CNNs and CRFs together extract more accurate water boundaries at both large and small scales than CNNs alone.

3.2.3. ANN, MLP, DNN, and Other DL-Based Methods for Water Body Detection

An NN architecture called a local excitatory globally inhibitory oscillator network (LEGION) is used in [28], where the authors compared the results of LEGION trained on NDWI and spectral information, respectively. In addition, they employed object-wise classification, instead of pixel-based classification used in most other work. The authors reported that the network is very computationally expensive.

Different methods of water body extraction work (or do not work) in different areas/terrain types. Each needs subjective thresholds and/or hand-crafted features. In addition, generating large sets of labeled data is difficult and expensive, as high-dimension RS data is difficult to analyze. Objects such as shadows, clouds, and buildings are hard to distinguish from water bodies. In [29], the authors used an autoencoder for unsupervised training and concluded that their results are more accurate than for an SVM and traditional NN.

Huang et al., 2015 [30] pointed out that not many people have focused on water body detection in urban settings. This is a problem because water bodies often look similar to shadows due to buildings at certain times of the day in optical imagery. The authors employed an extreme learning machine (ELM), an SVM, a tree bagger (TB), and an RF to detect water bodies. The authors reported that the RF and TB performed much better than the SVM and ELM. However, their method depends on optical imagery with subjective thresholds set through trial and error. Specifically, their method depends on subjective threshold values in NDWI, normalized difference vegetation index (NDVI), and morphological shadow index (MSI).

Ref. [10] compared MLP, NDWI, and a maximum likelihood model for water body classification and showed that MLP performed the best. However, the maximum likelihood model could not recognize small bodies of water and thin rivers, whereas NDWI was not able to distinguish seawater from land. The MLP could identify small bodies of water better, but the analysis depended on visual assessment.

3.2.4. "Shallow" ML-Based Water Body Detection

Although most of the recent methods for water body detection used DL and/or deeper neural networks (Sections 3.2.1–3.2.3), a few studies used only "shallow" ML methods (e.g., RF and SVM). In [35], the authors used band methods (where slope, NDVI, and NDWI were added as three secondary bands to integrate extra information into ML training), and then applied an SVM, a decision tree (DT), and an RF to analyze multiband RS data for water body extraction in the Himalayas. However, while their models worked well for flat and hilly terrain, they had to parse out high elevations and snow in this method (which involves extra preprocessing and limits when/where their method can work with optical data). The authors ran different experiments to analyze which input bands (NDWI vs. individual input bands from Landsat data) worked the best but could only compare results visually. The authors concluded that adding single secondary bands is better than adding multiple in most ML algorithms except for NNs.

Sentinel-1 data and four different ML models (K-nearest neighbors classifier (KNN), fuzzy-rules classification, Haralick's textural features of dissimilarity, Otsu valley-emphasis) were employed to classify water bodies in [54]. It involved many different ML methods in tandem (i.e., the output of one ML model was fed into other processing steps), which complicates interpretability. This method did not have very high accuracy and did not work well in flooded regions, near buildings, and in the presence of aquatic vegetation. However, it was an important attempt to use synthetic aperture radar (SAR) data, which is rare in water body detection literature.

3.3. *Recent Advances in Water Quality Monitoring Using AI*

From Table 3, we identify the following trends in the use of AI for water quality monitoring research: (1) Water quality monitoring differs from water body detection in that it is formulated as both a classification and a regression task. Because of this, recurrent neural networks (RNNs), long short-term memory (LSTMs), and gated recurrent units (GRUs) are much more prevalent in the water quality literature. (2) Accuracy, precision, and recall are common metrics, as are some variations of mean squared error (MSE) and R^2. (3) It is important to note that while water body detection papers describe integrating multiple data sources into one analysis, this practice is much more common in water quality monitoring research. This primarily takes the form of trying to match up water quality parameters from time series data or water samples to optical satellite RS imagery. In water quality monitoring, it is much more common to utilize Internet of Things (IoT) sensors, smaller probes such as unmanned aerial vehicle (UAVs) and stationary hyperspectral imagers, as well as government and private water quality time series data. (4) Some studies do not compare their model to any other models (detailed in Table 3), making it difficult to fully assess their methodologies.

Below, we provide a more detailed review of the methods used for water quality detection and monitoring. As our manual investigation (see Table 3) and machine summary (word cloud, see Figure 1) indicate, the dominant methods used in water quality detection with AI are LSTMs (Section 3.3.1) and ANNs, MLPs, DNNs, and other DL methods (Section 3.3.5). Beyond LSTM and ANN-based methods, there are other methods including LSTM hybrids (Section 3.3.2), CNN-based methods (Section 3.3.3), and "shallow" ML-based methods (Section 3.3.4).

3.3.1. LSTM-Based Water Quality Detection and Monitoring

Algal blooms cause serious harm to human and animal health and can damage both environments and economies. Various factors lead to algal blooms and gathering the data necessary to predict them is time- and cost-intensive. ML models can provide advanced warning for these events by taking into account time series data of basic water quality parameters. A linear regression model was compared with an MLP, an RNN, and an LSTM to predict harmful algae blooms in dammed pools from several rivers [57]. While the LSTM model was the most accurate overall, for several of the dammed pools that the authors tested, a least-squares regression model outperformed the LSTM. This casts doubt as to how the LSTM model generalizes and if it is worth the added complexity.

Water pollution is becoming an increasing problem because of rapid rates of development and urbanization. Large amounts of water quality parameters can be taken via IoT sensors, and DL techniques are well suited to finding patterns in the large quantity of data. An LSTM was used to predict future values of different water quality parameters [60]. Most importantly, the authors only used single-dimensional inputs and outputs (i.e., a 1D time series of dissolved oxygen as an input to predict dissolved oxygen at some time in the future). While the results were good, the authors noted that the architecture would benefit from training on multiple time series at the same time. The authors reported that long-term predictions on the order of 6 months into the future did not work well. Beyond monitoring water for different levels of pollutants, it is also important to find the sources of pollutants when they are identified. Cross-correlation was used to map pollutants to different water quality parameters [58]. They then used an LSTM to match pollutants to nearby polluting industries using the highly correlated water quality parameters.

Similar to LSTMs, RNNs have been demonstrated to be accurate for times series prediction but are also often criticized for being difficult to interpret. Meanwhile, process-based ecological models, although deterministic, fail to capture patterns at longer time scales. A process-based model was integrated with an RNN to better align predictions of phosphorus levels in lakes to eliminate outlier predictions. Constraining NN output with physics-based models better aligns their predictions with ecological principles [68].

Rapid development has led to decreased water quality. In [70], water quality parameters can be used to both classify the current water quality index and predict future water quality index states. However, the authors separately compared DL models for water quality prediction and ML models for water quality classification, making the methods not directly comparable. A nonlinear autoregressive neural network (NARNET), a type of ANN, performed better than an LSTM at predicting the water quality index, while an SVM performed better than other traditional ML models for classification.

3.3.2. LSTM Hybrids Water Quality Detection and Monitoring

To further improve model performance, a few recent studies have integrated other models with LSTMs. Water scarcity and drought are increasingly significant environmental challenges. Increased development is leading to worsening water pollution. Predicting the water quality from time series data is essential, but traditional ML models fail to capture long-term temporal patterns. This causes them to make false predictions in water quality monitoring applications. An RNN–Dempster–Shafer (RNN–DS) evidence theory hybrid model was used to make sense of multiple input time series of different time scales [63]. While evidence theory did make the predictions more stable, longer-term predictions did not work very well, even with the improvements to the model. The authors pointed out one possible reason might have been not taking spatial correlations between water quality parameters into account.

Economic development and urban growth have posed water quality issues. Wavelet domain threshold denoising (WDTD) and wavelet mean fusion (WMF) were used to analyze the output of LSTM predictions for multiple water quality parameters [65]. While multiple wavelet basis functions were used to analyze predictions, the LSTM was not compared to any other models in this analysis. The authors noted that not having enough observations was a limitation while training their LSTM model.

Mangrove wetlands provide habitats for many different types of animal species in addition to preventing coastal erosion. More recent research has focused on monitoring the water quality in these environments to assess the health of coastal ecosystems. Using water quality and meteorological time series data, three different submodels were used for each water quality parameter at different time intervals and fused their output predictions [66]. The authors tested this setup with a DNN, a gated recurrent unit (GRU), and an LSTM model. While the LSTM performed the best, the authors reported that the model is not very reusable or user-friendly.

Collecting and analyzing water samples is expensive, time-consuming, and labor-intensive. Thus, many researchers choose to use sensors to remotely monitor water quality parameters, but the number of parameters they can record are often limited. Ref. [69] used a submerged multiprobe sensor to monitor several important water quality parameters over the course of 1 year. They found that a CNN–LSTM model performs better than standalone DL models and traditional ML methods for predicting water quality parameter values; however, the authors did not use a validation set during NN training and the hybrid model was able to quickly learn the training and testing set data distributions.

3.3.3. CNN-Based Water Quality Detection and Monitoring

CNNs are the dominant architecture for water body detection (Sections 3.3.1 and 3.3.2) but are not used as widely for water quality. Here, we review two very interesting but effective CNN-based methods. In situ water quality measurements work really well but are very expensive. In addition, things such as total nitrogen and phosphorus, biological oxygen demand, and dissolved oxygen are hard to measure from satellites because they have weak optical properties. A CNN was used in [59] and showed that TL beats out traditional ML models when classifying water quality from RS imagery. However, their dataset was very small, and their focus was narrow (specifically, only two lakes in China, no rivers or coastal waters covered). Water bodies are often polluted, or their quality is affected from far away and thus it is difficult to identify and report on water quality. Methods for estimating water quality at scale are essential. Turbidity can be a proxy for total suspended solids (TSS) and suspended sediment concentration (SSC), so [72] used image detection and then applied edge detectors to UAV images of water. They employed CNNs to detect changes in water color and utilized this to approximate quality. They showed that image-based turbidity detection is as accurate as actual turbidity meters, but more importantly represents a very promising method for monitoring water quality at greater spatial scales.

3.3.4. "Shallow" ML-Based Water Quality Detection and Monitoring

Remote water bodies are hard to monitor for water quality. A simple NN architecture was designed to estimate several water quality parameters (i.e., chlorophyll-a, turbidity, phosphorus) both before and after an ecosystem restoration project during both the dry and wet seasons [55]. Importantly, their predictions, using seven different input bands for training the NN, were very close to the actual values.

Finding what data to input into an ML model for water quality monitoring is neither easy nor straightforward. Different indices are sensitive to different areas and varying weather and lighting conditions. To address this problem, [71] first correlated water quality parameters to different RS bands. These correlations were then used to test four ML models and their ability to predict a water quality index. Their R^2 statistics were not high, though.

3.3.5. ANN, MLP, DNN, and Other DL-Based Methods for Water Quality Detection and Monitoring

Climate change is making droughts and water shortages increasingly worse in arid regions. It is thus important to develop methods and systems for intelligent and efficient monitoring of the water resources in those regions. A water quality index for arid regions was proposed in [56] and attempted to find which bands and spectral indices are related to that water quality index. In situ water quality sampling is labor- and cost-intensive and often suffers from low temporal resolution. As bodies of water around the world are changing rapidly due to global warming, it is more important than ever to model their spatial variation through time. A point-centered regression CNN (PSRCNN) was used in [73] to analyze lake reflectance data to model water transparency. The authors concluded that their model outperformed different band ratios and traditional ML models (KNN, RF, SVM), although at the cost of generalization. The PSRCNN did not make stable predictions due to too little data.

There is currently not enough paired RS imagery and in situ water measurement to meaningfully create robust water quality monitoring applications. The generation of a synthetic dataset of atmospheric reflectances and its suitability for water quality monitoring were investigated in [76]. The synthetic dataset is physics-based and attempts to capture the natural variability in inland water reflectances and chlorophyll-a concentrations. An ANN outperforms several traditional ML models (KNN, RF, XGBoost) in predicting actual water quality parameter values when trained on the synthetic dataset, although only the ANN is validated against unseen data. Still, synthetic data generation is a promising research direction for water body and water quality detection. Without RS imagery, many water quality monitoring programs will suffer from lack of spatial coverage due to labor, time, and cost constraints. Yet while RS is a useful tool for monitoring water quality parameters, it has not been meaningfully integrated into operational water quality monitoring programs. Existing water quality time series data were used in [75] and assessed the effectiveness of multiple RS data platforms and ML models in estimating various water quality parameters. The authors showed that some sensors are poorly correlated with water quality parameters, while others are more suitable for water quality monitoring tasks. They concluded that more research needs to be carried out for assessing the suitability of paired RS imagery and in situ field data.

Current water quality monitoring systems are labor-, time-, and cost-intensive to operate. IoT sensors can monitor water quality parameters in near real time, allowing for much more data to be recorded with much higher temporal resolution. A wireless sensor network made up in part of IoT sensors was used in [61], and used an MLP to classify water quality as either good or bad. The authors utilized the MLP predictions to notify water quality managers via SMS if the water quality drops below a certain threshold value. However, because of the cost to deploy and run the network, the authors were not able to include additional water quality parameters from more types of bodies of water other than rivers. Water quality monitoring data collection is expensive and time consuming, and there are usually tradeoffs between spatial and temporal resolution when implementing data collection programs. In addition, several key water quality parameters (pH, turbidity, temperature) can be estimated directly from optical and infrared RS imagery. Randrianianina et al. [64] used RS imagery and DNNs to model water quality parameters directly, after which they extend their analysis to map the distributions of water quality parameters to an entire lake, but they only focused on one lake and did not test their methods on other bodies of water.

As bodies of water are exposed to increased nutrient loads, harmful algal blooms can occur, leading to eutrophication. This process can create dead zones that would kill wildlife and lead to negative economic impacts. Thus, it is important to monitor chlorophyll-a levels in water bodies and predict algal blooms before they happen. Zhao et al. [74] attempted to address this need by comparing DL models to traditional ML and curve-fitting methods to predict chlorophyll-a levels using time series measurements paired with RS imagery. The authors did not have much data as they limited the data collection process to one lake. Thus, the DL models did not perform well. Additionally, the ML models used in this paper needed more data and computing than simpler models in order to perform well.

It is often difficult to monitor inland water bodies for quality because of low signal-to-noise ratios and limitations in resolution. A proximal hyperspectral imager was used in [77] with high spectral and temporal time series data for continuous water quality observations. The authors found that index-based methods of water quality detection were difficult to calibrate as thresholding values are subjective, while ML and DL models performed much better. However, the authors show that their models do not generalize well to other water bodies with different water quality parameter distributions.

Anthropogenic activities have currently threatened largely coastal ecosystems. Coastal ecosystems are complex bodies of water but monitoring them is very important. The performance of an ANN was compared to traditional ML models in [62] for predicting various water quality parameters. In some cases, traditional ML methods outperform the ANN. More importantly, the authors conducted an analysis of relative variable importance to show which sets of input data helped the ML models to learn the most. While the relative variable importance analysis is critically important, the authors only test their method in cloud-free RS imagery, limiting its utility. Additionally, while biophysical and chemical water quality parameters were analyzed, little work was carried out with bio-optical data due to issues with data availability.

While recent advances in RS capabilities for water quality detection are substantial in the literature, few papers have collected and synthesized the resources available to researchers. In a paper reviewing recent trends in RS imagery, cloud computing, and ML methods, [67] used time series data from hundreds of water quality parameters and water samples and combined them with proximal imagery, hyperspectral imagery, and two sets of data from different satellite data platforms. They showed that DNNs outperform many other traditional processing and ML techniques for assessing water quality. The authors conclude that anomaly detection using multisensor data is the most promising method for algal bloom detection. As is sometimes the case in the water body detection and water quality monitoring literature, the authors did not have a third holdout set (necessary for DL projects so that the data is not memorized).

4. Challenges and Opportunities

In this section, we first provide a brief summary and discussion of the key themes and overall insights (Section 4.1) derived from reviewing the range of research discussed above. In Section 4.2, we provide and discuss some of the major challenges we identified through our systematic survey. Specifically, those challenges shared in both domains are detailed in Section 4.2.1, those specific only to water body extraction in Section 4.2.2, and those specific to water quality monitoring in Section 4.2.3. Finally, we discuss possible research directions and related opportunities for water body detection and water quality monitoring using RS and AI in Section 4.3.

4.1. Summary and Discussion

After introducing the essential terms in AI and RS (Appendix B) and commonly used evaluation metrics in ML and DL for classification, regression, and segmentation tasks

(Appendix C), we reviewed recent and influential research for water body detection and water quality monitoring using RS and AI (Section 3).

While the research investigated in Section 3 has demonstrated the power of using RS and AI to detect water bodies and monitor water quality, very few studies thus far performed integrative research of water body and water quality using the power of RS and AI. In addition, most existing RS and AI-based work on water bodies and water quality repeat the same (or very similar) methods in a different research location or on a different (usually small) dataset. However, real intelligent water resource management applications will require serious development that goes beyond this type of research. Before operational applications can be deployed, AI models (especially DL models) need to be trained on large and representative benchmark datasets with a focus on making models generalizable and interpretable.

We noticed that most work does not include hardware specifications (e.g., what CPU/GPU the authors used to run their models) and/or processing time. To make models comparable and for the sake of replicability and reproducibility, it is essential to report such information. This is even true for index-based methods and more traditional ML models so that researchers can fully evaluate the trade-offs between runtime, accuracy, and ease of implementation. We hope our review will provide a useful guide to make future research more replicable and reproducible. From our interactive web app (the web app tool URL and its brief demo video link are provided in Appendix A), we also noticed that while most papers have an open access PDF/HTML version of their manuscripts, a sizable portion of manuscripts (16 out of 56 of reviewed articles) do not. We suggest authors provide an open access version (e.g., posting the proofreading version after acceptance to ResearchGate/arXiv) in order to increase the visibility of their research and thus to accelerate the advancement of scientific knowledge.

4.2. Identified Major Challenges

Below, we provide the most commonly posed challenges for water body and water quality research in the literature we reviewed. Those challenges shared in both domains are outlined in Section 4.2.1 and those specific to each domain are detailed in Sections 4.2.2 and 4.2.3, respectively. Here are some specific issues to water body detection and water quality monitoring.

4.2.1. Shared Common Challenges in Both Domains

A summary of the shared common challenges and identified problems in water body extraction and water quality monitoring using RS and AI are provided below.

- Methods for water body detection and water quality monitoring need to be able to work quickly and reliably on large spatial and temporal scales, and yet high-resolution RS imagery is very complex. Index methods rely on subjective threshold values that can change over time and space depending on weather conditions. Shallow ML models are more accurate, but do not work at scale. DL models are complex, require very large datasets to train on, and are very computationally expensive; also, the hyperparameter tuning process is very tedious and difficult.
- It is difficult to know exactly what data to feed to ML and DL models, and it is difficult to know what to make of the output predictions. This often requires integrative expertise and/or interdisciplinary collaboration of RS, hydrology, biology, and CV/ML expertise.
- NNs generally perform the best in water quality and water body detection tasks but are often the least stable models (i.e., they do not generalize well). This is not surprising, as the datasets used in RS problem settings are often not large enough to allow NN models (too many parameters compared with shallow ML models) to overcome overfitting (see Appendix B). Table 4 summarizes the relatively few existing datasets we identified through our systematic review.

- Both domains over-rely on optical RS imagery, and thus clouds and shadows are a persistent problem and heavily skew the results towards working only in cloud-free conditions.

Table 4. Existing datasets for waterbody extraction and water quality monitoring.

Datasets	Source	Category	Link to the Dataset	Notes
DeepWaterMap v2	[40]	Water body	https://github.com/isikdogan/deepwatermap, accessed on 15 December 2021	>1 TB of Landsat-7 imagery paired with Global Inland Water dataset labels.
2020 GaoFen Challenge	/	Water body	https://github.com/AICyberTeam/2020Gaofen, accessed on 15 December 2021	Dataset containing both 2500 optical and 1200 SAR satellite images with pixel level labels for water body segmentation.
GID-15	[79]	Water body	https://captain-whu.github.io/GID15/, accessed on 15 December 2021	150 pixel-level annotated GaoFen-2 images for semantic segmentation tasks.
LandCover.ai	[80]	Water body	https://landcover.ai/, accessed on 15 December 2021	A dataset from 2015–2018 of 10,674 annotated tiles of RGB imagery with labeled water bodies.
SEN12MS	[81]	Water body	https://mediatum.ub.tum.de/1474000, accessed on 15 December 2021	A curated dataset of 180, 662 georeferenced multispectral Sentinel-1 and -2 imagery with MODIS land cover labels.
AquaSat	[82]	Water quality	https://github.com/GlobalHydrologyLab/AquaSat, accessed on 15 December 2021	600,000 data matchups between satellite imagery and water quality measurements from 1984–2019.
Forel–Ule Index	[83]	Water quality	https://doi.org/10.6084/m9.figshare.13014299, accessed on 15 December 2021	151 data matchups between satellite imagery and water quality measurements from 2000–2018.

4.2.2. Additional Challenges in Water Body Extraction

The specific challenges and problems identified for water body extraction are summarized below.

- The majority of reviewed research focused on inland bodies of water, where only a few papers discussed applications for coastal waters (not including oceans). Moreover, many papers focus solely on only one type of water body, for example, only on lakes or rivers in a specific area. As a result, water bodies from different landscapes (e.g., inland, coastal tidal flats, urban, wetlands) are difficult to recognize with one unified method (i.e., methods do not generalize). The same applies to water bodies of different colors, especially when distinguishing them from rock, ice, snow, clouds, and shadows.
- There are very few benchmark datasets. In contrast, there are huge volumes of unlabeled data not being fully leveraged.
- CNNs blur output boundaries during the segmentation process.

4.2.3. Additional Challenges in Water Quality Monitoring

The specific challenges and problems identified for water quality monitoring are summarized below.

- Collecting in situ water quality data is very time- and labor-intensive and financially expensive; also, it often does not have adequate temporal or spatial resolution.
- RS imagery and existing corresponding field samples are often not stored together. Allowing water quality researchers to easily retrieve and locate two or more sources

of data at the same location is critical, as computational methods require such data to verify their model performance in order to generalize to new water bodies.

- Remote water bodies are difficult to monitor.
- Urbanization, pollution, and drought are having serious effects on the economy, wildlife, and human health as they deteriorate water quality.
- Ecosystems are complex and their nutrient and pollution budgets are not well understood.
- Some studies do not use a training, validation, and testing set for DL projects (all three are necessary) or do not use nearly enough data to achieve good results with DL models.

4.3. Research Directions and Opportunities

Here, we provide five research directions, each along with its promising opportunities, from our investigation and based on the posed challenges discussed in Section 4.2 above.

4.3.1. Urgent Need of Large and Comprehensive Benchmark Datasets

Large representative, balanced, and open-access benchmark datasets are critical for any domain to let AI meaningfully shine [84–86]. In computer science, especially for its branches CV and DL, there are very comprehensive, large, and open-source databases (e.g., ImageNet [87] for image classification tasks, and Microsoft COCO [88] for object detection and segmentation tasks). The availability of big and open-source image repositories has dramatically boosted recent advances in novel and robust algorithms in DL and CV, as computer science researchers do not need to worry about collecting datasets. Instead, they can focus on developing new algorithms and/or methods.

In our systematic review, we identified an urgent need for more curated, labeled datasets for intelligent water body extraction and water quality monitoring. We found some of the few available open-source datasets with water body boundary labels through our literature review, but also sought out additional datasets. We identified datasets that were not used in our literature review but contain water body labels, or datasets that were used for water body detection or water quality monitoring that did not use ML/DL/CV but would be useful for benchmarking tasks. Our search results are summarized in Table 4 above. Below, we list a few opportunities in this direction.

(1) *More public data and code: currently, most authors do not share their code and/or datasets.* See the two quoted pieces below from [25]: (a) "Lack of deep learning-ready datasets within the water field [...] The main problem caused by this absence of many datasets is that the research community does not build upon previous work in terms of constructing better neural network architectures and moving the state of art to the next iteration [...]"; (b) "[...] many papers are published that achieve the same task with almost identical methods but different data.". Part of this issue is a replication crisis in the water body detection and water quality monitoring literature, but it stems more broadly from the lack of public codebases and datasets.

(2) Some promising ways to generate large datasets of good quality

- AI/ML/DL models need large datasets with good quality to guarantee meaningful (unbiased and generalize well) good to great performance, thus work on obtaining large but better subsets of data. Quality > quantity is critical and in urgent demand. See one piece of such evidence reported in [44], "[...] site-specific models improved as more training data was sampled from the area to be mapped, with the best models created from the maximum training datasets studied: [...] However, performance did not improve consistently for sites at the intermediate training data thresholds. This outcome exemplifies that model improvement is an issue of not only increasing the quantity of training data, but also the quality".
- Generating synthetic data as in [76] (detailed in the second paragraph in Section 3.3.5).
- Downloading RS images from Google Earth Engine (GEE) and annotating accordingly, or, even better, developing user-friendly interactive interfaces with GEE as a backend to directly allow researchers (or even citizen science volunteers) to contribute to the

annotation of RS imagery available on GEE. To our knowledge, no RS datasets for water body detection and water quality monitoring are downloaded from GEE and then annotated, let alone interfaces for directly annotating RS imagery on GEE.

- Obtaining RS imagery from Google Earth (GE) manually or with the help of code scripts, then annotating accordingly (see [34,42,49] for examples). For instance, the following two datasets generated and used in [34,49] are both from GE, but are not shared publicly.
    - "The first dataset was collected from the Google Earth service using the BIGEMAP software (http://www.bigemap.com, accessed on 15 December 2021). We named it as the GE-Water dataset. The GE-Water dataset contains 9000 images covering water bodies of different types, varying shapes and sizes, and diverse surface and environmental conditions all around the world. These images were mainly captured by the QuickBird and Land remote-sensing satellite (Landsat) 7 systems." [49].
    - "We constructed a new water-body data set of visible spectrum Google Earth images, which consists of RGB pan-sharpened images of a 0.5 m resolution, no infrared bands, or digital elevation models are provided. All images are taken from Suzhou and Wuhan, China, with rural areas as primary. The positive annotations include lakes, reservoirs, rivers, ponds, paddies, and ditches, while all other pixels are treated as negative. These images were then divided into patches with no overlap, which provided us with 9000 images [...]" [34].

4.3.2. Generalization

It is important to be able to obtain a good accuracy score when training an ML/DL model, but perhaps more important is that model's ability to generalize to unseen data. The ultimate goal of ML/DL is to develop predictive models through finding statistical patterns in a training set which then generalize well to new, previously unseen data outside the training set [89]. Ideally, this is achieved by training on large and representative datasets that capture nearly all variations in the data actual distribution of values [86,89]. A model's ability to generalize is critical to the success of a model. An ML/DL model with good generalization capability will have the best trade-off between underfitting and overfitting so that a trained model obtains the best performance (See "Generalization, overfitting, underfitting and regularization" entry in Appendix B for details). Below, we outline a few ways to make AI systems more generalizable for water body detection and water quality monitoring tasks.

(1) Create robust AI methods for tiny water body detection. Depending on resolution, tiny water bodies such as ponds or small lakes in desert cities are difficult to identify yet may play a more critical role than we think.

(2) Develop NN architectures and comprehensive datasets (see Section 4.3.1) that are able to recognize water bodies not just from

- One type of body of water (e.g., ponds, lakes, rivers);
- One color (e.g., different levels of sediment, aquatic vegetation and algae, nutrients, pollutants);
- One size: Water bodies present in RS imagery come with different sizes (large and small water bodies) and various shapes. Many studies reported that it is not an easy task to correctly classify small water bodies and/or water bodies with different shapes.
- One environment setting (e.g., desert, urban, inland, coastal).

(3) Utilize data from multiple sources to train ML/DL models. From our comprehensive investigation, most of the current AI methods are only able to deal with water quality and/or water body detection data from one specific type of RS imagery. This should be improved and indicates a promising new research direction. Specifically, it will be important to focus on using data from multiple data platforms or resolutions, from varying weather conditions, and regions which have different ecosystem and terrain types. We humans can recognize water bodies in different RS imagery with different weather conditions. We

expect that machines should be able to mimic humans to perform this task well if we have robust AI algorithms and comprehensive datasets. See some example research below:

- Extraction of water bodies at multiple resolutions and scales using CNNs [49,53];
- Evaluation of CNN performance on multisensor data from multiple RS platforms [52];
- Integration of data from multiple sources (e.g., SAR, UAV, smaller sensors, water quality time series);
- Data fusion of Landsat-8 and Sentinel-2 RS imagery for water quality estimation [67]. "Virtual constellation" learning introduced in [67] could be a future direction for both water body detection and water quality estimation. A virtual constellation is constructed by using multiple RS platforms to "shorten" the revisit time and improve the spatial coverage of individual satellites. This entails fusing data sources from separate RS platforms with potentially different resolutions.

(4) Propose new frameworks for improving generalizability. Generalization is one of the fundamental unsolved problems in DL. The goal of a generalization theory in supervised learning is to understand when and why trained ML/DL models have small test errors [90]. The recently proposed deep bootstrap framework [90] provides a new lens for understanding generalization in DL. This new framework has the potential to advance our understanding of water domain research empowered by RS and AI by highlighting important design choices when processing RS imagery with DL.

4.3.3. Addressing Interpretability

DL has achieved significant advances with great performance in many tasks in a variety of domains, including some water domain tasks (detailed in Section 3). In the literature we reviewed for this paper, DL models have produced results comparable to, and in some scenarios even superior to, human experts. Improving predictive accuracy is important; however, improving the interpretability of ML/DL models is more important, especially through visualization techniques of ML/DL model output for later analysis by humans [18]. Interpretability is one of the primary weaknesses of DL techniques and raises wide concerns and attention in DL [91]. Due to the overparameterized and black-box nature of DL models, it is often difficult to understand the prediction results of DL models [92,93]. Understanding and explaining their black-box behaviors remains challenging due to their hierarchical, nonlinear nature. The lack of interpretability raises major concerns across several domains; for example, in high-stakes prediction applications, such as autonomous driving, healthcare, and financial services [94], the trust of DL models is critical. While many interpretation tools (e.g., image perturbation and occlusion [95], visualizing NN activation weights and class activation mapping [96,97] or attention mechanisms [98,99], feature inversion [100], local interpretable model-agnostic explanations or "LIME" [101]) have been proposed to interpret how DL models make decisions, either from a scientific perspective or a social angle, explaining the behaviors of DL models is still in progress [92]. For water domains, we list some specific potential opportunities in terms of interpretability we identified below.

- More ablation studies are needed (see Appendix B for an introduction) to investigate the role of each DL component in terms of model performance contribution and ultimately which component(s) control the model performance.
- Exploring the output of hidden layers to obtain some information to help investigate whether the model works as expected.
- Hybrid models for analyzing NN output and improving an NN's decision-making process through post-processing, for example, CNN–LR hybrids [32], CNN–CRF hybrids [36,38], CNN–SVM hybrids [39], RNN–DS hybrids [63], and CNN-LSTM hybrids [69].
- More research needs to be carried out on analyzing the importance of input data to output predictions. See examples in [62,75], each detailed below.
    - The authors in [62] systematically analyzed relative variable importance to show which sets of input data contributed to the ML models' performance. See

the quoted text below: "Relative variable importance was also conducted to investigate the consistency between in situ reflectance data and satellite data, and results show that both datasets are similar. The red band (wavelength ≈ 0.665 µm) and the product of red and green band (wavelength ≈ 0.560 µm) were influential inputs in both reflectance data sets for estimating SS and turbidity, and the ratio between red and blue band (wavelength ≈ 0.490 µm) as well as the ratio between infrared (wavelength ≈ 0.865 µm) and blue band and green band proved to be more useful for the estimation of Chl-a concentration, due to their sensitivity to high turbidity in the coastal waters".

 ○ The authors in [75] utilized existing water quality time series data and assessed the effectiveness of multiple RS data platforms and ML models in estimating various water quality parameters. One of their interesting findings is that some sensors are poorly correlated with water quality parameters, while others are more suitable for water quality monitoring tasks. They suggested that more research needs to be carried out for assessing the suitability of paired RS imagery and in situ field data. See the quoted text below: "[. . .] assess the efficacy of available sensors to complement the often limited field measurements from such programs and build models that support monitoring tasks [. . .] We observed that OLCI Level-2 Products are poorly correlated with the RNMCA data and it is not feasible to rely only on them to support monitoring operations. However, OLCI atmospherically corrected data is useful to develop accurate models using an ELM, particularly for Turbidity ($R^2 = 0.7$)." (RNMCA is the acronym for the Mexican national water quality monitoring system).

- Water quality monitoring will benefit from more research exploring how well a certain ML/DL model contributes to which water quality parameter(s). See an example in [67], where the authors investigated how well DNNs could predict certain water quality parameters.
- Physics-constrained or process-based ML/DL predictions as demonstrated in [68,69].
- The need for automatic and visually-based model evaluation metrics that are better than current visual assessment as an evaluation metric. For example, automatic assessment of how DL methods are performing in large and complex RS imagery (e.g., specifically, Bayesian DL, and Gaussian DL/ML for uncertainty measurement and visualization).

4.3.4. Ease of Use

As emphasized in [13,14], one of the major current challenges for water resource management is the integration of water quality data and indices from multiple sources into usable and meaningful insights for actionable management decisions. Geovisualization, also known as geographic visualization, uses the visual representations of geospatial data and the use of cartographic techniques to facilitate thinking, understanding, knowledge construction, and decision support about human and physical environments at geographic scales of measurement [102,103]. Geovisualization is widely utilized in different domains (e.g., public health [104], crisis management [105,106], environmental analysis [107–109], and climate change strategies [110]) for the exploration and analysis of spatiotemporal data. To the best of our knowledge, very little research has leveraged geovisualization in this way for water resources management. The only piece of work similar to this we noticed is in [111], where a web interface powered by GEE allows their expert system, combined with visual analytics, to be run on any Landsat 5, 7, or 8 imagery to draw boundaries for water bodies. Geovisualization through interactive web applications provides a promising solution to the posed challenge of integrating water quality data and indices from multiple sources [112–115]. We provide a few suggested research opportunities in this direction below.

- Simply applying (or with minor modifications) existing AI/ML/CV/DL algorithms/methods to RS big data imagery-based problems is still very far away from producing real-world applications that meet water management professionals' and policymakers' needs. As echoed in [13], "[...] realizing the full application potential of emerging technologies requires solutions for merging various measurement techniques and platforms into useful information for actionable management decisions, requiring effective communication between data providers and water resource managers" [116]. Much more multidisciplinary and integrative collaboration in terms of depth and breadth are in high demand. Those scholars and practitioners who have an interdisciplinary background will play a major role in this in-depth and in-breadth integration. For example, researchers who have expertise in RS but also know how to utilize AI, through collaboration with domain expertise such as water resources management officers, will significantly advance this research direction. Intuitive interactive web apps that are powered by both geovisualization and AI/ML/DL/CV will definitely make interdisciplinary collaboration much more seamless and thus easier.

  - ○ Interactive web portal empowered by geovisualization for integration of various water quality data sources. As noted in [117], it is natural and intuitive in many studies to use "space" as the organizing paradigm.
  - ○ More smart and responsive water management systems through the development of interactive web apps/libraries that integrate ML/DL backends and intuitive, user-friendly front ends are needed. Such systems would allow collaboration between technical experts and domain experts, including stakeholders, and even community volunteers, from anywhere at any time.
  - ○ This requires very close collaboration and thus very integrative research from researchers in many domains (e.g., computer science, cognitive science, informatics, RS, and water-related sub-domains). We reinforce that geovisualization will be the ideal tool to make the collaboration smooth, productive, and insightful.
  - ○ There is one recent work [118] that takes a small step in this direction, but much more work and efforts are in demand.

- Resource hubs for standardized AI/ML/DL/CV models and easy-to-follow and understandable tutorials for how to use them are needed.
- More data "matchups" as demonstrated in [82,83]. When more in situ measurements come in, they should be matched up and stored with satellite data for easy calibration studies.

4.3.5. Shifting Focus

From our investigation, it is clear that with enough annotated data and allocated computing, DL models are more accurate than traditional ML models, which are in turn more accurate than index-based methods for water body detection and water quality monitoring tasks. Increasing the accuracy of models by fractions of a percent should be given much less focus and attention moving forward. Water body detection methods are unlikely to improve upon the high rates of accuracy already reported in the literature without very high-resolution, very large, labeled datasets or the use of UAVs to detect small water bodies. Instead, we suggest that future research should focus more on reducing model parameters and making model training less computationally expensive in terms of time (e.g., designing neural networks to use constant memory at inference time [40], or by using TL [37,59]). Below, we outline some additional potential research directions we identified through our systematic review.

- As noted in Section 4.3.1, the lack of large benchmark datasets is a bottleneck in water body detection and water quality monitoring research utilizing RS imagery and AI. The dominant methods in both water domains are supervised learning, which often requires very large, labeled datasets to train on, thus, there is a clear, urgent need for semi-supervised and unsupervised learning methods [15].
  - ○ Unsupervised learning methods are able to learn from big sets of *unlabeled* data, as demonstrated in [29,46].

    - Semi-supervised learning methods are able to learn from limited good-quality labeled samples. DL models do not require feature engineering, and they are also much better at discovering intricate patterns hidden in big data. However, pure supervised DL is impractical in some situations, such as those for which the labeling tasks require domain knowledge from experts. Very few domain experts have the time and are willing to label very large sets of RS images [84]. An active learning-enabled DL approach that uses a visualization interface and methods to iteratively collect modest amounts of input from domain experts and uses that input to refine the DL classifiers [84] provides a promising direction to produce well-performing DL models with limited good-quality datasets.
- From our systematic review, we can easily see that current work on water body extraction and water quality monitoring using AI and RS are, in general, carried out separately. We call for a closer integration of water body detection and water quality monitoring research and more attention focusing on handling massive datasets that may include information in a variety of formats, of varying quality, and from diverse sources. This integration is critical as it will provide the essential foundation for developing real, intelligent water monitoring systems using RS and AI capable of producing insights used for actionable decision making.
- GEE + AI: as noted in [18], GEE is a good solution to address computational costs and overcome technical challenges of processing RS big data. However, online DL functionality is still not supported on GEE. To the best of our knowledge, the only piece of research integration of the Google AI platform with GEE is performed in [119]; however, as the authors reported, "data migration and computational demands are among the main present constraints in deploying these technologies in an operational setting". Thus, the ideal solution is to develop DL models directly on the GEE platform.
- Most current ML/DL-based RS research focuses on borrowing or slightly improving ML/DL/CV models from computer science [79,120]. Compared with natural scene images, RS data are multiresolution, multitemporal, multispectral, multiview, and multitarget [15]. Slight modifications of ML/DL/CV models simply cannot cope with the special challenges posed in RS big data. New ML/DL models specialized for RS big data are thus urgently needed [15,18]. We hope our review will draw the attention of researchers who have a multidisciplinary background to this issue. Looking deep into the mechanisms of RS and land surface processes, studying the characteristics of RS imagery would guide the design of specialized ML/DL models for RS big data and thus further improve RS applications using AI in breadth and depth [15].

5. Conclusions

Building intelligent and synoptic water monitoring systems requires automation of water body extent detection using RS imagery, from which volume can be computed, and also automation of their corresponding water quality, eventually linking the two to allow synoptic water quality monitoring. Yet, to date, water body detection and water quality monitoring research has been historically separate. Our systematic investigation indicates the following trends: deep learning is much more commonly used in water body detection, the dominant data source of which is RS imagery, whereas water quality literature often involves other types of data sources (e.g., in situ sensors, smaller RS devices that are not satellites). The trends relate to the scale of projects in the two domains: water body extraction is usually undertaken across large spatial scales, whereas the water quality monitoring literature is still only focused on smaller, often individual, bodies of water. This points to one of the future research directions in the water quality literature that we touch on above in Section 4.3; that is, we need to scale up water quality estimation using RS imagery through matching it with ground-truth water quality measurements.

Overall, based on the systematic review above, we contend that RS integrated with AI/ML/DL/CV methods, along with geovisualization, have great potential to provide smart and intelligent support for water resources monitoring and management. Thus, this integration has considerable potential to address major scientific and societal challenges, such as climate change and natural hazards risk management.

Author Contributions: All authors have contributed to this review paper. L.Y. initiated the review, contributed to writing and overall organization, identified selected research to include in the review, supervised the web app design and development, and coordinated input from other authors. J.D. took the lead on identifying relevant literature, contributed to writing and editing the text, and provided the data for the accompanying interactive web app. S.S. contributed to the web app design and development, word clouds visualization, and editing. Q.W., C.D.L. and M.M. have contributed to editing and M.M. also contributed to writing part of the introduction section along with identifying some relevant literature. All authors have revised the manuscript. All authors have read and agreed to the published version of the manuscript.

Funding: This research received no external funding.

Institutional Review Board Statement: Not applicable.

Informed Consent Statement: Not applicable.

Data Availability Statement: Not applicable.

Acknowledgments: This material is partly based upon work supported by the funding support from the College of Arts and Sciences at University of New Mexico. The authors are also grateful to the three reviewers for their useful suggestions.

Conflicts of Interest: The authors declare no conflict of interest.

Abbreviations

The following abbreviations (in alphabetical order) are used in this manuscript:

AE	Autoencoder
AI	Artificial Intelligence
ANN	Artificial Neural Network
ARIMA	Autoregressive Integrated Moving Average
BOA	Boundary Overall Accuracy
CA	Class Accuracy
CART	Classification and Regression Trees
CB	Cubist Regression
CE	Commission Error
CNN	Convolutional Neural Network
COCO	Common Objects in Context
CPU	Central Processing Unit
CRF	Conditional Random Field
CV	Computer Vision
DL	Deep Learning
DNN	Dense Neural Network
DS	Dempster–Shafer Evidence Theory
DT	Decision Tree
DEM	Digital Elevation Model
ECE	Edge Commission Error
ELM	Extreme Learning Machine
ELR	Extreme Learning Regression
ESA	European Space Agency
EOE	Edge Omission Error
EOA	Edge Overall Accuracy
FN	False Negative
FP	False Positive
FWIoU	Frequency Weighted Intersection over Union
GA	Global Accuracy

GAN	Generative Adversarial Network
GBM	Gradient Boosted Machine
GE	Google Earth
GEE	Google Earth Engine
GPR	Gaussian Process Regression
GPU	Graphics Processing Unit
GRU	Gated Recurrent Unit
IoT	Internet of Things
IoU	Intersection over Union
Kappa	Kappa Coefficient
KNN	K-Nearest Neighbors Classifier
LORSAL	Logistic Regression via Variable Splitting and Augmented Lagrangian
LSTM	Long Short-Term Memory
MA	Mapping Accuracy
MAE	Mean Absolute Error
MAPE	Mean Absolute Percentage Error
mIoU	Mean Intersection over Union
MK	Mann–Kendall
ML	Machine Learning
MLC	Maximum-Likelihood Classifier
MLP	Multilayer Perceptron
MLR	Multiple Linear Regression
MNDWI	Modified Normalized Difference Water Index
MPC	Microsoft Planetary Computer
MRE	Mean Relative Error
MSE	Mean Squared Error
MSI	Morphological Shadow Index
NB	Naive Bayes Classifier
NDMI	Normalized Difference Moisture Index
NDVI	Normalized Difference Vegetation Index
NDWI	Normalized Difference Water Index
NIR	Near-Infrared
NN	Neural Network
NSEC	Nash–Sutcliffe Efficiency Coefficient
OA	Overall Accuracy
OE	Omission Error
PA	Producer's Accuracy
PCC	Percent Classified Correctly
RBFNN	Radial Basis Function Neural Network
R-CNN	Region Based Convolutional Neural Network
RF	Random Forests
RMSE	Root Mean Squared Error
RMSLE	Root Mean Squared Log Error (referred to in Table 3 as RMSELE by the authors)
RNN	Recurrent Neural Network
RPART	Recursive Partitioning And Regression Trees
RPD	Relative Percent Difference
RS	Remote Sensing
SAR	Synthetic Aperture Radar
SRN	Simple Recurrent Network (same abbreviation given for Elman Neural Network)
SOTA	State-of-the-Art
SVM	Support Vector Machine
SVR	Support Vector Regression
SWIR	Short Wave Infrared

TB	Tree Bagger
TL	Transfer Learning
TN	True Negative
TP	True Positive
VHR	Very High Resolution
UA	User's Accuracy
UAV	Unmanned Aerial Vehicle

Appendix A. The Accompanying Interactive Web App Tool for the Literature of Intelligent Water Information Extraction Using AI

In Section 1.1, we provided a brief map and graphic summary of the papers covered in this review. To allow readers to obtain more useful and dynamic information and insights from the papers reviewed, we have developed an interactive web app. Through the web app, readers can keep track of the major researchers and access an up-to-date list of publications in the reviewed topics. Updated publications are accessible through (1) a researcher's public academic profile on Google Scholar or ResearchGate (see Figure A1a for an example), and (2) a continuously updated citations count of the papers that we reviewed in this paper (see Figure A1b for an example: the cited by as of 10 November 2021 is 47, which is when we first entered the data in our data file when we reviewed the paper, and then before this paper submission, when we clicked on the cited by URL, the page shows that the up-to-date citation number is 49). The web app can be accessed publicly, *free of charge* at

- Web app tool: https://geoair-lab.github.io/WaterFeatureAI-WebApp/index.html, accessed on 28 February 2022.
- Brief web app demo video (about 6 min duration): the video link is accessible at the web app page.

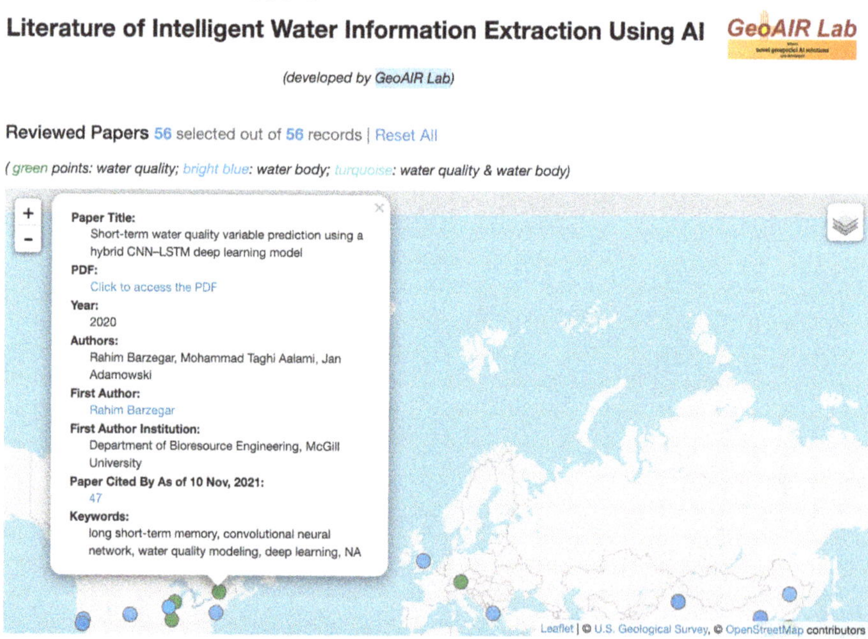

(a)

Figure A1. *Cont.*

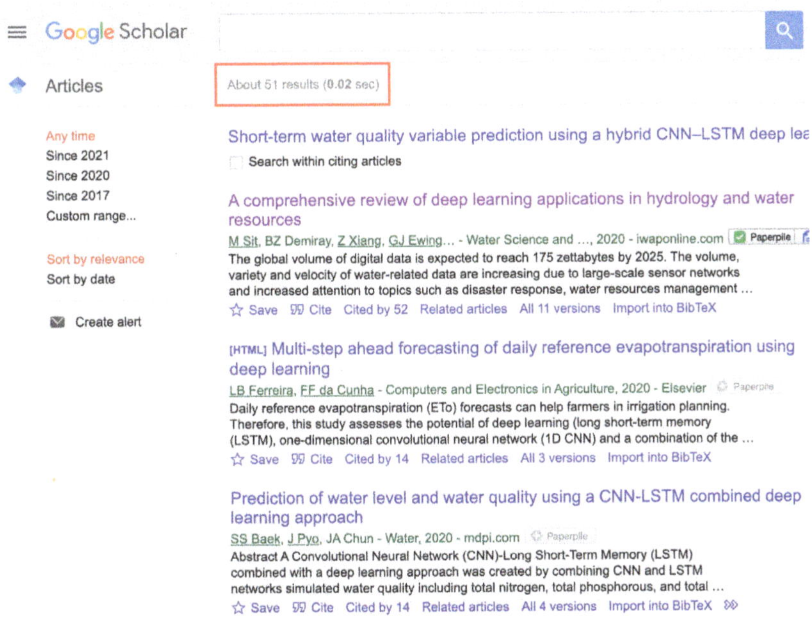

(b)

Figure A1. Our highly interactive web app (accessible publicly at: https://geoair-lab.github.io/WaterFeatureAI-WebApp/index.html, accessed on 5 December 2021) provides the track of scholars and publications with just a few clicks. See an example on the pop-up. Our readers can access (1) a direct link to the PDF file of the paper (note that if there is no free, publicly available version of the paper, we link directly to the journal page of the paper so our readers can obtain the paper if their institution purchases the journal database), (2) the scholar profile (Google Scholar/ResearchGate URL) of the first author, and (3) "Cited by" Google Scholar page. (**a**) Water body and quality AI literature map pop-up. (**b**) "Cited by" Google Scholar page corresponding to the paper shown in (**a**).

Appendix B. Essential AI/ML/DL/CV Terms

In this appendix, we provide brief definitions to some essential terms (ordered alphabetically) in ML/DL/RS in our review. For readability, we group some related concepts together.

Ablation Studies: In AI, particularly in ML and DL, ablation is the removal of a component of an AI system. Ablation studies are crucial for AI, especially for DL research. An ablation study investigates the performance of an AI system by removing certain components to understand the contribution of the component to the overall system. The term is analogous to ablation in biology (removal of components of an organism). Note that ablation studies require that the systems exhibit *graceful degradation* (i.e., they continue to function even when certain components are missing or degraded). The motivation was that, while individual components are engineered, the contribution of an individual component to the overall system performance is not clear; removing components allows this analysis. Simpler is better: if we can obtain the same performance with two models, we prefer the simpler one.

Convolution, kernel (i.e., filter), and feature map [121–123]:

Convolutional layers are the major building blocks in CNNs. A convolution is the simple application of a filter (i.e., kernel) to an input that results in an activation. Repeated application of the same filter to an input results in a map of activations called a feature map, indicating the locations and strength of a detected feature in an input (e.g., an image).

Convolution: Convolution is one of the most important operations in signal and image processing. Convolution is a mathematical operation to merge two sets of information. Convolution provides a way of multiplying together two arrays of numbers, generally of different sizes, but of the same dimensionality, to produce a third array of numbers of the same dimensionality. This can be used in image processing to implement operators whose output pixel values are simple linear combinations of certain input pixel values.

A convolutional filter (i.e., kernel) is a weight matrix (vector for one-dimensional and *cube* for three-dimensional data) which operates through a sliding window on input data. The convolution is performed by determining the value of a central pixel through adding the weighted pixel values of all its neighbors together. Specifically, it is carried out by sliding the kernel over the input image, generally starting at the top left corner, so as to move the kernel through all the positions where the kernel fits entirely within the boundaries of the input image. Each kernel position corresponds to a single output pixel, the value of which is calculated by multiplying together the kernel value and the underlying image pixel value for each of the cells in the kernel, and then adding all these numbers together. The output is a new modified filtered image. Convolution is a general purpose filter effect for images. Depending on the kernel structure, the operation enhances some features of the input data (e.g., blurring, sharpening, and edge detection).

In the context of a CNN, a convolution is a linear operation that involves the multiplication of a set of weights with the input. Given that the technique was designed for two-dimensional input, the multiplication is performed between an array of input data and a two-dimensional array of weights (i.e., a filter or a kernel). Technically, note that in CNNs, although it is referred to as a "convolution" operation, it is actually a "cross-correlation". That is, in CNNs, the filter is not flipped as is required in typical image convolutions; except for this flip, both operations are identical.

Kernel (i.e., filter): A kernel is a small matrix used in image convolution, which slides over the input image from left to right and top to bottom. Differently sized kernels, which contain different patterns of numbers, produce different results through convolution operation. The size of a kernel is arbitrary, but 3×3 or 5×5 is often used. Think of a filter similar to a membrane that allows only the desired qualities of the input to pass through it.

Feature map: The feature maps of a CNN capture the application result of the filters to an input image (i.e., at each layer, the feature map is the output of that layer). Think of it as (higher level) representations of the input. The feature map(s) is/are the output image(s) of each convolutional layer(s). The resultant number of feature maps equals the number of filters.

Data augmentation (DA) [124]:

ML (especially DL) model performance often improves with an increase in the amount of data. The common case in most ML/DL applications, especially in image classification tasks, is that obtaining new training data is not easy. Thus, we need to make good use of the existing (relatively small) training set. DA is one technique to expand the training dataset from existing training data in order to improve the performance and generalizability of DL models. DA enriches (i.e., "augments") the training data by creating new examples through random transformation of existing ones. This way, we artificially boost the size of the training set, reducing overfitting. Thus, to some extent, DA can also be viewed as a regularization technique.

Image DA is perhaps the most well-known type of DA and involves creating transformed versions of images in the training dataset that belong to the same class as the original image. The ultimate goal is to expand the training dataset with new, plausible examples (i.e., variations of the training set images that are most likely to be seen by DL models). For example, a horizontal flip of a bike photo may make sense, because the photo could be taken from the left or right. A vertical flip of a bike image does not make sense and would probably not be appropriate as the model is very unlikely to see a picture of an upside down bike. Transformations for image DA include a range of operations from the field of image manipulation (e.g., rotation, shifting, resizing, flipping, zooming, exposure adjustment, contrast change, and much more). This way, a lot of new samples can be generated from a single training example.

Note that image DA is typically only applied to the training dataset, and NOT to the validation or test dataset. This is different from data preparation such as image resizing and pixel scaling; those must be performed consistently across all datasets that interact with the model. The choice of the specific DA techniques used for a training dataset must be chosen carefully and within the context of the training dataset and knowledge of the problem domain. It can be useful to experiment with DA methods in isolation and in concert to see if they result in a measurable improvement to model performance, perhaps with a small prototype dataset, model, and training run.

DeepLabV3+ [125]: DeepLabV3 was firstly proposed to enable deep CNNs to segment features in images *at multiple scales*. ResNet-50 and ResNet-101, two variations on the popular residual network (ResNet) architecture, are the tested backbones for DeepLabV3. Through the use of residual blocks, atrous convolution, and a spatial pyramid pooling module, the authors showed that their new architecture achieved comparable performance to other SOTA models in image segmentation tasks without the need for further post-processing. The authors further improved DeepLabV3 and named the new version DeepLabV3+ [126], which combines atrous spatial pyramid pooling modules with an encoder–decoder module. This further improved the performance of DeepLabV3 while sharpening predicted feature boundaries. The DeepLabV3+ architecture is very popular in the water body extraction literature.

Generative adversarial network (GAN): GAN is a class of unsupervised DL frameworks in which two neural networks compete with each other. One network, the generator, tries to create synthetic or false images which fool the discriminator network. The discriminator, in turn, attempts to discern which images coming from the generator are actual vs. synthetic images [127]. GANs use a cooperative zero-sum game framework to learn. Among many variants of GAN, cycleGAN [128] is a technique for training unsupervised image translation models using the GAN architecture and unpaired collections of images from two different domains. CycleGAN has been demonstrated on a wide range of applications, including season translation, object transfiguration, style transfer, and generating photos from paintings.

Generalization, overfitting, underfitting and regularization (referenced [123,129,130]):

The prediction results of an ML/DL model sit somewhere between (a) low-bias, low-variance, (b) low-bias, high-variance, (c) high-bias, low-variance, and (d) high-bias, high-variance. A low-biased, high-variance model is called overfit and a high-biased, low-variance model is called underfit. A trained model achieves the best performance, through generalization, when the best trade-off between underfitting and overfitting is found. Learning with good accuracy is good, but generalization is what matters most. A good model is supposed to have both low bias and low variance. Overfitting and underfitting should both be avoided, where regularization may help.

Generalization: In ML/DL, generalization refers to the ability of a trained ML/DL model to react to new (i.e., previously unseen) data, drawn from the same distribution as the training data used to create the model. That is, after being trained on a training set, an ML/DL model can digest new data and make accurate predictions. The generalizability of an ML/DL model is central to the success of that model.

Overfitting vs. underfitting: Variance and bias are two important terms in ML. Variance refers to the variety of predicted values made by an ML model (target function). Bias means the distance of the predictions from the actual (true) target values. A high-biased model means its prediction values (average) are far from the actual values. In addition, high-variance prediction means the prediction values are highly varied.

If an ML/DL model has been trained too well on training data, it will be unable to generalize. It will make inaccurate predictions when given new data, making the model useless even though it is able to make accurate predictions for the training data. This is called overfitting. Underfitting happens when a model has not been trained enough on the data. Underfitting models are not useful either, as they are not capable of making accurate predictions, even with the training data.

Low error rates and a high variance are good indicators of overfitting. To avoid overfitting, part of the training dataset is typically set aside as the "test set" to check whether a trained model is overfitting. If the training data has a low error rate and the test data has a high error rate, it signals overfitting. An overfit model would have very low training error on seen training data but very high error from unseen datasets (e.g., testing dataset and new datasets beyond training and testing data). This is because the model maps the training set perfectly and any deviation from the training set would result in errors. An underfit model has high training error in training data and testing error in testing data and thus in new unseen data. This is because the model cannot generalize the training data correctly. Thus, the model will have a very high training error.

Regularization (also known as shrinkage): When an ML/DL model becomes too complex, it is most likely to suffer from *overfitting*. To avoid overfitting, regularization is a collection of methods to constrain and make an ML/DL model simpler and less flexible. Specifically, regularization methods are used to avoid high variance (i.e., bias/underfitting) and overfitting and thus to increase generalization. Intuitively, it follows that the function the model represents is simpler, less unsteady. Thus, predictions are smoother, and overfitting is less likely. Certain approaches are applied to different ML algorithms, for example, pruning for DT, dropout techniques for NN, and adding a penalty parameter to the cost function in regression.

Google Earth (GE): GE is a computer software, formerly known as Keyhole EarthViewer, that renders a 3D representation of Earth based primarily on satellite imagery. It has a web version at https://earth.google.com/web/, accessed on 2 January 2022. Since GE version 4.3, Google fully integrated Street View into Google Earth. Street View displays 360° panoramic street-level photos of select cities and their surroundings. The photos were taken by cameras mounted on automobiles, can be viewed at different scales and from many angles, and are navigable by arrow icons imposed on them.

Google Earth Engine (GEE) and Microsoft Planetary Computer (MPC):

GEE and MPC share similar goals (e.g., cloud storage and computing support for geospatial datasets), but have their own primary focus. For example, GEE is the pioneer in the area of RS cloud computing (launched in 2010, has 495 datasets in total as of 22 December 2021), and MPC, launched in 2020 (contains 17 datasets in total as of 22 December 2021), with a primary focus on climate change and sustainable environmental studies.

GEE [131,132]: GEE is a cloud-based platform for planetary-scale geospatial analysis, launched in 2010 by Google. GEE combines a multipetabyte catalog of satellite imagery and geospatial datasets with planetary-scale analysis capabilities. Scientists, researchers, and developers use GEE to detect changes, map trends, and quantify differences on the Earth's surface. GEE brings Google's massive computational capabilities to bear a variety of high-impact societal problems (e.g., deforestation, drought, disaster, disease, food security, water management, climate monitoring, and environmental protection). GEE has been available for commercial use from 2021 and remains free for academic and research use.

MPC [133,134]: The world lacks comprehensive, global environmental data. Microsoft Chief Environmental Officer (CEO), Dr. Lucas Joppa, imagines an international database that would provide the world with "information about every tree, every species, all of our natural resources". Microsoft President Brad Smith further emphasized that "it should be as easy for anyone in the world to search the state of the planet as it is to search the internet for driving directions or dining options", and Microsoft believes technology and AI is the key to get there, in hopes that this information will allow people to "come together and solve some of the greatest environmental and sustainability challenges we face today".

To support sustainability decision-making with the power of cloud computing and AI, similar to GEE, since December 2020, Microsoft is using ML and computing power to aggregate global environmental data (contributed by individuals around the world coupled with machinery placed in water, space, land, and air environments) into a planetary computer for a sustainable future. MPC, described as a "global portfolio of applications connecting trillions of data points", is designed to use AI to synthesize environmental data into practical information regarding the Earth's current ecosystems. For the first time, there will be a concise and comprehensive compendium of international ecosystem data. Not only will this allow for essential environmental information to be readily available to individuals across the world, but the planetary computer will predict future environmental trends through ML. In short, MPC integrates a multipetabyte catalog of global environmental data with APIs, a flexible scientific computing environment that allows people to answer global questions about that data, and applications that place those answers in the hands of conservation stakeholders.

Image classification: The concept of image classification in RS and ML/DL settings has different meanings. In RS research, the image classification is at pixel level (this is what semantic segmentation does in CV, ML, and DL settings; see the concept definition below). In contrast, in an ML and DL setting, image classification does not refer to assigning each individual pixel to a class (e.g., vegetation, water), but rather to assign the entire image to a specific class (e.g., flooded vs. not flooded) [135].

Instance segmentation: Unlike semantic segmentation, instance segmentation identifies each object instance of each pixel for every known object within an image. Thus, labels are instance-aware. Instance segmentation is essential to tasks such as counting the number of objects and reasoning about occlusion.

Normalized difference moisture index (NDMI) [136,137]: Normalized difference moisture index (NDMI) is a satellite-derived index from the near-infrared (NIR) and short wave infrared (SWIR) channels of RS imagery (note that some literature used NDMI interchangeably with NDWI; check the NDWI entry in this Appendix B for clarification).

NDMI is sensitive to the moisture levels in vegetation, and thus used to determine vegetation water content. It can be used to monitor droughts as well as monitor fuel levels in fire-prone areas. NDMI uses NIR and SWIR bands to create a ratio designed to mitigate illumination and atmospheric effects. It is calculated as a ratio between the NIR and SWIR values from RS imagery, see the formula below. For example, in Landsat 4–7, NDMI = (Band 4 − Band 5)/(Band 4 + Band 5). In Landsat 8, NDMI = (Band 5 − Band 6)/(Band 5 + Band 6). Delivered NDMI is a single band image. Similar to NDVI, NDMI values are between −1 and 1.

$$\text{NDMI} = (\text{NIR} - \text{SWIR})/(\text{NIR} + \text{SWIR})$$

Normalized difference vegetation index (NDVI) [138]: NDVI is a pixel-wise mathematical calculation rendered on an image. It is an indicator of plant health, calculated by comparing the values of absorption and reflection of red and near-infrared (NIR) light. A single NDVI value can be determined for every pixel in an image, ranging from an individual leaf to a 500-acre wheat field, depending on the RS imagery resolution.

$$\text{NDVI} = (\text{NIR} - \text{Red})/(\text{NIR} + \text{Red})$$

NDVI values always fall between −1 and 1. Values between −1 and 0 indicate dead plants, or inorganic objects (e.g., water surfaces, manmade structures such as houses, stones/rocks, roads, clouds, snow). Bare soil usually falls within 0.1–0.2 range; and plants will always have positive values between 0.2 and 1 (1 being the healthiest plants). Healthy, dense vegetation canopy should be above 0.5, and sparse vegetation will most likely fall within 0.2 to 0.5. However, it is only a rule of thumb and we should always take into account the season, type of plant, and regional peculiarities to meaningfully interpret NDVI values.

Normalized difference water index (NDWI) and modified NDWI (MNDWI) [139–141]: The NDWI is an RS-based indicator sensitive to the change in the water content of leaves or water content in water bodies (detailed below). There are two versions of NDWI.

One was defined to monitor changes in water content of leaves, using near-infrared (NIR) and short-wave infrared (SWIR) wavelengths, proposed by Gao in 1996 [139] (to avoid confusion of the two versions of NDWI, this version is also called NDMI, see NDMI entry in this Appendix B).

$$\text{NDWI} = (\text{NIR} - \text{SWIR})/(\text{NIR} + \text{SWIR})$$

The other version of NDWI, proposed by McFeeters in 1996, was defined to monitor changes related to water content in water bodies, using green and NIR wavelengths [140]. The calculation formula is given below. It is obvious that the NDWI in the papers we reviewed in this article is the version of water content in water bodies. Modification of normalized difference water index (MNDWI) was proposed [141] for improved detection of open water by replacing NIR spectral band with SWIR.

$$\text{NDWI} = (\text{Green} - \text{NIR})/(\text{Green} + \text{NIR})$$

PyTorch [142]: PyTorch is an open-source deep learning framework developed and maintained by Facebook Artificial Intelligence Research (FAIR). At its core, PyTorch is a mathematical library that performs efficient computation and automatic differentiation on graph-based models. Achieving this directly is challenging, although thankfully, the modern PyTorch API provides classes and methods that allow you to easily develop a suite of deep learning models.

Random forest (RF): It is an ML (particularly, ensemble learning) algorithm that can be used for both continuous (regression) and categorical (classification) tasks [143]. RF is widely accepted as an efficient ensemble approach for land cover classification using RS data. It handles imbalanced data, missing values, and outliers well [144].

Semantic segmentation: In contrast to instance segmentation, semantic segmentation aims to predict categorical labels for each pixel for every known object within an image, without differentiating object instances [145]. Thus, its labels are class-aware.

Support vector machine (SVM): SVM is a (supervised) machine learning algorithm that provides solutions for both classification and regression problems. The support-vector clustering [146] algorithm applies the statistics of support vectors (developed in the support vector machine algorithm) to categorize unlabeled data and is one of the most widely used clustering algorithms in many applications.

TensorFlow: TensorFlow is an open-source deep learning framework developed and maintained by Google. Although using TensorFlow directly can be challenging, the modern tf.keras API brings the simplicity and ease of use of Keras to the TensorFlow project.

Transfer learning (TL): TL is one powerful technique that makes learning in (deep) ML transferable. TL was initially proposed in [147] and recently received considerable attention due to recent significant advances in DL [123,148–152]. Inspired by humans' capabilities to transfer knowledge across domains (e.g., the knowledge gained while learning violin can be helpful to learn piano faster), TL aims to leverage learned knowledge from a related domain to achieve a desirable learning performance with minimized number of labeled samples in a target domain [151]. The main idea behind TL is that it is more efficient to take a DL model trained on an (unrelated) massive image dataset (e.g., ImageNet [87]) in one domain, and transfer its knowledge to a smaller dataset in another domain instead of training a DL classifier from scratch [153], as there are universal, low-level features shared between images for different problems.

U-Net: CNNs gave decent results in easier image segmentation problems but have not made any good progress on complex ones. This is where UNet comes in. UNet was first designed especially for medical image segmentation in [154]. It demonstrated such good results that it was used in many other fields afterwards. UNet is an improved architecture developed for biomedical image segmentation [154]. The UNet architecture stems from a fully convolutional network (FCN) first proposed by Long and Shelhamer in [155] and its architecture was modified and extended to work with fewer training images and to yield more precise segmentations. The architecture of UNet resembles a "U", which justifies its name.

The UNet architecture includes three sections: the contraction, the bottleneck, and the expansion section. The bottommost layer mediates between the contraction layer and the expansion layer. The number of expansion blocks is the same as the number of contraction blocks. Most importantly, UNet uses a novel loss weighting scheme for each pixel such that there is a higher weight at the border of segmented objects. Specifically, all pixel-wise softmax applied on the resultant image is followed by a cross-entropy loss function. Each pixel is classified into one of the classes. The idea is that even in segmentation, every pixel has to lie in some predefined category. Thus, a segmentation problem was converted into a multiclass classification and it performed very well compared to the traditional loss functions.

Appendix C. Common Evaluation Metrics in AI/ML/DL/CV Classification and Regression, and Segmentation Tasks

Many evaluation criteria have been proposed and are frequently used to assess the performance of AI/ML/DL/CV models. No single evaluation metric can tell a full story of a trained model. To better select appropriate evaluation metrics for certain domain problems and tasks, in this appendix, we provide brief definitions to some commonly used evaluation metrics (ordered alphabetically; referenced [123,129,130,156,157]) in AI/ML/DL/CV for classification, regression, and segmentation tasks in our review (i.e., those listed in the field of "Evaluation metrics" in Tables 2 and 3). For readability, we group some related metrics together. In the following formulas, TP refers to true positive, FP to false positive, FN to false negative, and TN to true negative. TP samples are those that are in the positive category and are correctly predicted as positive. FPs are not annotated as the positive category but are incorrectly predicted as positive. TNs are correctly predicted as negative, while FNs are predicted as negative when they are actually labeled as positive.

Accuracy, overall accuracy (OA), commission error (CE), omission error (OE), producer's accuracy (PA), user's accuracy (UA), and pixel accuracy (PixA) [31,156,158–161]:

To better understand the metrics in this group, let us use the same confusion matrix shown below in Figure A2 to calculate the accuracy metrics in this group. Confusion matrix, also called error matrix, is a table that allows us to visualize the performance of a classification algorithm by comparing the predicted value of the target variable with its actual value [162].

		Reference Data			
		Water	Forest	Urban	Total
Classified Data	Water	21	6	0	**27**
	Forest	5	31	1	**37**
	Urban	7	2	22	**31**
	Total	**33**	**39**	**23**	**95**

Figure A2. Example confusion matrix. The classified data indicate the ML/DL model predicted results and the reference data refer to the actual manually annotated data (image source: [161]).

(Average) Accuracy: Classification accuracy is the number of correct predictions made as a ratio of all predictions made. Accuracy with a binary classifier is measured as the following:

$$\text{Accuracy (for binary classifier)} = (TP + TN)/(TP + TN + FP + FN)$$

Note, however, that (average) accuracy for a multiclass classifier is calculated as the average of each accuracy per category (i.e., sum of accuracy for each category/number of categories) (see the definition and examples of binary classification and multiclass classification in Appendix A4 in [84]). For the example confusion matrix shown in Figure A2 (it is a multiclass classification problem), the (average) accuracy is calculated as follows:

$$\text{(average) accuracy} = (21/27 + 31/37 + 22/31)/3 = 77.5\%$$

Accuracy is perhaps the most common evaluation metric for classification problems, and it is also the most misused. It is really only meaningful and appropriate when there are an equal number of observations in each category and that all predictions and prediction errors are equally important, which is often not the case. Accuracy alone cannot tell a full meaningful story of the ML/DL models, especially when a dataset encounters a severe data imbalance problem (detailed in [86]); other metrics, such as F-score, need to tell whether an ML/DL is not suffering from overfitting when the trained model has very high accuracy.

OA: It essentially tells us out of all of the samples what proportion were classified correctly. OA is usually expressed as a percent, with 100% accuracy being a perfect classification where all samples were classified correctly. OA is the easiest to calculate and understand but ultimately only provides very basic accuracy information. OA is formally defined as follows, where N is the number of total samples. OA calculation from the example confusion matrix in Figure A2 is (21 + 31+ 22)/95 = 74/95 = 77.9%

$$OA = \textit{Number of correctly classified samples/N} = (TP + TN)/N$$

OE [31]: Errors of omission refer to samples that were left out (or omitted, as its name implies) from the correct category in the classified results. An example of OE is when pixels of a certain thing (such as maple trees), are not classified as maple trees.

OE is sometimes also referred to as a type II error (false negative). An OE in one category will be counted as a CE in another category. OEs are calculated by reviewing the reference sites for incorrect classifications. In the example confusion matrix shown in Figure A2, this is carried out by going down the columns for each category and adding together the incorrect classifications and dividing them by the total number of samples for each category. A separate OE is generally calculated for each category, as this will allow us to evaluate the classification accuracy and error for each category. OE is the inverse of the PA (i.e., OE = 1 − PA).

OE example based on the confusion matrix shown in Figure A2:

Water: Incorrectly classified reference sites: 5 + 7 = 12. Total # of reference sites = 33.

$$OE = 12/33 = 36\%$$

Forest: Incorrectly classified reference sites: 6 + 2 = 8. Total # of reference sites = 39.

$$OE = 8/39 = 20\%$$

Urban: Incorrectly classified reference sites: 0 + 1 = 1. Total # of reference sites = 23.

$$OE = 1/23 = 4\%$$

CE [31]: Errors of commission are in relation to the classified results. An example of an CE is when a pixel predicts the presence of a feature (such as trees) and, in reality, it is absent (no trees are actually present). CE is sometimes also referred to as a type I error (false positive). CEs are calculated by reviewing the classified sites for incorrect classifications. This is performed by going across the rows for each class and adding together the incorrect classifications and dividing them by the total number of classified sites for each class. CE is the inverse of the UA (i.e., CE = 1 − UA). This makes sense and is easy to interpret, as when the predicted results are very reliable (with high UA score), the classification error would be low.

CE example based on the confusion matrix shown in Figure A2:

Water: Incorrectly classified sites: 6 + 0 = 6. Total # of classified sites = 27.

$$CE = 6/27 = 22\%$$

Forest: Incorrectly classified sites: 5 + 1 = 6. Total # of classified sites = 37.

$$CE = 6/37 = 16\%$$

Urban: Incorrectly classified sites: 7 + 2 = 9. Total # of classified sites = 31.

$$CE = 9/31 = 29\%$$

PA: Similar to UA, PA is category-level-based accuracy. PA is the accuracy from the point of view of the "producer". PA tells us how often real features in the ground truth are correctly shown in the classified results, or the probability that a certain ground truth category is classified as such. PA is formally defined as the following and is complement of the omission error (OE). PA = 100% − OE.

PA = Number of correctly classified reference samples for a particular category/Number of samples from reference (i.e., annotated) data for that category = 1 − omission error

PA example based on the example confusion matrix in Figure A2:

PA for water category = Correctly classified reference sites for water category/Total # of reference sites for water category = 21/33 = 64%.

PA for forest category = Correctly classified reference sites for forest category/Total # of reference sites for water category = 31/39 = 80%.

PA for urban category = Correctly classified reference sites for urban category/Total # of reference sites for uran category = 22/23 = 96%.

UA: Similar to PA, UA is category-level-based accuracy. UA is the accuracy from the point of view of a "user", not the "producer". UA essentially tells us how often the classified category will actually align with the ground truth. This is referred to as reliability (memory tip: users often care about reliability). The UA is a complement of the commission error (i.e., UA = 100% − Commission Error). UA is defined as the following:

UA = *Number of correctly classified samples for a particular category/Number of samples classified (i.e., predicted) to that category* = 1 − *commission error.*

UA example based on the example confusion matrix in Figure A2:

UA for *water* category = *Correctly classified sites for water category/Total # of classified sites for water category* = 21/27 = 78%.

UA for *forest* category = *Correctly classified sites for forest category/Total # of classified sites for water category* = 31/37 = 84%.

UA for *urban* category = *Correctly classified sites for urban category/Total # of classified sites for uran category* = 22/31 = 70%.

PixA [158]: Pixel accuracy is perhaps the easiest to understand metric conceptually. It is the percent of pixels in the image that are classified correctly. It is the simplest metric, simply computing a ratio between the amount of properly classified pixels and the total number of pixels. See the PixA calculation formula below, where N represents the total number of pixels in the assessment image, which equals *TP* + *TN* + *FP* + *FN*. *TP* denotes the number of target-pixels that were correctly detected, *FN* denotes the number of water body pixels not classified, *FP* is the number of nontarget pixels classified, and *TN* is the number of nontarget pixels classified as nontarget pixels. This metric can sometimes provide misleading results when the category representation is small within the image, as the measure will be biased in mainly reporting how well the classifier identifies negative category (i.e., where the category we care about, such as the water body category, is not present).

$$\text{PixA} = (TP + TN)/N$$

Edge overall accuracy (EOA), edge commission error (ECE), and edge omission error [33]:

The authors in [33] defined a few evaluation metrics for water edge pixel extraction accuracy. See the following steps for how these metrics are computed.

1. Manually draw the boundary of water body.
2. Apply morphological expansion to the water body boundary from step (1) to create a buffer zone, which is centered on the boundary line (radius = three pixels).
3. Finally, the pixels in the buffer area are judged.

Let the total number of pixels in the buffer area be M, the number of correctly classified pixels be M_R, the number of missing pixels be M_O, and the number of false alarm pixels be M_C. EOA, EOE, and ECE are defined as below:

$$\text{EOA} = \text{M}_\text{R}/\text{M} \times 100\%$$

$$\text{EOE} = \text{M}_\text{o}/\text{M} \times 100\%$$

$$\text{ECE} = \text{M}_\text{c}/\text{M} \times 100\%$$

Intersection over union (IoU), mean intersection over union (mIoU), and frequency weighted intersection over union (FWIoU):

In the formal definitions below, TP, TN, FP, and FN are the number of true positive, true negative, false positive, and false negative samples, respectively.

IoU [163,164]: It is the most popular and simple evaluation metric for object detection and image segmentation used to measure the overlap between any two shapes such as two bounding boxes or masks (e.g., ground-truth and predicted bounding boxes). Values of IoU lie between 0 and 1, where 0 means two boxes do not intersect and 1 indicates two boxes completely overlap. If the prediction is completely correct, IoU = 1. The lower the IoU, the worse the prediction.

mIoU [43]: It is a common evaluation metric for semantic image segmentation, which first computes the IOU for each semantic class and then computes the average over classes. The formula is given below.

$$\text{mIoU} = TP/(TP + FP + FN)$$

FWIoU [46,158]: It is an improvement over mIoU. As its name implies, it weights each class importance depending on their appearance frequency. The formal definition of FWIoU is given below, where n is the number of categories.

$$\text{FWIoU} = \frac{1}{n+1}\sum\nolimits_{i=0}^{n}\left(\frac{TP_i}{TP_i + TN_i + FN_i} \times \frac{TP_i + FN_i}{TP_i + FP_i + TN_i + FN_i}\right)$$

Kappa statistic [156,159,165–172]:

Kappa (aka Cohen's kappa) statistic, a statistic that is frequently used to measure inter-annotator reliability (i.e., agreement) and also intra-annotator reliability for qualitative (i.e., categorical) items, is a very useful, but underutilized, metric. The importance of rater reliability is important because it represents the extent to which the data collected in a study are correct representations of measured variables. Note that this measure is to compare labeling by different human annotators, not a classifier versus a ground truth.

Cohen's kappa statistic is a very good measure that can handle both multiclass and imbalanced class problems very well. In ML, for a multiclass classification problem (see Appendix A4.2 in [84] for the definition and other types of classification tasks), measures such as accuracy, precision, or recall do not provide the complete picture of the performance of a classifier. In addition, for imbalanced class problems (see section II.D Imbalanced data in [86] for details about data imbalance), measures such as accuracy are misleading, so measures such as precision and recall are used. There are different ways to combine the two, such as the F-measure, but the F-measure does not have a very good intuitive explanation, other than it being the harmonic mean of precision and recall.

The kappa statistic can be calculated by the following formula, where Pr(a) represents the actual observed agreement, and Pr(e) represents expected (i.e., estimated) chance agreement). Thus, Pr(a) = OA.

$$\text{Kappa Statistic} = (\Pr(\text{a}) - \Pr(\text{e}))/(1 - \Pr(\text{e}))$$

Note that the sample size consists of the number of observations made across which raters are compared. Cohen specifically discussed two raters in his papers. The kappa is based on the chi-square table, and the Pr(e) is obtained through the following formula [166], where: $\text{cm}^1, \text{cm}^2, \text{rm}^1, \text{rm}^2$ represent column 1 marginal, column 2 marginal, row 1 marginal, row 2 marginal, respectively, and n represents the number of observations (not the number of raters).

$$\text{Expected (Chance) Agreement} = \frac{\left(\frac{\text{cm}^1 \times \text{rm}^1}{n}\right) + \left(\frac{\text{cm}^2 \times \text{rm}^2}{n}\right)}{n}$$

Similar to most correlation statistics, the kappa score can range from −1 to +1. Scores above 0.8 are generally considered good agreement; zero or lower mean no agreement (practically random labels). According to the scheme of [165], a value of <0 indicates no agreement, 0–0.20 is slight, 0.21–0.40 is fair, 0.41–0.60 is moderate, 0.61–0.80 is substantial, and 0.81–1 is almost perfect agreement.

Kappa is one of the most commonly used statistics to test interrater reliability, but it has limitations. Judgments about what level of kappa should be acceptable for health research are questioned. Cohen's suggested interpretation may be too lenient for health-related studies because it implies that a score as low as 0.41 might be acceptable [166]. Additional measures have been proposed to make use of the kappa framework.

For example, in [159], the authors advocate against the use of kappa and proposed the alternative measures of quantity and allocation disagreement. Quantity disagreement (QD) is the disagreement between the classification and reference data resulting from a difference in proportion of categories. Allocation disagreement (AD) assesses a difference in the spatial location of categories. The two measures (i.e., QD and AD) sum to overall error (i.e., 1–OA).

Mean Absolute Error (MAE) and Mean Absolute Percentage Error (MAPE) [123, 129,130,173]:

MAE: also called mean absolute deviation, MAE finds the average of the absolute differences between actual and predicted values. It gives an idea of how wrong the predictions were. MAE measure gives an idea of the magnitude of the error, but no idea of the direction (e.g., over- or underpredicting). MAE is defined as below [174], where y_i is the actual true value, and $\hat{y}_i$ is the predicted value. MAE value lies between 0 to ∞. Small value indicates a better model, and a value of 0 indicates no error, or perfect predictions.

$$\text{MAE} = \frac{1}{N}\sum_{i=1}^{N}|y_i - \hat{y}_i|$$

MAE is more robust to the outliers than MSE, as it is not sensitive to outliers. MAE treats larger and small errors equally. The main reason is that in MSE, through squaring the errors, the outliers, which usually have higher errors than other samples, obtain more attention and dominance in the final error and thus impact the model parameters. In addition, there is an intuitive maximum likelihood (MLE) interpretation behind MSE and MAE metrics. If we assume a linear dependence between features and targets, then MSE and MAE correspond to the MLE on the model parameters by assuming Gaussian and Laplace priors on the model errors, respectively.

MAPE [175]: MAPE, also known as mean absolute percentage deviation (MAPD), is the mean or average of the absolute percentage errors of forecasts. Error is defined as actual value (i.e., observed value) minus the forecasted value. Percentage errors are summed without regard to sign to compute MAPE. It is the most common measure used to forecast error and works best if there are no extremes to the data (and no zeros). Because absolute percentage errors are used, it avoids the problem of canceling positive and negative errors. The formula is given below, where M is mean absolute percentage error, n is number of times the summation iteration happens, At is the actual value, and Ft is the forecast value. The smaller the MAPE, the better the forecast.

$$M = \frac{1}{n}\sum_{t=1}^{n}\left|\frac{A_t - F_t}{A_t}\right|$$

Precision, recall, sensitivity, specificity, and F-score [156]:

Each measure in this group is a set-based measure [176]. The values of those measures are all from 0 to 1, with the best value at 1 and the worst score at 0.

Precision: The precision is mathematically defined by the following formula. Precision attempts to answer the question What proportion of positive identifications was actually correct? Precision refers to the proportion of the samples that is correctly classified amongst the samples predicted to be positive and is equivalent to user's accuracy (UA) for the positive category, which is also equivalent to 1 − commission error.

$$\text{Precision} = TP/(TP + FP)$$

Recall (also called sensitivity or true positive rate): it refers to the proportion of the reference data for the positive category that is correctly classified and is equivalent to producer's accuracy (also equivalent to 1 − omission error) for the positive category. It is calculated by the following formula. Recall attempts to answer the following question: What proportion of actual positives was identified correctly?

$$\text{Recall} = TP/(TP + FN)$$

Specificity (also called true negative rate): it refers to the proportion of negative samples that is correctly predicted and is equivalent to the producer's accuracy (PA) for the negative category [177].

$$\text{Specificity} = TN/(TN + FP)$$

F-score (also called F1-score, F measure): Depending on the application domain, we may need to give a higher priority to recall or precision, but there are many applications where both recall and precision are important. Thus, it is natural to think of a way to combine these two metrics into a single one. One popular metric that combines precision and recall is called F1-score. The F1-score can be interpreted as a weighted harmonic mean of the precision and recall and is formally defined as below. There is always a trade-off between precision and recall of a model; if making the precision too high, we would see a drop in the recall rate, and vice versa.

$$\text{F1-score} = (2 \times Precision \times Recall)/(Precision + Recall)$$

The generalized version of F-score is defined as follows. F1-score is a special case of F_β when β = 1.

$$F_\beta = \left(1+\beta^2\right) \times \frac{precision \times recall}{\beta^2 \times precision + recall}$$

R^2, mean squared error (MSE), root mean squared error (RMSE), and root mean squared logarithmic error (RMSLE) [123,129,130,173]:

R^2 is based on correlation between actual and predicted value; MAE is based on absolute value of error; MSE and RMSE are both based on square of error.

R^2: R-squared, also known as the coefficient of determination, is a value between 0 and 1 that measures how well a regression line fits the data (i.e., indication of the goodness of fit of a set of predictions to the actual values in a regression model). The value range of R^2 lies between 0 and 1 for no-fit and perfect fit, respectively. R^2 is not sensitive to outliers.

The R-squared formula compares our fitted regression line to a baseline model. This baseline model is considered the "worst" model. The baseline model is a flat line that predicts that every value of y will be the mean value of y. R-squared checks to see if our fitted regression line will predict y better than the mean.

$$\mathrm{R}^2 = 1 - \frac{SS_{RES}}{SS_{TOT}} = 1 - \frac{\sum_i (y_i - \hat{y}_i)^2}{\sum_i (y_i - \overline{y})^2}$$

SS_{RES} refers to the residual sum of squared errors of the regression model; y_i is the actual value, and $\hat{y}_i$ is the predicted value through the regression model. For example, if the actual y value was 58 but we had predicted it would be 47 then the residual squared error would be 121 and we would add that to the rest of the residual squared errors for the model.

SS_{TOT} is the total sum of squared errors. This compares the actual y values to the baseline model (i.e., the mean). We square the difference between all the actual y values and the mean $\overline{y}$ and add them together.

MSE: MSE is perhaps the most popular metric used for regression problems. It essentially finds the mean (i.e., average) of the square of the difference (i.e., squared error) between actual and estimated values. Similar to MAE, MSE provides a gross idea of the magnitude of error. Let us assume we have a regression model that predicts the price of houses in the Boston area and let us say for each house we also have the actual price the house was sold for. The MSE can be calculated as the following, where N is the number of samples, y_i is the actual house price, and $\hat{y}_i$ is the predicted value through the regression model. MSE value lies between 0 to ∞. Small value indicates a better model. Sensitive to outliers, it punishes larger errors more. MSE incorporates both the variance and the bias of the predicting model.

$$\mathrm{MSE} = \frac{1}{N}\sum_{i=1}^{N}(y_i - \hat{y}_i)^2$$

MSE measures how far the data are from the model's predicted values, whereas R^2 measures how far the data are from the model's predicted values compared to how far the data are from the mean. The difference between how far the data are from the model's predicted values and how far the data are from the mean is the improvement in prediction from the regression model.

RMSE: very straightforward, RMSE is the square root of MSE. Sometimes people use RMSE to have a metric with scale as the target values. Taking the house pricing prediction example, RMSE essentially shows what is the average deviation in your model predicted house prices from the target values (the prices the houses are sold for). Similar to MSE, RMSE value lies between 0 to ∞, with a small value indicating a better model. Similar to MSE, RMSE is sensitive to outliers and punishes larger errors more. The value of RMSE is always greater than or equal to MAE (RMSE >= MAE). The greater difference between them indicates greater variance in individual errors in the sample.

RMSLE: both RMSE and RMSLE are the techniques to find out the difference between the actual values and the predicted values by an ML/DL model. RMSLE is the root mean squared error of the log-transformed predicted and log-transformed actual values. RMSLE is formally defined as follows, where X denotes the predicted value and Y denotes the actual value, and n is the number of samples. Note that RMSLE adds 1 to both actual and predicted values before taking the natural logarithm to avoid taking the natural log of possible 0 (zero) values.

$$\text{RMLSE} = \sqrt{\frac{1}{n}\sum_{i=1}^{n}(log(x_i+1) - \log(\,y_i+1))^2}$$

RMSLE is very robust to outliers. When we compare the formula of the RMSE and RMSLE, the only difference is the log function. Basically, what changes is the variance measured. This small difference makes RMSLE much more robust to outliers than RMSE. In RMSE, outliers can explode the error term to a very high value, but in RMLSE, the outliers are drastically scaled down, therefore nullifying their effect.

RMSLE is often used when we do not want to penalize huge differences in the predicted and the actual values when both predicted and true values are huge numbers. (1) If both predicted and actual values are small: RMSE and RMSLE is same. (2) If either predicted or the actual value is big: RMSE > RMSLE. (3) If both predicted and actual values are big: RMSE > RMSLE (RMSLE becomes almost negligible).

References

1. UN Water. Climate Change Adaptation: The Pivotal Role of Water (2010). UN Water. 2010. Available online: https://www.unwater.org/publications/climate-change-adaptation-pivotal-role-water/#:~{}:text=Higher%20temperatures%20and%20changes%20in,likely%20to%20be%20adversely%20affected (accessed on 15 December 2021).
2. U.S. Bureau of Reclamation California-Great Basin Area Office Water Facts—Worldwide Water Supply. Available online: https://www.usbr.gov/mp/arwec/water-facts-ww-water-sup.html (accessed on 3 December 2021).
3. Reidmiller, D.R.; Avery, C.W.; Easterling, D.R.; Kunkel, K.E.; Lewis, K.L.M.; Maycock, T.K.; Stewart, B.C. *Impacts, Risks, and Adaptation in the United States: Fourth National Climate Assessment*; U.S. Global Change Research Program: Washington, DC, USA, 2018; Volume II. [CrossRef]
4. IPCC (Intergovernmental Panel on Climate Change). *Climate Change 2014–Impacts, Adaptation and Vulnerability: Part A: Global and Sectoral Aspects: Working Group II Contribution to the IPCC Fifth Assessment Report*; Cambridge University Press: Cambridge, UK, 2014; ISBN 9781107058071.
5. Steffen, W.; Richardson, K.; Rockström, J.; Cornell, S.E.; Fetzer, I.; Bennett, E.M.; Biggs, R.; Carpenter, S.R.; de Vries, W.; de Wit, C.A.; et al. Planetary Boundaries: Guiding Human Development on a Changing Planet. *Science* **2015**, *347*, 1259855. [CrossRef] [PubMed]
6. Rockström, J.; Steffen, W.; Noone, K.; Persson, A.; Chapin, F.S., 3rd; Lambin, E.F.; Lenton, T.M.; Scheffer, M.; Folke, C.; Schellnhuber, H.J.; et al. A Safe Operating Space for Humanity. *Nature* **2009**, *461*, 472–475. [CrossRef] [PubMed]
7. Walker, B.; Salt, D. *Resilience Thinking: Sustaining Ecosystems and People in a Changing World*; Island Press: Washington, DC, USA, 2006; ISBN 9781597266222.
8. Jiménez Cisneros, B.E.; Oki, N.W.; Arnell, G.; Benito, J.G.; Cogley, P.; Döll, T.; Jiang, S.S. Mwakalila Freshwater Resources. In *Climate Change 2014: Impacts, Adaptation, and Vulnerability. Part A: Global and Sectoral Aspects. Contribution of Working Group II to the Fifth Assessment Report of the Intergovernmental Panel on Climate Change*; Field, C.B., Barros, V.R., Dokken, D.J., Mach, K.J., Mastrandrea, M.D., Bilir, T.E., Chatterjee, M., Ebi, K.L., Estrada, Y.O., Genova, R.C., et al., Eds.; Cambridge University Press: Cambridge, UK; New York, NY, USA, 2014; pp. 229–269, ISBN 9781107058163.
9. Yamazaki, D.; Trigg, M.A.; Ikeshima, D. Development of a Global ~90m Water Body Map Using Multi-Temporal Landsat Images. *Remote Sens. Environ.* **2015**, *171*, 337–351. [CrossRef]
10. Jiang, W.; He, G.; Long, T.; Ni, Y. Detecting Water Bodies In Landsat8 Oli Image Using Deep Learning. *Int. Arch. Photogramm. Remote Sens. Spat. Inf. Sci.* **2018**, *XLII-3*, 669–672. [CrossRef]
11. Shao, Z.; Fu, H.; Li, D.; Altan, O.; Cheng, T. Remote Sensing Monitoring of Multi-Scale Watersheds Impermeability for Urban Hydrological Evaluation. *Remote Sens. Environ.* **2019**, *232*, 111338. [CrossRef]
12. Wang, X.; Xie, H. A Review on Applications of Remote Sensing and Geographic Information Systems (GIS) in Water Resources and Flood Risk Management. *Water* **2018**, *10*, 608. [CrossRef]
13. El Serafy, G.Y.H.; Schaeffer, B.A.; Neely, M.-B.; Spinosa, A.; Odermatt, D.; Weathers, K.C.; Baracchini, T.; Bouffard, D.; Carvalho, L.; Conmy, R.N.; et al. Integrating Inland and Coastal Water Quality Data for Actionable Knowledge. *Remote Sens.* **2021**, *13*, 2899. [CrossRef]

14. Brown, C.M.; Lund, J.R.; Cai, X.; Reed, P.M.; Zagona, E.A.; Ostfeld, A.; Hall, J.; Characklis, G.W.; Yu, W.; Brekke, L. The Future of Water Resources Systems Analysis: Toward a Scientific Framework for Sustainable Water Management. *Water Resour. Res.* **2015**, *51*, 6110–6124. [CrossRef]
15. Zhang, X.; Zhou, Y.; Luo, J. Deep Learning for Processing and Analysis of Remote Sensing Big Data: A Technical Review. *Big Earth Data* **2021**, 1–34. [CrossRef]
16. Zhu, X.X.; Tuia, D.; Mou, L.; Xia, G.; Zhang, L.; Xu, F.; Fraundorfer, F. Deep Learning in Remote Sensing: A Comprehensive Review and List of Resources. *IEEE Geosci. Remote Sens. Mag.* **2017**, *5*, 8–36. [CrossRef]
17. Hoeser, T.; Kuenzer, C. Object Detection and Image Segmentation with Deep Learning on Earth Observation Data: A Review-Part I: Evolution and Recent Trends. *Remote Sens.* **2020**, *12*, 1667. [CrossRef]
18. Reichstein, M.; Camps-Valls, G.; Stevens, B.; Jung, M.; Denzler, J.; Carvalhais, N. Prabhat Deep Learning and Process Understanding for Data-Driven Earth System Science. *Nature* **2019**, *566*, 195–204. [CrossRef] [PubMed]
19. Boyd, C.E. *Water Quality: An Introduction*; Springer Nature: Cham, Switzerland, 2019; ISBN 9783030233358.
20. Ahuja, S. *Monitoring Water Quality: Pollution Assessment, Analysis, and Remediation*; Newnes: London, UK, 2013; ISBN 9780444594044.
21. Ramadas, M.; Samantaray, A.K. Applications of Remote Sensing and GIS in Water Quality Monitoring and Remediation: A State-of-the-Art Review. In *Water Remediation*; Bhattacharya, S., Gupta, A.B., Gupta, A., Pandey, A., Eds.; Springer: Singapore, 2018; pp. 225–246, ISBN 9789811075513.
22. Bijeesh, T.V.; Narasimhamurthy, K.N. Surface Water Detection and Delineation Using Remote Sensing Images: A Review of Methods and Algorithms. *Sustain. Water Resour. Manag.* **2020**, *6*, 68. [CrossRef]
23. Gholizadeh, M.H.; Melesse, A.M.; Reddi, L. A Comprehensive Review on Water Quality Parameters Estimation Using Remote Sensing Techniques. *Sensors* **2016**, *16*, 1298. [CrossRef]
24. Sibanda, M.; Mutanga, O.; Chimonyo, V.G.P.; Clulow, A.D.; Shoko, C.; Mazvimavi, D.; Dube, T.; Mabhaudhi, T. Application of Drone Technologies in Surface Water Resources Monitoring and Assessment: A Systematic Review of Progress, Challenges, and Opportunities in the Global South. *Drones* **2021**, *5*, 84. [CrossRef]
25. Sit, M.; Demiray, B.Z.; Xiang, Z.; Ewing, G.J.; Sermet, Y.; Demir, I. A Comprehensive Review of Deep Learning Applications in Hydrology and Water Resources. *Water Sci. Technol.* **2020**, *82*, 2635–2670. [CrossRef]
26. Doorn, N. Artificial Intelligence in the Water Domain: Opportunities for Responsible Use. *Sci. Total Environ.* **2021**, *755*, 142561. [CrossRef]
27. Hassan, N.; Woo, C.S. Machine Learning Application in Water Quality Using Satellite Data. *IOP Conf. Ser. Earth Environ. Sci.* **2021**, *842*, 012018. [CrossRef]
28. Li, M.; Xu, L.; Tang, M. An Extraction Method for Water Body of Remote Sensing Image Based on Oscillatory Network. *J. Multimed.* **2011**, *6*, 252–260. [CrossRef]
29. Yang, L.; Tian, S.; Yu, L.; Ye, F.; Qian, J.; Qian, Y. Deep Learning for Extracting Water Body from Landsat Imagery. *Int. J. Innov. Comput. Inf. Control* **2015**, *11*, 1913–1929.
30. Huang, X.; Xie, C.; Fang, X.; Zhang, L. Combining Pixel- and Object-Based Machine Learning for Identification of Water-Body Types From Urban High-Resolution Remote-Sensing Imagery. *IEEE J. Sel. Top. Appl. Earth Obs. Remote Sens.* **2015**, *8*, 2097–2110. [CrossRef]
31. Isikdogan, F.; Bovik, A.C.; Passalacqua, P. Surface Water Mapping by Deep Learning. *IEEE J. Sel. Top. Appl. Earth Obs. Remote Sens.* **2017**, *10*, 4909–4918. [CrossRef]
32. Yu, L.; Wang, Z.; Tian, S.; Ye, F.; Ding, J.; Kong, J. Convolutional Neural Networks for Water Body Extraction from Landsat Imagery. *Int. J. Comput. Intell. Appl.* **2017**, *16*, 1750001. [CrossRef]
33. Chen, Y.; Fan, R.; Yang, X.; Wang, J.; Latif, A. Extraction of Urban Water Bodies from High-Resolution Remote-Sensing Imagery Using Deep Learning. *Water* **2018**, *10*, 585. [CrossRef]
34. Miao, Z.; Fu, K.; Sun, H.; Sun, X.; Yan, M. Automatic Water-Body Segmentation From High-Resolution Satellite Images via Deep Networks. *IEEE Geosci. Remote Sens. Lett.* **2018**, *15*, 602–606. [CrossRef]
35. Acharya, T.D.; Subedi, A.; Lee, D.H. Evaluation of Machine Learning Algorithms for Surface Water Extraction in a Landsat 8 Scene of Nepal. *Sensors* **2019**, *19*, 2769. [CrossRef] [PubMed]
36. Feng, W.; Sui, H.; Huang, W.; Xu, C.; An, K. Water Body Extraction From Very High-Resolution Remote Sensing Imagery Using Deep U-Net and a Superpixel-Based Conditional Random Field Model. *IEEE Geosci. Remote Sens. Lett.* **2019**, *16*, 618–622. [CrossRef]
37. Li, L.; Yan, Z.; Shen, Q.; Cheng, G.; Gao, L.; Zhang, B. Water Body Extraction from Very High Spatial Resolution Remote Sensing Data Based on Fully Convolutional Networks. *Remote Sens.* **2019**, *11*, 1162. [CrossRef]
38. Li, Z.; Wang, R.; Zhang, W.; Hu, F.; Meng, L. Multiscale Features Supported DeepLabV3+ Optimization Scheme for Accurate Water Semantic Segmentation. *IEEE Access* **2019**, *7*, 155787–155804. [CrossRef]
39. Meng, X.; Zhang, S.; Zang, S. Lake Wetland Classification Based on an SVM-CNN Composite Classifier and High-Resolution Images Using Wudalianchi as an Example. *J. Coast. Res.* **2019**, *93*, 153–162. [CrossRef]
40. Isikdogan, L.F.; Bovik, A.; Passalacqua, P. Seeing Through the Clouds With DeepWaterMap. *IEEE Geosci. Remote Sens. Lett.* **2020**, *17*, 1662–1666. [CrossRef]

41. Song, S.; Liu, J.; Liu, Y.; Feng, G.; Han, H.; Yao, Y.; Du, M. Intelligent Object Recognition of Urban Water Bodies Based on Deep Learning for Multi-Source and Multi-Temporal High Spatial Resolution Remote Sensing Imagery. *Sensors* **2020**, *20*, 397. [CrossRef]
42. Yang, F.; Feng, T.; Xu, G.; Chen, Y. Applied Method for Water-Body Segmentation Based on Mask R-CNN. *JARS* **2020**, *14*, 014502. [CrossRef]
43. Wang, G.; Wu, M.; Wei, X.; Song, H. Water Identification from High-Resolution Remote Sensing Images Based on Multidimensional Densely Connected Convolutional Neural Networks. *Remote Sens.* **2020**, *12*, 795. [CrossRef]
44. O'Neil, G.L.; Goodall, J.L.; Behl, M.; Saby, L. Deep Learning Using Physically-Informed Input Data for Wetland Identification. *Environ. Model. Softw.* **2020**, *126*, 104665. [CrossRef]
45. Chen, Y.; Tang, L.; Kan, Z.; Bilal, M.; Li, Q. A Novel Water Body Extraction Neural Network (WBE-NN) for Optical High-Resolution Multispectral Imagery. *J. Hydrol.* **2020**, *588*, 125092. [CrossRef]
46. Dang, B.; Li, Y. MSResNet: Multiscale Residual Network via Self-Supervised Learning for Water-Body Detection in Remote Sensing Imagery. *Remote Sens.* **2021**, *13*, 3122. [CrossRef]
47. Yuan, K.; Zhuang, X.; Schaefer, G.; Feng, J.; Guan, L.; Fang, H. Deep-Learning-Based Multispectral Satellite Image Segmentation for Water Body Detection. *IEEE J. Sel. Top. Appl. Earth Obs. Remote Sens.* **2021**, *14*, 7422–7434. [CrossRef]
48. Tambe, R.G.; Talbar, S.N.; Chavan, S.S. Deep Multi-Feature Learning Architecture for Water Body Segmentation from Satellite Images. *J. Vis. Commun. Image Represent.* **2021**, *77*, 103141. [CrossRef]
49. Yu, Y.; Yao, Y.; Guan, H.; Li, D.; Liu, Z.; Wang, L.; Yu, C.; Xiao, S.; Wang, W.; Chang, L. A Self-Attention Capsule Feature Pyramid Network for Water Body Extraction from Remote Sensing Imagery. *Int. J. Remote Sens.* **2021**, *42*, 1801–1822. [CrossRef]
50. Li, W.; Li, Y.; Gong, J.; Feng, Q.; Zhou, J.; Sun, J.; Shi, C.; Hu, W. Urban Water Extraction with UAV High-Resolution Remote Sensing Data Based on an Improved U-Net Model. *Remote Sens.* **2021**, *13*, 3165. [CrossRef]
51. Zhang, L.; Fan, Y.; Yan, R.; Shao, Y.; Wang, G.; Wu, J. Fine-Grained Tidal Flat Waterbody Extraction Method (FYOLOv3) for High-Resolution Remote Sensing Images. *Remote Sens.* **2021**, *13*, 2594. [CrossRef]
52. Li, M.; Wu, P.; Wang, B.; Park, H.; Yang, H.; Wu, Y. A Deep Learning Method of Water Body Extraction From High Resolution Remote Sensing Images With Multisensors. *IEEE J. Sel. Top. Appl. Earth Obs. Remote Sens.* **2021**, *14*, 3120–3132. [CrossRef]
53. Su, H.; Peng, Y.; Xu, C.; Feng, A.; Liu, T. Using Improved DeepLabv3+ Network Integrated with Normalized Difference Water Index to Extract Water Bodies in Sentinel-2A Urban Remote Sensing Images. *JARS* **2021**, *15*, 018504. [CrossRef]
54. Ovakoglou, G.; Cherif, I.; Alexandridis, T.K.; Pantazi, X.-E.; Tamouridou, A.-A.; Moshou, D.; Tseni, X.; Raptis, I.; Kalaitzopoulou, S.; Mourelatos, S. Automatic Detection of Surface-Water Bodies from Sentinel-1 Images for Effective Mosquito Larvae Control. *JARS* **2021**, *15*, 014507. [CrossRef]
55. Chebud, Y.; Naja, G.M.; Rivero, R.G.; Melesse, A.M. Water Quality Monitoring Using Remote Sensing and an Artificial Neural Network. *Water Air Soil Pollut.* **2012**, *223*, 4875–4887. [CrossRef]
56. Wang, X.; Zhang, F.; Ding, J. Evaluation of Water Quality Based on a Machine Learning Algorithm and Water Quality Index for the Ebinur Lake Watershed, China. *Sci. Rep.* **2017**, *7*, 12858. [CrossRef]
57. Lee, S.; Lee, D. Improved Prediction of Harmful Algal Blooms in Four Major South Korea's Rivers Using Deep Learning Models. *Int. J. Environ. Res. Public Health* **2018**, *15*, 1322. [CrossRef]
58. Wang, P.; Yao, J.; Wang, G.; Hao, F.; Shrestha, S.; Xue, B.; Xie, G.; Peng, Y. Exploring the Application of Artificial Intelligence Technology for Identification of Water Pollution Characteristics and Tracing the Source of Water Quality Pollutants. *Sci. Total Environ.* **2019**, *693*, 133440. [CrossRef]
59. Pu, F.; Ding, C.; Chao, Z.; Yu, Y.; Xu, X. Water-Quality Classification of Inland Lakes Using Landsat8 Images by Convolutional Neural Networks. *Remote Sens.* **2019**, *11*, 1674. [CrossRef]
60. Liu, P.; Wang, J.; Sangaiah, A.; Xie, Y.; Yin, X. Analysis and Prediction of Water Quality Using LSTM Deep Neural Networks in IoT Environment. *Sustainability* **2019**, *11*, 2058. [CrossRef]
61. Chowdury, M.S.U.; Emran, T.B.; Ghosh, S.; Pathak, A.; Alam, M.M.; Absar, N.; Andersson, K.; Hossain, M.S. IoT Based Real-Time River Water Quality Monitoring System. *Procedia Comput. Sci.* **2019**, *155*, 161–168. [CrossRef]
62. Hafeez, S.; Wong, M.S.; Ho, H.C.; Nazeer, M.; Nichol, J.; Abbas, S.; Tang, D.; Lee, K.H.; Pun, L. Comparison of Machine Learning Algorithms for Retrieval of Water Quality Indicators in Case-II Waters: A Case Study of Hong Kong. *Remote Sens.* **2019**, *11*, 617. [CrossRef]
63. Li, L.; Jiang, P.; Xu, H.; Lin, G.; Guo, D.; Wu, H. Water Quality Prediction Based on Recurrent Neural Network and Improved Evidence Theory: A Case Study of Qiantang River, China. *Environ. Sci. Pollut. Res. Int.* **2019**, *26*, 19879–19896. [CrossRef]
64. Randrianiaina, J.J.C.; Rakotonirina, R.I.; Ratiarimanana, J.R.; Fils, L.R. Modelling of Lake Water Quality Parameters by Deep Learning Using Remote Sensing Data. *Am. J. Geogr. Inf. Syst.* **2019**, *8*, 221–227.
65. Yu, Z.; Yang, K.; Luo, Y.; Shang, C. Spatial-Temporal Process Simulation and Prediction of Chlorophyll-a Concentration in Dianchi Lake Based on Wavelet Analysis and Long-Short Term Memory Network. *J. Hydrol.* **2020**, *582*, 124488. [CrossRef]
66. Zou, Q.; Xiong, Q.; Li, Q.; Yi, H.; Yu, Y.; Wu, C. A Water Quality Prediction Method Based on the Multi-Time Scale Bidirectional Long Short-Term Memory Network. *Environ. Sci. Pollut. Res. Int.* **2020**, *27*, 16853–16864. [CrossRef]
67. Peterson, K.T.; Sagan, V.; Sloan, J.J. Deep Learning-Based Water Quality Estimation and Anomaly Detection Using Landsat-8/Sentinel-2 Virtual Constellation and Cloud Computing. *GISci. Remote Sens.* **2020**, *57*, 510–525. [CrossRef]

68. Hanson, P.C.; Stillman, A.B.; Jia, X.; Karpatne, A.; Dugan, H.A.; Carey, C.C.; Stachelek, J.; Ward, N.K.; Zhang, Y.; Read, J.S.; et al. Predicting Lake Surface Water Phosphorus Dynamics Using Process-Guided Machine Learning. *Ecol. Modell.* **2020**, *430*, 109136. [CrossRef]
69. Barzegar, R.; Aalami, M.T.; Adamowski, J. Short-Term Water Quality Variable Prediction Using a Hybrid CNN–LSTM Deep Learning Model. *Stoch. Environ. Res. Risk Assess.* **2020**, *34*, 415–433. [CrossRef]
70. Aldhyani, T.H.H.; Al-Yaari, M.; Alkahtani, H.; Maashi, M. Water Quality Prediction Using Artificial Intelligence Algorithms. *Appl. Bionics Biomech.* **2020**, *2020*, 6659314. [CrossRef]
71. Li, X.; Ding, J.; Ilyas, N. Machine Learning Method for Quick Identification of Water Quality Index (WQI) Based on Sentinel-2 MSI Data: Ebinur Lake Case Study. *Water Sci. Technol. Water Supply* **2021**, *21*, 1291–1312. [CrossRef]
72. Sharma, C.; Isha, I.; Vashisht, V. Water Quality Estimation Using Computer Vision in UAV. In Proceedings of the 2021 11th International Conference on Cloud Computing, Data Science Engineering (Confluence), Noida, India, 28–29 January 2021; pp. 448–453.
73. Cui, Y.; Yan, Z.; Wang, J.; Hao, S.; Liu, Y. Deep Learning-Based Remote Sensing Estimation of Water Transparency in Shallow Lakes by Combining Landsat 8 and Sentinel 2 Images. *Environ. Sci. Pollut. Res. Int.* **2022**, *29*, 4401–4413. [CrossRef] [PubMed]
74. Zhao, X.; Xu, H.; Ding, Z.; Wang, D.; Deng, Z.; Wang, Y.; Wu, T.; Li, W.; Lu, Z.; Wang, G. Comparing Deep Learning with Several Typical Methods in Prediction of Assessing Chlorophyll-a by Remote Sensing: A Case Study in Taihu Lake, China. *Water Supply* **2021**, *21*, 3710–3724. [CrossRef]
75. Arias-Rodriguez, L.F.; Duan, Z.; de Díaz-Torres, J.; Basilio Hazas, M.; Huang, J.; Kumar, B.U.; Tuo, Y.; Disse, M. Integration of Remote Sensing and Mexican Water Quality Monitoring System Using an Extreme Learning Machine. *Sensors* **2021**, *21*, 4118. [CrossRef]
76. Kravitz, J.; Matthews, M.; Lain, L.; Fawcett, S.; Bernard, S. Potential for High Fidelity Global Mapping of Common Inland Water Quality Products at High Spatial and Temporal Resolutions Based on a Synthetic Data and Machine Learning Approach. *Front. Environ. Sci.* **2021**, *19*. [CrossRef]
77. Sun, X.; Zhang, Y.; Shi, K.; Zhang, Y.; Li, N.; Wang, W.; Huang, X.; Qin, B. Monitoring Water Quality Using Proximal Remote Sensing Technology. *Sci. Total Environ.* **2021**, *803*, 149805. [CrossRef] [PubMed]
78. Chen, Y.; Fan, R.; Bilal, M.; Yang, X.; Wang, J.; Li, W. Multilevel Cloud Detection for High-Resolution Remote Sensing Imagery Using Multiple Convolutional Neural Networks. *ISPRS Int. J. Geo-Inf.* **2018**, *7*, 181. [CrossRef]
79. Tong, X.-Y.; Xia, G.-S.; Lu, Q.; Shen, H.; Li, S.; You, S.; Zhang, L. Land-Cover Classification with High-Resolution Remote Sensing Images Using Transferable Deep Models. *Remote Sens. Environ.* **2020**, *237*, 111322. [CrossRef]
80. Boguszewski, A.; Batorski, D.; Ziemba-Jankowska, N.; Dziedzic, T.; Zambrzycka, A. LandCover. Ai: Dataset for Automatic Mapping of Buildings, Woodlands, Water and Roads From Aerial Imagery. In Proceedings of the IEEE/CVF Conference on Computer Vision and Pattern Recognition, Virtual, 19–25 June 2021; pp. 1102–1110.
81. Schmitt, M.; Hughes, L.H.; Qiu, C.; Zhu, X.X. SEN12MS—A Curated Dataset of Georeferenced Multi-Spectral Sentinel-1/2 Imagery for Deep Learning and Data Fusion. *arXiv* **2019**, arXiv:1906.07789. [CrossRef]
82. Ross, M.R.V.; Topp, S.N.; Appling, A.P.; Yang, X.; Kuhn, C.; Butman, D.; Simard, M.; Pavelsky, T.M. AquaSat: A Data Set to Enable Remote Sensing of Water Quality for Inland Waters. *Water Resour. Res.* **2019**, *55*, 10012–10025. [CrossRef]
83. Wang, S.; Li, J.; Zhang, W.; Cao, C.; Zhang, F.; Shen, Q.; Zhang, X.; Zhang, B. A Dataset of Remote-Sensed Forel-Ule Index for Global Inland Waters during 2000–2018. *Sci. Data* **2021**, *8*, 26. [CrossRef]
84. Yang, L.; MacEachren, A.M.; Mitra, P.; Onorati, T. Visually-Enabled Active Deep Learning for (Geo) Text and Image Classification: A Review. *ISPRS Int. J. Geo-Inf.* **2018**, *7*, 65. [CrossRef]
85. Yang, L.; Gong, M.; Asari, V.K. Diagram Image Retrieval and Analysis: Challenges and Opportunities. In Proceedings of the IEEE/CVF Conference on Computer Vision and Pattern Recognition Workshops, Virtual, 14–19 June 2020; pp. 180–181.
86. Yang, L.; MacEachren, A.M.; Mitra, P. Geographical Feature Classification from Text Using (active) Convolutional Neural Networks. In Proceedings of the 2020 19th IEEE International Conference on Machine Learning and Applications (ICMLA), Virtual, 14–17 December 2020; pp. 1182–1198.
87. Deng, J.; Dong, W.; Socher, R.; Li, L.; Li, K.; Fei-Fei, L. ImageNet: A Large-Scale Hierarchical Image Database. In Proceedings of the IEEE Conference on Computer Vision and Pattern Recognition, Miami, FL, USA, 20–25 June 2009; pp. 248–255.
88. Lin, T.-Y.; Maire, M.; Belongie, S.; Hays, J.; Perona, P.; Ramanan, D.; Dollár, P.; Zitnick, C.L. Microsoft COCO: Common Objects in Context. In Proceedings of the European Conference on Computer Vision (ECCV), Zurich, Switzerland, 6–12 September 2014; pp. 740–755.
89. Maaten, L.; Chen, M.; Tyree, S.; Weinberger, K. Learning with Marginalized Corrupted Features. In Proceedings of the International Conference on Machine Learning (ICML), Atlanta, GA, USA, 16–21 June 2013; pp. 410–418.
90. Nakkiran, P.; Neyshabur, B.; Sedghi, H. The Deep Bootstrap Framework: Good Online Learners Are Good Offline Generalizers. *arXiv* **2020**, arXiv:2010.08127.
91. Montavon, G.; Samek, W.; Müller, K.-R. Methods for Interpreting and Understanding Deep Neural Networks. *Digit. Signal Process.* **2018**, *73*, 1–15. [CrossRef]
92. Li, X.; Xiong, H.; Li, X.; Wu, X.; Zhang, X.; Liu, J.; Bian, J.; Dou, D. Interpretable Deep Learning: Interpretations, Interpretability, Trustworthiness, and beyond. *arXiv* **2021**, arXiv:2103.10689.
93. Doshi-Velez, F.; Kim, B. Towards A Rigorous Science of Interpretable Machine Learning. *arXiv* **2017**, arXiv:1702.08608.

94. Carvalho, D.V.; Pereira, E.M.; Cardoso, J.S. Machine Learning Interpretability: A Survey on Methods and Metrics. *Electronics* **2019**, *8*, 832. [CrossRef]
95. Fong, R.C.; Vedaldi, A. Interpretable Explanations of Black Boxes by Meaningful Perturbation. In Proceedings of the IEEE International Conference on Computer Vision, Venice, Italy, 22–29 October 2017; pp. 3429–3437.
96. Zeiler, M.D.; Fergus, R. Visualizing and Understanding Convolutional Networks. In Proceedings of the European Conference on Computer Vision, Zurich, Switzerland, 6–12 September 2014; pp. 818–833.
97. Zhou, B.; Khosla, A.; Lapedriza, A.; Oliva, A.; Torralba, A. Learning Deep Features for Discriminative Localization. In Proceedings of the IEEE Conference on Computer Vision and Pattern Recognition, Las Vegas, NV, USA, 27–30 June 2016; pp. 2921–2929.
98. Xu, K.; Ba, J.; Kiros, R.; Cho, K.; Courville, A.; Salakhudinov, R.; Zemel, R.; Bengio, Y. Show, Attend and Tell: Neural Image Caption Generation with Visual Attention. In Proceedings of the International Conference on Machine Learning, Lille, France, 6–11 July 2015; Bach, F., Blei, D., Eds.; pp. 2048–2057.
99. Woo, S.; Park, J.; Lee, J.-Y.; Kweon, I.S. Cbam: Convolutional Block Attention Module. In Proceedings of the European Conference on Computer Vision (ECCV), Munich, Germany, 8–14 September 2018; pp. 3–19.
100. Mahendran, A.; Vedaldi, A. Understanding Deep Image Representations by Inverting Them. In Proceedings of the IEEE Conference on Computer Vision and Pattern Recognition, Boston, MA, USA, 7–12 June 2015; pp. 5188–5196.
101. Ribeiro, M.T.; Singh, S.; Guestrin, C. "Why Should I Trust You?": Explaining the Predictions of Any Classifier. *arXiv* **2016**, arXiv:1602.04938.
102. Laurini, R. *Geographic Knowledge Infrastructure: Applications to Territorial Intelligence and Smart Cities*; Elsevier: Oxford, UK, 2017.
103. MacEachren, A.M.; Gahegan, M.; Pike, W.; Brewer, I.; Cai, G.; Lengerich, E.; Hardisty, F. Geovisualization for Knowledge Construction and Decision Support. *IEEE Comput. Graph. Appl.* **2004**, *24*, 13–17. [CrossRef]
104. Lan, Y.; Desjardins, M.R.; Hohl, A.; Delmelle, E. Geovisualization of COVID-19: State of the Art and Opportunities. *Cartographica* **2021**, *56*, 2–13. [CrossRef]
105. MacEachren, A.M.; Cai, G. Supporting Group Work in Crisis Management: Visually Mediated Human—GIS—Human Dialogue. *Environ. Plann. B Plann. Des.* **2006**, *33*, 435–456. [CrossRef]
106. Tomaszewski, B.; MacEachren, A.M. Geovisual Analytics to Support Crisis Management: Information Foraging for Geo-Historical Context. *Inf. Vis.* **2012**, *11*, 339–359. [CrossRef]
107. Harrower, M.; MacEachren, A.; Griffin, A.L. Developing a Geographic Visualization Tool to Support Earth Science Learning. *Cartogr. Geogr. Inf. Sci.* **2000**, *27*, 279–293. [CrossRef]
108. Cova, T.J.; Dennison, P.E.; Kim, T.H.; Moritz, M.A. Setting Wildfire Evacuation Trigger Points Using Fire Spread Modeling and GIS. *Trans. GIS* **2005**, *9*, 603–617. [CrossRef]
109. Cliburn, D.C.; Feddema, J.J.; Miller, J.R.; Slocum, T.A. Design and Evaluation of a Decision Support System in a Water Balance Application. *Comput. Graph.* **2002**, *26*, 931–949. [CrossRef]
110. Kiss, E.; Zichar, M.; Fazekas, I.; Karancsi, G.; Balla, D. Categorization and Geovisualization of Climate Change Strategies Using an Open-Access WebGIS Tool. *Infocommun. J.* **2020**, *12*, 32–37. [CrossRef]
111. Pekel, J.-F.; Cottam, A.; Gorelick, N.; Belward, A.S. High-Resolution Mapping of Global Surface Water and Its Long-Term Changes. *Nature* **2016**, *540*, 418–422. [CrossRef]
112. Brodlie, K.; Fairbairn, D.; Kemp, Z.; Schroeder, M. Connecting People, Data and Resources—distributed Geovisualization. In *Exploring Geovisualization*; Elsevier: Amsterdam, The Netherlands, 2005; pp. 423–443.
113. Robinson, A.C. *Design for Synthesis in Geovisualization*; The Pennsylvania State University: Ann Arbor, MI, USA, 2008.
114. Robinson, A.C. Supporting Synthesis in Geovisualization. *Int. J. Geogr. Inf. Sci.* **2011**, *25*, 211–227. [CrossRef]
115. Andrienko, G.; Andrienko, N.; Jankowski, P.; Keim, D.; Kraak, M.-J.; MacEachren, A.; Wrobel, S. Geovisual Analytics for Spatial Decision Support: Setting the Research Agenda. *Int. J. Geogr. Inf. Sci.* **2007**, *21*, 839–857. [CrossRef]
116. Schaeffer, B.A.; Schaeffer, K.G.; Keith, D.; Lunetta, R.S.; Conmy, R.; Gould, R.W. Barriers to Adopting Satellite Remote Sensing for Water Quality Management. *Int. J. Remote Sens.* **2013**, *34*, 7534–7544. [CrossRef]
117. Smith, M.J.; Hillier, J.K.; Otto, J.-C.; Geilhausen, M. Geovisualization. In *Treatise on Geomorphology*; Shroder, J.F., Ed.; Academic Press: Cambridge, MA, USA, 2013; Volume 3, pp. 299–325.
118. Sit, M.; Sermet, Y.; Demir, I. Optimized Watershed Delineation Library for Server-Side and Client-Side Web Applications. *Open Geospat. Data Softw. Stand.* **2019**, *4*, 8. [CrossRef]
119. Mayer, T.; Poortinga, A.; Bhandari, B.; Nicolau, A.P.; Markert, K.; Thwal, N.S.; Markert, A.; Haag, A.; Kilbride, J.; Chishtie, F.; et al. Deep Learning Approach for Sentinel-1 Surface Water Mapping Leveraging Google Earth Engine. *ISPRS Open J. Photogramm. Remote Sens.* **2021**, *2*, 100005. [CrossRef]
120. Cheng, G.; Zhou, P.; Han, J. Learning Rotation-Invariant Convolutional Neural Networks for Object Detection in VHR Optical Remote Sensing Images. *IEEE Trans. Geosci. Remote Sens.* **2016**, *54*, 7405–7415. [CrossRef]
121. CS231n: Convolutional Neural Networks for Visual Recognition. Available online: http://cs231n.stanford.edu/ (accessed on 8 December 2021).
122. Mather, P.M.; Koch, M. *Computer Processing of Remotely-Sensed Images: An Introduction*; John Wiley & Sons: Hoboken, NJ, USA, 2011; ISBN 9781119956402.
123. Goodfellow, I.; Bengio, Y.; Courville, A. *Deep Learning*; MIT Press: Cambridge, MA, USA, 2016.
124. Chollet, F. *Deep Learning with Python*, 2nd ed.; Manning Publications Co.: Shelter Island, NY, USA, 2021; ISBN 9781617296864.

125. Chen, L.-C.; Papandreou, G.; Schroff, F.; Adam, H. Rethinking Atrous Convolution for Semantic Image Segmentation. *arXiv* **2017**, arXiv:1706.05587.
126. Chen, L.-C.; Zhu, Y.; Papandreou, G.; Schroff, F.; Adam, H. Encoder-Decoder with Atrous Separable Convolution for Semantic Image Segmentation. In Proceedings of the European Conference on Computer Vision (ECCV), Munich, Germany, 8–14 September 2018; pp. 801–818.
127. Goodfellow, I.; Pouget-Abadie, J.; Mirza, M.; Xu, B.; Warde-Farley, D.; Ozair, S.; Courville, A.; Bengio, Y. Generative Adversarial Nets. In Proceedings of the Advances in Neural Information Processing Systems, Montreal, QC, Canada, 8–13 December 2014; pp. 2672–2680.
128. Zhu, J.-Y.; Park, T.; Isola, P.; Efros, A.A. Unpaired Image-to-Image Translation Using Cycle-Consistent Adversarial Networks. In Proceedings of the IEEE International Conference on Computer Vision, Venice, Italy, 22–29 October 2017; pp. 2223–2232.
129. Bishop, C.M. *Pattern Recognition and Machine Learning*; Springer: New York, NY, USA, 2006.
130. Hastie, T.; Friedman, J.; Tibshirani, R. *The Elements of Statistical Learning: Data Mining, Inference, and Prediction*; Springer: New York, NY, USA, 2001.
131. Gorelick, N.; Hancher, M.; Dixon, M.; Ilyushchenko, S.; Thau, D.; Moore, R. Google Earth Engine: Planetary-Scale Geospatial Analysis for Everyone. *Remote Sens. Environ.* **2017**, *202*, 18–27. [CrossRef]
132. Google Earth Engine. Available online: https://earthengine.google.com/ (accessed on 15 December 2021).
133. Toadvine, H. Microsoft Launches a "Planetary Computer" To Support Global Sustainability. Available online: https://earth.org/planetary-computer/ (accessed on 22 December 2021).
134. Microsoft Planetary Computer. Available online: https://planetarycomputer.microsoft.com/ (accessed on 15 December 2021).
135. Yang, L.; Cervone, G. Analysis of Remote Sensing Imagery for Disaster Assessment Using Deep Learning: A Case Study of Flooding Event. *Soft Comput.* **2019**, *23*, 13393–13408. [CrossRef]
136. Wilson, E.H.; Sader, S.A. Detection of Forest Harvest Type Using Multiple Dates of Landsat TM Imagery. *Remote Sens. Environ.* **2002**, *80*, 385–396. [CrossRef]
137. Skakun, R.S.; Wulder, M.A.; Franklin, S.E. Sensitivity of the Thematic Mapper Enhanced Wetness Difference Index to Detect Mountain Pine Beetle Red-Attack Damage. *Remote Sens. Environ.* **2003**, *86*, 433–443. [CrossRef]
138. Measuring Vegetation (NDVI & EVI). Available online: https://earthobservatory.nasa.gov/features/MeasuringVegetation (accessed on 20 December 2021).
139. Gao, B.-C. NDWI—A Normalized Difference Water Index for Remote Sensing of Vegetation Liquid Water from Space. *Remote Sens. Environ.* **1996**, *58*, 257–266. [CrossRef]
140. McFEETERS, S.K. The Use of the Normalized Difference Water Index (NDWI) in the Delineation of Open Water Features. *Int. J. Remote Sens.* **1996**, *17*, 1425–1432. [CrossRef]
141. Xu, H. Modification of Normalised Difference Water Index (NDWI) to Enhance Open Water Features in Remotely Sensed Imagery. *Int. J. Remote Sens.* **2006**, *27*, 3025–3033. [CrossRef]
142. Paszke, A.; Gross, S.; Massa, F.; Lerer, A.; Bradbury, J.; Chanan, G.; Killeen, T.; Lin, Z.; Gimelshein, N.; Antiga, L.; et al. PyTorch: An Imperative Style, High-Performance Deep Learning Library. In Proceedings of the Advances in Neural Information Processing Systems, Vancouver, BC, Canada, 8–14 December 2019; pp. 8026–8037.
143. Breiman, L. Random Forests. *Mach. Learn.* **2001**, *45*, 5–32. [CrossRef]
144. Pal, M. Random Forest Classifier for Remote Sensing Classification. *Int. J. Remote Sens.* **2005**, *26*, 217–222. [CrossRef]
145. Wu, X.; Sahoo, D.; Hoi, S.C.H. Recent Advances in Deep Learning for Object Detection. *Neurocomputing* **2020**, *396*, 39–64. [CrossRef]
146. Ben-Hur, A.; Horn, D.; Siegelmann, H.T.; Vapnik, V. Support Vector Clustering. *J. Mach. Learn. Res.* **2001**, *2*, 125–137. [CrossRef]
147. Pratt, L.Y. Discriminability-Based Transfer between Neural Networks. In Proceedings of the Advances in Neural Information Processing Systems, Denver, CO, USA, 29 November–2 December 1993; pp. 204–211.
148. Pan, S.J.; Yang, Q. A Survey on Transfer Learning. *IEEE Trans. Knowl. Data Eng.* **2010**, *22*, 1345–1359. [CrossRef]
149. Weiss, K.; Khoshgoftaar, T.M.; Wang, D. A Survey of Transfer Learning. *J. Big Data* **2016**, *3*, 9. [CrossRef]
150. Tan, C.; Sun, F.; Kong, T.; Zhang, W.; Yang, C.; Liu, C. A Survey on Deep Transfer Learning. In Proceedings of the International Conference on Artificial Neural Networks, Rhodes, Greece, 4–7 October 2018; pp. 270–279.
151. Zhuang, F.; Qi, Z.; Duan, K.; Xi, D.; Zhu, Y.; Zhu, H.; Xiong, H.; He, Q. A Comprehensive Survey on Transfer Learning. *Proc. IEEE* **2019**, *109*, 43–76. [CrossRef]
152. Li, C.; Zhang, S.; Qin, Y.; Estupinan, E. A Systematic Review of Deep Transfer Learning for Machinery Fault Diagnosis. *Neurocomputing* **2020**, *407*, 121–135. [CrossRef]
153. Bar, Y.; Diamant, I.; Wolf, L.; Lieberman, S.; Konen, E.; Greenspan, H. Chest Pathology Detection Using Deep Learning with Non-Medical Training. In Proceedings of the 2015 IEEE 12th International Symposium on Biomedical Imaging (ISBI), Brooklyn, NY, USA, 16–19 April 2015; pp. 294–297.
154. Ronneberger, O.; Fischer, P.; Brox, T. U-Net: Convolutional Networks for Biomedical Image Segmentation. In Proceedings of the Medical Image Computing and Computer-Assisted Intervention (MICCAI), Munich, Germany, 5–9 October 2015; pp. 234–241.
155. Long, J.; Shelhamer, E.; Darrell, T. Fully Convolutional Networks for Semantic Segmentation. In Proceedings of the IEEE Conference on Computer Vision and Pattern Recognition, Boston, MA, USA, 7–12 June 2015; pp. 3431–3440.

156. Maxwell, A.E.; Warner, T.A. Thematic Classification Accuracy Assessment with Inherently Uncertain Boundaries: An Argument for Center-Weighted Accuracy Assessment Metrics. *Remote Sens.* **2020**, *12*, 1905. [CrossRef]
157. Sokolova, M.; Lapalme, G. A Systematic Analysis of Performance Measures for Classification Tasks. *Inf. Process. Manag.* **2009**, *45*, 427–437. [CrossRef]
158. Garcia-Garcia, A.; Orts-Escolano, S.; Oprea, S.; Villena-Martinez, V.; Martinez-Gonzalez, P.; Garcia-Rodriguez, J. A Survey on Deep Learning Techniques for Image and Video Semantic Segmentation. *Appl. Soft Comput.* **2018**, *70*, 41–65. [CrossRef]
159. Pontius, R.G.; Millones, M. Death to Kappa: Birth of Quantity Disagreement and Allocation Disagreement for Accuracy Assessment. *Int. J. Remote Sens.* **2011**, *32*, 4407–4429. [CrossRef]
160. Yang, X.; Zhao, S.; Qin, X.; Zhao, N.; Liang, L. Mapping of Urban Surface Water Bodies from Sentinel-2 MSI Imagery at 10 M Resolution via NDWI-Based Image Sharpening. *Remote Sens.* **2017**, *9*, 596. [CrossRef]
161. GSP216Online Accuracy Metrics. Available online: http://gsp.humboldt.edu/olm_2019/courses/GSP_216_Online/lesson6-2/metrics.html (accessed on 23 December 2021).
162. Stehman, S.V. Selecting and Interpreting Measures of Thematic Classification Accuracy. *Remote Sens. Environ.* **1997**, *62*, 77–89. [CrossRef]
163. Rezatofighi, H.; Tsoi, N.; Gwak, J.; Sadeghian, A.; Reid, I.; Savarese, S. Generalized Intersection over Union: A Metric and a Loss for Bounding Box Regression. In Proceedings of the IEEE/CVF Conference on Computer Vision and Pattern Recognition, Long Beach, CA, USA, 16–20 June 2019; pp. 658–666.
164. Padilla, R.; Netto, S.L.; da Silva, E.A.B. A Survey on Performance Metrics for Object-Detection Algorithms. In Proceedings of the 2020 International Conference on Systems, Signals and Image Processing (IWSSIP), Niteroi, Brazil, 1–3 July 2020; pp. 237–242.
165. Landis, J.R.; Koch, G.G. The Measurement of Observer Agreement for Categorical Data. *Biometrics* **1977**, *33*, 159–174. [CrossRef]
166. McHugh, M.L. Interrater Reliability: The Kappa Statistic. *Biochem. Med.* **2012**, *22*, 276–282. [CrossRef]
167. Congalton, R.G. A Review of Assessing the Accuracy of Classifications of Remotely Sensed Data. *Remote Sens. Environ.* **1991**, *37*, 35–46. [CrossRef]
168. Stehman, S.V.; Foody, G.M. Key Issues in Rigorous Accuracy Assessment of Land Cover Products. *Remote Sens. Environ.* **2019**, *231*, 111199. [CrossRef]
169. Stehman, S.V.; Wickham, J.D. Pixels, Blocks of Pixels, and Polygons: Choosing a Spatial Unit for Thematic Accuracy Assessment. *Remote Sens. Environ.* **2011**, *115*, 3044–3055. [CrossRef]
170. Foody, G.M. Status of Land Cover Classification Accuracy Assessment. *Remote Sens. Environ.* **2002**, *80*, 185–201. [CrossRef]
171. Congalton, R.G. Accuracy Assessment and Validation of Remotely Sensed and Other Spatial Information. *Int. J. Wildland Fire* **2001**, *10*, 321–328. [CrossRef]
172. Stehman, S.V. Sampling Designs for Accuracy Assessment of Land Cover. *Int. J. Remote Sens.* **2009**, *30*, 5243–5272. [CrossRef]
173. Strutz, T. Data Fitting and Uncertainty. Available online: https://link.springer.com/book/9783658114558 (accessed on 31 December 2021).
174. Willmott, C.J.; Matsuura, K. Advantages of the Mean Absolute Error (MAE) over the Root Mean Square Error (RMSE) in Assessing Average Model Performance. *Clim. Res.* **2005**, *30*, 79–82. [CrossRef]
175. Rosenblad, A. The Concise Encyclopedia of Statistics. *J. Appl. Stat.* **2011**, *38*, 867–868. [CrossRef]
176. Schütze, H.; Manning, C.D.; Raghavan, P. *Introduction to Information Retrieval*; Cambridge University Press: Cambridge, UK, 2008; Volume 39.
177. Goutte, C.; Gaussier, E. A Probabilistic Interpretation of Precision, Recall and F-Score, with Implication for Evaluation. In Proceedings of the Advances in Information Retrieval, Santiago de Compostela, Spain, 21–23 March 2005; pp. 345–359.

MDPI
St. Alban-Anlage 66
4052 Basel
Switzerland
www.mdpi.com

Sensors Editorial Office
E-mail: sensors@mdpi.com
www.mdpi.com/journal/sensors

www.ingramcontent.com/pod-product-compliance
Lightning Source LLC
Chambersburg PA
CBHW040805250926
51850CB00007BA/1087
9783036595948